An Identification Handbook to Angiosperm Families
and Genera of the East China Sea Coastal Islands

中国东海近陆岛屿被子植物科属图志

General Editor BinJie GE
主　编　葛斌杰

Henan Science and Technology Press
· Zheng zhou ·

河南科学技术出版社
· 郑州 ·

图书在版编目（CIP）数据

中国东海近陆岛屿被子植物科属图志 / 葛斌杰主编. —郑州：河南科学技术出版社, 2020.2

ISBN 978-7-5349-9406-7

Ⅰ. ①中… Ⅱ. ①葛… Ⅲ. ①东海—岛—被子植物—植物志—中国 Ⅳ. ①Q949.708

中国版本图书馆CIP数据核字(2018)第300760号

出版发行：河南科学技术出版社
　　　　　地址：郑州市郑东新区祥盛街27号　邮编：450016
　　　　　电话：（0371）65737028　65788613
　　　　　网址：www.hnstp.cn

策划编辑：陈淑芹

责任编辑：陈淑芹

责任校对：司丽艳

装帧设计：张德琛

责任印制：张艳芳

印　　刷：北京盛通印刷股份有限公司

经　　销：全国新华书店

开　　本：889 mm×1 194 mm　1/16　印张：25.25　字数：800千字

版　　次：2020年2月第1版　2020年2月第1次印刷

定　　价：470.00元

如发现印、装质量问题，影响阅读，请与出版社联系并调换。

Authors of *An Identification Handbook to Angiosperm Families and Genera of the East China Sea Coastal Islands*

General Editor BinJie GE

Editors BinJie GE, SiYue XIAO, MinYu CHEN, Xin ZHONG, James W. BYNG

Photographer BinJie GE, Bin CHEN, Xin ZHONG, ZhengWei WANG, Yuan WANG, Bin SHEN, MinYu CHEN, WeiLiang MA, Qi TIAN, Yuan XU

Proofreader Bing LIU, XiYang YE, HongQing LI, YuKe BI, Yuan WANG, Pan LI, HuaiZhen TIAN, Chao HU

This book was supported by grants from Shanghai Municipal Administration of Forestation and City Appearances (F112424, F132438, G192426), and Specific Project for Strategic Biological Resources and Technology Supporting System from Chinese Academy of Sciences (ZSZY-001) and National Wild Plant Germplasm Resource Center

《中国东海近陆岛屿被子植物科属图志》编写人员

主　编	葛斌杰
编写人员	葛斌杰　肖斯悦　陈敏愉　钟　鑫　James W. BYNG
摄　影	葛斌杰　陈　彬　钟　鑫　王正伟　汪　远　沈　彬 陈敏愉　马炜梁　田　旗　许　源
校　对	刘　冰　叶喜阳　李宏庆　毕玉科　汪　远　李　攀 田怀珍　胡　超

本书得到了上海市绿化和市容管理局科学技术项目 (F112424, F132438, G192426) 和中国科学院战略生物资源科技支撑体系运行专项植物园运行补助经费 (ZSZY-001) 和国家重要野生植物种质资源库的资助

主编简介

BinJie GE, born in Hangzhou, Zhejiang Province in 1983. In 2007, he graduated from College of Science and Technology, Zhejiang Normal University. In 2010, he graduated from School of Life Sciences, East China Normal University with a Master degree in botany and entered Shanghai Chenshan Botanical Garden in the same year. He mainly interested in plant diversity on coastal islands of East China Sea, construction of herbarium digitalization, and fine structure and morphology of plants in spare time. He had published 18 papers/books.

葛斌杰，1983年生于浙江杭州，2007年毕业于浙江师范大学理工学院，2010年毕业于华东师范大学生命科学学院，获植物学硕士，同年进入上海辰山植物园工作。主要从事华东沿海岛屿植物多样性研究，标本馆数字化建设，业余从事植物精细解剖工作。已发表论著18篇/部。

Preface

The coastal islands in the East China Sea, from near the Sheshan Island outside the Yangtze River estuary to the Dongjia - Xiaori islands in Taiwan Strait, belong to the coastal Mesozoic volcanic area of the eastern part of China, located in Zhejiang-Fujian Sea uplift. The plant species of this area account for about 40% of the total vascular plants of East China, including many endemic coastal taxa. The eastern economic coastal zone of China is an area of active industrialization changing the ecology of the islands. The rapid increase of artificial coastline has brought great destruction to the ecosystem of coastal areas. Biological invasion, a decrease in biodiversity and other problems have since followed. The geological history of the coastal islands in the East China Sea is relatively young, the area of most of islands is small, the soil is barren, the vegetation type is simple, and the ecosystem is relatively fragile, which is easily disturbed by the outside world, and once destroyed, is difficult to recover naturally. In recent years, the development of tourism in the several of the coastal islands has brought significant changes on the native vegetation of those islands.

After four thorough surveys in China since the 1980s, the vegetation of the coastal islands of China have been studied and data has been accumulated. However, with a rapid development and re-construction, some of the islands vegetation have been severely disturbed, and at present there are no national nature reserves for vascular plants in the East China Sea. Some areas are protected by traditional means, such as temples and Fengshui forests, but the general lack of monitoring and protection makes conservation of the island's vegetation very difficult. Therefore, in 2011, with support from Shanghai Municipal Administration of Forestation and City Appearances, Shanghai Chenshan Plant Science Research Center, Chinese Academy Sciences/ Shanghai Chenshan Herbarium cooperated with East China Normal University, and Zhejiang Agriculture and Forestry University, with the purpose to survey the quality and diversity of vegetation on the coastal islands in the East China Sea.

There are more than 2,000 islands in the East China Sea. Based on previous studies, four geological units were established namely the Zhoushan Islands and Hangzhou Bay, Coast of central and north Zhejiang Province, Coast of south Zhejiang Province and northern coast of Fujian. Considering the type of vegetation, species richness and accessibility conditions, 63 of the islands were selected for further investigation and field studies. After six years of research, as of 2016, the expedition team collected 9,372 plant specimens, from which 1,727 angiosperm taxa were identified. More than 3,000 silica gel samples which belong to 1,100 taxa were collected, more than 120 taxa of coastal plants were introduced to Shanghai Chenshan Botanical Garden, and more than 15,000 pictures of the living plants in the field were taken. In addition, 1,034 species were geo-referenced from the region.

The work, *An Identification Handbook to Angiosperm Families and Genera of the East China Sea Coastal Islands* aims to cover native species, exotic species, naturalized species, and also some species for ecological restoration, wind prevention and sand fixation. This book contains flowering plants in 151 families and 665 genera distributed in the coastal islands of the East China Sea.

The majority of pictures in this book were taken by the project team, and the preparation of this book has been supported by our our numerous friends and colleagues. With their help, problems such as for some species and replacing poor quality ones, were solved. In addition, we would like to thank WeiLiang MA, Bing LIU, Pan LI, HuaiZhen TIAN, Chao HU, KaiWen JIANG and Yuan WANG for their suggestions, and identification of specimens.

This is the second book of the plant diversity and conservation series of East China Flora, the first of which was *The Catalogue of Vascular Plants Biodiversity of Eastern China*. (2014, Science Press). After reviewing the genera and families of angiosperms in these areas, it may help readers to understand the diversity and particularity of plants on the coastal islands in the East China Sea. This book will be useful for plant enthusiasts, researchers, teaching staff and various personnel for reference

Editors
September, 2019

前 言

中国东海近陆岛屿北起长江口外的佘山岛，南到台湾海峡的东甲列岛至小日列岛，属于中国东部沿海地区的中部中生代火山岩区，位于浙闽隆起之上。本海域岛屿维管束植物种类约占整个华东植物区系维管束植物的40%，且分布有大量的滨海特有种。东海作为中国沿海经济带最具活力的地区之一，沿海经济的快速崛起一直引人注目。然而海岸线的人工化比例急剧上升，给沿海地区的生态环境带来了巨大的压力，物种入侵、生物多样性降低等问题接踵而至。东海近陆岛屿成岛时间短，大部分岛屿面积小，土层瘠薄，植被类型简单，生态系统较脆弱，极易受到外界干扰，并且一旦遭到破坏，几乎无法自然恢复。近些年来的岛屿旅游热、开发热等活动，更是对岛屿自然分布的原生植被产生了严重的影响。

自20世纪80年代以来中国沿海岛屿的植被资源经过4次全国性的大规模考察之后，有了一定的数据基础，但期间伴随着快速的开发建设，部分岛屿的植被面貌受到强烈的干扰，且目前东海海域还未有专门以维管束植物为保护对象的国家级自然保护区，除个别区域因庙宇、风水林等传统习俗而得以保留原生植被外，其余地区极易遭到破坏。对岛屿植被的跟踪科考和保护尚十分缺乏，使得岛屿植被的保护形势严峻。2011年，在上海市绿化和市容管理局的支持下，中国科学院上海辰山植物科学研究中心/上海辰山植物标本馆与华东师范大学、浙江农林大学合作，启动了中国东海近陆岛屿植物多样性调查与编目项目，并对东海近陆岛屿植物的现状进行详细调查。

东海海域分布着2 000多个岛屿，项目组成员通过前期调研和分析，最终划定了4个地理单元，即舟山群岛及杭州湾、浙江中北部沿岸、浙江南部沿岸和福建中北部沿岸，综合考虑植被类型、物种丰富程度以及交通情况，共选择了其中的63个岛屿进行重点考察和采集。经过6年的调查，截至2016年，课题组成员共采集到植物标本9 372号，已鉴定出被子植物1 727种，收集了1 100个物种3 000余份分子材料（叶片干燥后保存于硅胶中），引种栽培滨海特色植物120余种，拍摄野外图片逾1.5万张，经统计，有准确GPS定位的植物共计1 034种。

《中国东海近陆岛屿被子植物科属图志》尽可能多地覆盖东海近陆岛屿被子植物种类，包括野生、外来逸生、归化（入侵）植物，同时也收录一些用于生态修复、防风固沙的常见造林树种等。全书共收录中国东海近陆岛屿被子植物151科665属。

本书主要图片来自项目组的第一手资料，并包括合作单位所拍摄的照片。本书编写过程中得到了许多人的帮助，才使得一些问题，如图片缺失或质量欠佳等得以解决，在此一并致谢。同时，要特别感谢马炜梁、刘冰、李攀、田怀珍、胡超、蒋凯文、汪远等在植物鉴定及审稿过程中给予的建议和帮助，最终才使得本书顺利出版。

本书是继《华东植物区系维管束植物多样性编目》（2014，科学出版社）以来，华东植物区系植物多样性保护系列丛书的第二部，较全面系统地梳理了东海近陆岛屿被子植物的科属组成，希望通过本书能让更多的读者了解并认识中国东海近陆岛屿植物的多样性和特殊性。本书适合植物爱好者、植物相关科研和教学人员及各类管理人员参考使用。

编者
2019年9月

About this book

This book includes descriptions and details for 151 families and 665 genera of angiosperms distributed in the coastal islands of the East China Sea. The family arrangement in the book follows the APG IV linear sequence (Angiosperm Phylogeny Group, 2016). The main features and detailed characteristics of the families were written by SiYue XIAO and BinJie GE, and mainly refer to *Flora of China*, and *The Flowering Plants Handbook: A practical guide to families and genera of the world* (J.W. Byng, 2015. eBook version), *Plant Systematics: A Phylogenetic Approach* (Third Edition) (W.S. JUDD, 2008), *Plant Systematics: A Phylogenetic Approach* (Third Edition) (Chnese Edition) (DeZhu Li, 2012). The number of genera and species in each family present in the world and China, and their distributions follow *A Dictionary of the Families and Genera of Chinese Vascular Plants* (DeZhu LI, 2018). A few exceptions are marked in the book. The keys to genera were compiled by MinYu CHEN and BinJie GE, and refer mainly to *Flora of China*, *Handbook of Seed Plants in East China* (HongQing LI, 2010), *Handbook for Identification of Seed Plants in Zhejiang* (ChaoZong ZHENG, 2005).

All fine structure images of plants and pictures with their characteristics in this book were prepared and taken by BinJie Ge unless noted. The English portion of this book was compiled by Xin Zhong.

Example family treatment

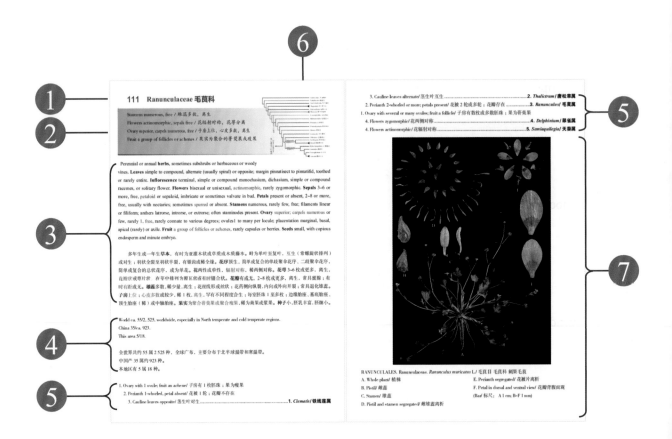

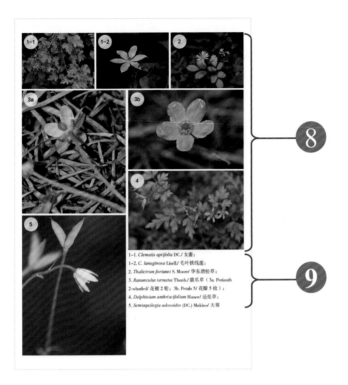

① The family number in the APG IV linear sequence, the scientific name, Chinese name.

② Important characteristics of the family, both in Chinese and English.

③ Detailed description of the family, both in Chinese and English.

④ The number of genera and taxa of the family in the world, and their global distribution; the number of genera and taxa of this family in China & the Coastal Islands of the East China Sea area.

⑤ The key to genera of the family distributed in the Coastal Islands of the East China Sea.

⑥ The family location in the angiosperm phylogenetic tree

⑦ Fine anatomy structure illustration and annotation.

⑧ Notable diagnostic characters figure numbers correspond to the serial number in the key table: a and b represents different parts of the same species, 1-1 and 1-2 represents different species in the same genus

⑨ Figures legend.

编写说明

全书共收录了分布于东海近陆岛屿的被子植物 151 科 665 属，按照 APG IV（APG 为"被子植物系统发育研究组"的英文 Angiosperm Phylogeny Group 的缩写，APG IV 即为被子植物系统发育研究组建立的被子植物分类系统的第 4 版）编排。科的重要识别特征及详细特征描述由肖斯悦、葛斌杰编写，主要参考了 *Flora of China* [《中国植物志》（英文版）， 1994–2013]、*The Flowering Plants Handbook: A practical guide to families and genera of the world* (J. W. BYNG, 2015. eBook version)；《植物系统学》（第三版）（中文版，李德铢，2012）；科下的属数和种数及分布区域，参考了《中国维管植物科属词典》（李德铢，2018），个别例外情况在文中另行标注了出处。属级检索表由陈敏愉、葛斌杰编写，主要参考了 *Flora of China*、《华东种子植物检索手册》（李宏庆，2010)、《浙江种子植物检索鉴定手册》(郑朝宗，2005)。

文中出现的植物精细解剖图片和各属特征图片除特别标注外，均由葛斌杰拍摄。全书英文由钟鑫校订。

内文说明

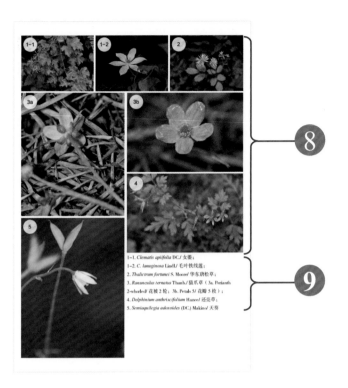

❶ APG IV 系统下，科的线性排列序号，及科的拉丁名和中文名。
❷ 本科的重要特征，含中英文对照。
❸ 本科详细特征，含中英文对照。
❹ 本科在全世界的属数与种数，主要分布区；本科在中国和东海近陆岛屿的属数与种数。
❺ 本科在东海近陆岛屿有分布的属级检索表。
❻ 本科在被子植物系统发育树上的位置。
❼ 本科代表植物的精细解剖图版及注释。
❽ 各属重要识别特征图，图序号与检索表中属序号对应，a、b 表示同种的不同部位图，1–1、1–2 表示同属中不同的种。
❾ 图注。

目 录 Index

007	Schisandraceae 五味子科	1	124	Hamamelidaceae 金缕梅科	119
010	Saururaceae 三白草科	4	126	Daphniphyllaceae 虎皮楠科	122
011	Piperaceae 胡椒科	6	128	Grossulariaceae 茶藨子科	123
012	Aristolochiaceae 马兜铃科	9	129	Saxifragaceae 虎耳草科	124
014	Magnoliaceae 木兰科	12	130	Crassulaceae 景天科	126
019	Calycanthaceae 蜡梅科	15	134	Haloragaceae 小二仙草科	128
025	Lauraceae 樟科	17	136	Vitaceae 葡萄科	129
026	Chloranthaceae 金粟兰科	21	138	Zygophyllaceae 蒺藜科	131
027	Acoraceae 菖蒲科	22	140	Fabaceae 豆科	134
028	Araceae 天南星科	23	142	Polygalaceae 远志科	150
030	Alismataceae 泽泻科	26	143	Rosaceae 蔷薇科	153
032	Hydrocharitaceae 水鳖科	28	146	Elaeagnaceae 胡颓子科	160
038	Potamogetonaceae 眼子菜科	31	147	Rhamnaceae 鼠李科	163
043	Nartheciaceae 沼金花科	32	148	Ulmaceae 榆科	166
045	Dioscoreaceae 薯蓣科	33	149	Cannabaceae 大麻科	167
053	Melanthiaceae 藜芦科	36	150	Moraceae 桑科	169
059	Smilacaceae 菝葜科	37	151	Urticaceae 荨麻科	171
060	Liliaceae 百合科	38	153	Fagaceae 壳斗科	173
061	Orchidaceae 兰科	41	154	Myricaceae 杨梅科	176
066	Hypoxidaceae 仙茅科	45	155	Juglandaceae 胡桃科	177
070	Iridaceae 鸢尾科	46	156	Casuarinaceae 木麻黄科	179
072	Asphodelaceae 阿福花科	49	158	Betulaceae 桦木科	181
073	Amaryllidaceae 石蒜科	52	163	Cucurbitaceae 葫芦科	182
074	Asparagaceae 天门冬科	55	166	Begoniaceae 秋海棠科	185
076	Arecaceae 棕榈科	60	168	Celastraceae 卫矛科	187
078	Commelinaceae 鸭跖草科	62	171	Oxalidaceae 酢浆草科	190
079	Philydraceae 田葱科	65	173	Elaeocarpaceae 杜英科	192
080	Pontederiaceae 雨久花科	66	179	Rhizophoraceae 红树科	193
085	Musaceae 芭蕉科	68	186	Hypericaceae 金丝桃科	194
089	Zingiberaceae 姜科	70	200	Violaceae 堇菜科	196
090	Typhaceae 香蒲科	72	202	Passifloraceae 西番莲科	198
094	Eriocaulaceae 谷精草科	73	204	Salicaceae 杨柳科	199
097	Juncaceae 灯芯草科	74	207	Euphorbiaceae 大戟科	201
098	Cyperaceae 莎草科	76	211	Phyllanthaceae 叶下珠科	203
103	Poaceae 禾本科	80	212	Geraniaceae 牻牛儿苗科	205
104	Ceratophyllaceae 金鱼藻科	100	215	Lythraceae 千屈菜科	207
106	Papaveraceae 罂粟科	101	216	Onagraceae 柳叶菜科	209
108	Lardizabalaceae 木通科	102	218	Myrtaceae 桃金娘科	212
109	Menispermaceae 防己科	104	219	Melastomataceae 野牡丹科	214
110	Berberidaceae 小檗科	106	226	Staphyleaceae 省沽油科	216
111	Ranunculaceae 毛茛科	108	239	Anacardiaceae 漆树科	217
112	Sabiaceae 清风藤科	111	240	Sapindaceae 无患子科	219
113	Nelumbonaceae 莲科	113	241	Rutaceae 芸香科	221
115	Proteaceae 山龙眼科	114	242	Simaroubaceae 苦木科	223
117	Buxaceae 黄杨科	116	243	Meliaceae 楝科	225
123	Altingiaceae 蕈树科	118	247	Malvaceae 锦葵科	227

页码	科名	页码
249	Thymelaeaceae 瑞香科	230
268	Capparaceae 山柑科	232
270	Brassicaceae 十字花科	233
276	Santalaceae 檀香科	235
282	Plumbaginaceae 白花丹科	238
283	Polygonaceae 蓼科	240
284	Droseraceae 茅膏菜科	242
295	Caryophyllaceae 石竹科	243
297	Amaranthaceae 苋科	246
304	Aizoaceae 番杏科	249
305	Phytolaccaceae 商陆科	251
308	Nyctaginaceae 紫茉莉科	252
309	Molluginaceae 粟米草科	254
312	Basellaceae 落葵科	255
314	Talinaceae 土人参科	256
315	Portulacaceae 马齿苋科	257
317	Cactaceae 仙人掌科	258
318	Nyssaceae 蓝果树科	259
320	Hydrangeaceae 绣球科	260
324	Cornaceae 山茱萸科	263
325	Balsaminaceae 凤仙花科	265
332	Pentaphylacaceae 五列木科	266
334	Ebenaceae 柿科	268
335	Primulaceae 报春花科	270
336	Theaceae 山茶科	274
337	Symplocaceae 山矾科	276
339	Styracaceae 安息香科	277
342	Actinidiaceae 猕猴桃科	279
345	Ericaceae 杜鹃花科	280
351	Garryaceae 丝缨花科	282
352	Rubiaceae 茜草科	283
353	Gentianaceae 龙胆科	287
354	Loganiaceae 马钱科	289
356	Apocynaceae 夹竹桃科	291
357	Boraginaceae 紫草科	294
359	Convolvulaceae 旋花科	297
360	Solanaceae 茄科	300
366	Oleaceae 木犀科	302
369	Gesneriaceae 苦苣苔科	304
370	Plantaginaceae 车前科	306
371	Scrophulariaceae 玄参科	308
373	Linderniaceae 母草科	310
376	Pedaliaceae 芝麻科	312
377	Acanthaceae 爵床科	313
379	Lentibulariaceae 狸藻科	315
382	Verbenaceae 马鞭草科	316
383	Lamiaceae 唇形科	318
384	Mazaceae 通泉草科	323
386	Paulowniaceae 泡桐科	325
387	Orobanchaceae 列当科	326
392	Aquifoliaceae 冬青科	329
394	Campanulaceae 桔梗科	331
401	Goodeniaceae 草海桐科	333
403	Asteraceae 菊科	334
408	Adoxaceae 五福花科	350
409	Caprifoliaceae 忍冬科	352
413	Pittosporaceae 海桐科	354
414	Araliaceae 五加科	355
416	Apiaceae 伞形科	357
	Index to Scientific Names 拉丁学名索引	361
	Index to Chinese Names 中文名称索引	377

007 Schisandraceae 五味子科

Leaves simple, alternate / 单叶互生
Leaves gland-dotted, aromatic / 叶芳香，具腺点
Carpels numerous, free / 心皮多数，分离
Perianth spirally arranged / 花被片螺旋状排列

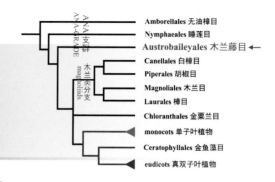

Woody climbers or **small trees** to **shrubs**. **Leaves** usually gland-dotted, aromatic (peppery in *Illicium*), simple, alternate (spiral); margins usually entire; petioles present; stipules absent. **Inflorescence** usually axillary solitary flowers or a few aggregated; sometimes cauliflorous. **Flowers** bisexual (*Illicium*) or unisexual (plants monoecious or dioecious: *Kadsura*, *Schisandra*), actinomorphic; often bracteolate. **Perianth** spirally arranged; outer whorl sepaloid, bract-like; inner whorl petaloid, white, yellow to red. **Stamen** filaments free (*Illicium*) or fused basally (e.g. *Schisandra*) or filaments ±fused (most *Kadsura*); anthers basifixed; often 3–22 staminodes. **Ovary** superior; carpels free; ovule(s) 1 (*Illicium*), 2–3 (*Schisandra*), 2–5(–11) (*Kadsura*); placentation marginal to basal. **Fruit** an aggregation of free carpels which are fleshy (red to yellow when mature) in *Kadsura* and *Schisandra* or star-shaped aggregated follicles (*Illicium*).

攀缘木质藤本或**小乔木**至**灌木**。叶常具腺点，具芳香味（八角属有辛辣味），单叶，互生（轮生）；通常全缘；具叶柄；托叶缺。花序常为单花，腋生，或少量簇生；有时具茎生花。花两性（八角属）或单性（雌雄同株或异株：南五味子属，五味子属），辐射对称；常具小苞片。花被片螺旋状排列；外轮萼片状，似苞片；内轮花瓣状，白色、黄色或红色。雄蕊花丝分离（八角属）或基部愈合（五味子属）或花丝多少愈合（大部分南五味子属）；花药基着；常具3~22枚退化雄蕊。子房上位，心皮分离；胚珠1枚（八角属），2~3枚（五味子属），2~5（~11）枚（南五味子属）；边缘胎座至基底胎座。果实由离生心皮构成的聚合果，南五味子属和五味子属为聚合浆果（成熟时红色至黄色），八角属为星状聚合蓇葖果。

World 3/ca. 70, Southeast Asia and Southeast North America.
China 3/54.
This area 2/3.

全世界共3属约70种，分布于亚洲东南部和北美东南部。
中国产3属54种。
本地区有2属3种。

1. Aggregated berry-like fruits are close together and often globose or ellipsoid/ 聚合浆果密集成球状或椭圆体状 .... ***1. Kadsura*/ 南五味子属**
1. Aggregated berry-like fruits widely separated, infructescence spike/ 聚合浆果分散，果序呈穗状 .... ***2. Schisandra*/ 五味子属**

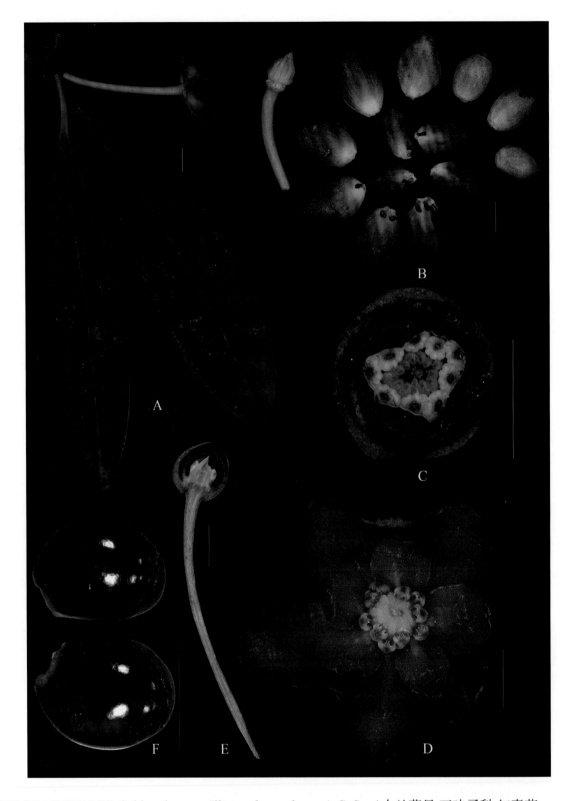

AUSTROBAILEYALES. Schisandraceae. *Illicium lanceolatum* A.C. Sm./ 木兰藤目 五味子科 红毒茴

A. Flowering branch/ 花枝
B. Perianth spirally arranged/ 花被片螺旋状排列
C. Cross section of flower bud/ 花蕾横切面
D. Frontal view of flower/ 花冠正面观
E. Vertical section of flower bud/ 花芽纵切面
F. Seeds/ 种子
(Scale/ 标尺：5 mm，The short vertical line in the figure represent for scale/ 图中短竖线为标尺)

1. *Kadsura longipedunculata* Finet & Gagnep./ 南五味子（1a. Infructescence/ 果序；1b. Male flower/ 雄花；1c. Female flower/ 雌花）；
2. *Schisandra sphenanthera* Rehder & E.H. Wilson/ 华中五味子

010　Saururaceae 三白草科

Stem jointed, often ridged / 茎节明显，常具脊
Inflorescence dense raceme or spike / 花序密集总状或穗状
Sometimes involucral petaloid bracts / 总苞苞片花瓣状
Perianth absent / 花被片缺失

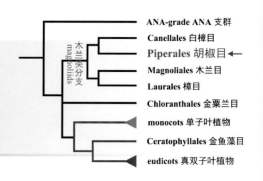

Perennial, rhizomatous or stoloniferous **herbs**; stems jointed, often ridged. **Leaves** aromatic, simple, alternate, leaf bases cordate; venation usually palmate; margin entire; petioles present; stipules fused to petiole. **Inflorescence** dense racemes or spikes; often involucral petaloid bracts, creating a false flower (e.g. *Anemopsis*, *Houttuynia*). **Flowers** small, bisexual, actinomorphic. **Perianth** absent. **Stamens** 3 (*Houttuynia*) or 6(–8), free (*Saururus*) or attached to the ovary base; anthers basifixed. **Ovary** superior to inferior; carpels fused or basally fused (*Saururus*); 1-loculed or many (*Saururus*); ovules 2 to many; placentation parietal. **Fruit** an indehiscent schizocarp (*Saururus*) or apically dehiscent capsule.

多年生**草本**，具根状茎或匍匐茎；茎节明显，常具脊。**叶**具芳香味，单叶互生，叶基心形，常为掌状脉；全缘；具叶柄；托叶与叶柄愈合。**花序**为密集的总状或穗状；总苞苞片常花瓣状，呈假花状（蕺菜属）。**花**小，两性，辐射对称。**花被片**缺。**雄蕊**3枚（蕺菜属）或6（~8）枚，离生（三白草属）或贴生于子房基部；花药基着。**子房**上位至下位；心皮合生或仅基部合生（三白草属），1室或多室（三白草属）；胚珠2至多枚；侧膜胎座。**果实**为不开裂的分果（三白草属）或顶端开裂的蒴果。

World 4/6, North America, East and South Asia.
China 3/4.
This area 2/2.

全世界共4属6种，分布于北美，亚洲的东部和南部。
中国产3属4种。
本地区有2属2种。

1. Inflorescence a dense spike, with 4 rarely 6 or 8, involucral petaloid bracts at base/ 花序呈密集穗状，基部具4片，极少为6或8片，花瓣状总苞片 .... **1. *Houttuynia*/ 蕺菜属**
1. Inflorescence a raceme, without involucral petaloid bracts/ 花序呈总状，基部无花瓣状总苞片 .... **2. *Saururus*/ 三白草属**

1. *Houttuynia cordata* Thunb./ 鱼腥草（蕺菜）；
2. *Saururus chinensis* (Lour.) Baill./ 三白草

011   Piperaceae 胡椒科

Branches with swollen nodes / 茎节膨大
Leaves peppery / 叶具辛辣味
Inflorescence dense spikes / 花序密集穗状
Flowers minute, perianth absent / 花微小，花被缺

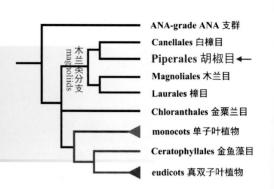

Herbs, **shrubs**, **small trees** or **woody climbers**, sometimes epiphytic (some *Peperomia*) or succulent; branches often with swollen nodes. **Leaves** aromatic (peppery), often gland-dotted, simple, alternate (spiral, rarely 2-ranked) or opposite (some *Peperomia*); venation palmate or pinnate; margin entire; petioles present; stipule-like sheath (some *Piper*) or absent. **Inflorescence** dense spikes on a fleshy axis or racemes; bracts peltate, triangular or umbrella-shaped. **Flowers** minute, unisexual (plants monoecious or dioecious) or bisexual, actinomorphic to zygomorphic. **Perianth** absent. **Stamens** 2 (*Peperomia*) or (2–)3–6(–7); sometimes staminodes. **Ovary** superior; carpel 1 (*Peperomia*), fused and 3 (*Piper*); 1-loculed; ovule 1; placentation basal. **Fruit** a small drupe or berry.

草本、灌木、小乔木或为**木质攀缘藤本**，有时为附生植物（部分草胡椒属）或多肉植物；茎节常膨大。**叶**具辛辣味，常具腺点，单叶，互生（螺旋状，极少呈 2 列）或对生（部分草胡椒属）；掌状脉或羽状脉；全缘；具叶柄；有托叶状鞘（部分胡椒属）或缺。**花序**为密集穗状，着生于肉质花序轴上或为总状；苞片盾形、三角形或伞形。**花**微小，单性（雌雄同株或异株）或两性，辐射对称至两侧对称。**花被片缺**。**雄蕊** 2 枚（草胡椒属）或（2~）3–6（~7）枚；有时具退化雄蕊。**子房**上位；心皮 1 枚（草胡椒属），心皮 3 枚合生（胡椒属）；子房 1 室；胚珠 1 枚；基底胎座。**果实**为小型核果或浆果。

World 5/3, 600+, tropics and warm subtropics, especially tropical America.
China 3/68.
This area 2/3.

全世界共 5 属 3 600 余种，分布于热带和亚热带温暖地区，尤其是热带美洲。
中国产 3 属 68 种。
本地区有 2 属 3 种。

1. Leaves often opposite or whorled in Chinese species; stigma 1, rarely 2-lobed/ 中国产种类叶常为对生或轮生；柱头 1，极少 2 裂 .... **1. *Peperomia*/ 草胡椒属**
1. Leaves alternate; stigmas 3–5, rarely 2/ 叶互生；柱头 3~5，极少为 2 .... **2. *Piper*/ 胡椒属**

PIPERALES. Piperaceae. *Piper hancei* Maxim./ 胡椒目 胡椒科 山蒟

A. Fruiting branch/ 果枝

B. Vertical section of infructescence/ 果序纵切面

C. Infructescence/ 果序

D. Fruits and seeds/ 果实与种子

(Scale/ 标尺：5 mm)

1. *Peperomia blanda* (Jacq.) Kunth/ 石蝉草；
2. *Piper* sp./ 胡椒属

012　Aristolochiaceae 马兜铃科

Leaves simple, alternate, venation often palmate / 单叶互生，常掌状脉
Leaves often aromatic / 叶常具芳香
Perianth tubes usually inflated at base / 花被管基部囊状膨大
Gynostemium / 具合蕊柱

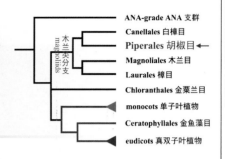

Perennial **herbs**, usually climbing, less often **subshrubs**, **shrubs** or **woody climbers**, **rarely root-holoparasitic herbs** (Hydnoroideae). **Leaves** often aromatic, sometimes with oil cells, simple, alternate; venation often palmate; margin usually entire, rarely 3–5-lobed; petioles present; stipules absent. **Inflorescence** solitary flower, a few aggregated, cymes, racemes or spikes; sometimes cauliflorous; bracteate (*Asarum*, *Saruma*). **Flowers** often bisexual, actinomorphic or zygomorphic (*Aristolochia*), with an inflated utricle, sometimes foetid (*Aristolochia*). **Perianth** usually 1-whorled, fused, petaloid, valvate or free with imbricate sepaloid (*Lactoris*) or 2-whorled (*Saruma*) with sepaloid and petaloid whorls. **Stamen** filaments adnate to the ovary (*Asarum*), filaments and anthers fused to the style column (*Aristolochia*); anthers extrorse. **Ovary** inferior (*Aristolochia*), part-inferior (*Saruma*) or inferior to part-inferior (*Asarum*) or superior (*Lactoris*); carpels free, fused or only basally fused (*Saruma*); locules 3–6(–9); ovules 4 to many per locule; placentation marginal, parietal or axile. **Fruit** an aggregation of follicles, a fleshy or dry capsule.

多年生**草本**，通常具攀缘性，少为**亚灌木、灌木或木质攀缘藤本**，极少为根全寄生草本（鞭寄生亚科）。**叶**常具芳香味，有时具油细胞，单叶，互生；常为掌状脉；常全缘，极少 3~5 裂；具叶柄；托叶缺或具与叶柄基部合生的杯状托叶（囊粉花属）。**花序**多样，单花，少数簇生，聚伞花序，总状花序或穗状花序；有时为茎生花；具苞片（细辛属、马蹄香属）。**花**常两性，辐射对称或两侧对称（马兜铃属），具囊，有时具恶臭（马兜铃属）。**花被片**常 1 轮，合生，花瓣状，或具萼片状和花瓣状 2 轮花被片（马蹄香属）。**雄蕊**花丝与子房合生（细辛属），或形成雌雄蕊合生的合蕊柱（马兜铃属）；花药外向。**子房**下位（马兜铃属），半下位（马蹄香属），下位至半下位（细辛属）或上位（囊粉花属）；心皮离生，合生或仅基部合生（马蹄香属）；子房 3~6（~9）室；每室胚珠 4 枚至多数；边缘胎座、侧膜胎座或中轴胎座。**果实**为聚合蓇葖果，肉质或干燥蒴果。

World 10/ca. 800 mainly in tropical to temperate regions.
China 4/86.
This area 2/2.

全世界共 10 属约 800 种，主要分布于热带至温带地区。
中国产 4 属 86 种。
本地区有 2 属 2 种。

1. Perianth zygomorphic; stamens 6; stems woody or herbaceous, usually climbing or twining / 花两侧对称；雄蕊 6 枚；茎木质或草质，常攀缘或缠绕 ································· **1. *Aristolochia* / 马兜铃属**

1. Perianth actinomorphic; stamens usually 12; stems herbaceous, rhizomatous / 花辐射对称；雄蕊常 12 枚；茎草质，具根状茎 .... **2. *Asarum* / 细辛属**

PIPERALES. Aristolochiaceae. *Asarum* L./ 胡椒目 马兜铃科 细辛属

A. Cross section of perianth tube/ 花被筒横切面
B. Lobes of perianth tube/ 花被筒裂片
C. Lateral view of flower/ 花侧面观
D. Stamen filaments fused to the ovary/ 雄蕊与子房合生
E. Cross section of ovary/ 子房横切面
F. Vertical section of ovary / 子房纵切面
G. Seeds/ 种子
H. Frontal view of flower, perianth tube partly removed/ 花正面观，除去部分花被筒
I. Lateral view of gynostemium/ 合蕊柱侧面观
(Scale/ 标尺：5 mm)

1. *Aristolochia dabieshanense* C.Y. Cheng & W. Yu/ 大别山马兜铃（1a. Flower/ 花；1b. Leaves/ 叶片；not distributed here/ 非本区域分布）；

2. *Asarum ichangense* C.Y. Cheng & C.S. Yang/ 小叶马蹄香（2a. Flower/ 花；2b. Leaves/ 叶片）

014 Magnoliaceae 木兰科

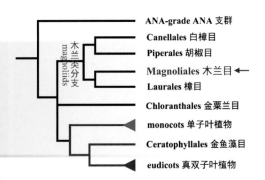

Stipules annular scar / 有托叶环痕
Tepals fleshy, 3-merous / 花被片肉质，3 基数
Pistils and stamens many, distinct / 雌雄蕊多数，离生
Aggregate of follicle or samaretum / 聚合蓇葖果或聚合翅果

Trees or **shrubs**. **Leaves** aromatic, often gland-dotted, often deciduous, simple, alternate (spiral); margin entire or lobed (4–10-lobed, *Liriodendron*); petioles present; stipules hooded, enclosing leaf buds, caducous but with a remaining annular scar on twig or petiole. **Inflorescence** usually terminal, solitary flower; bracts spathe-like, deciduous leaving a scar. **Flowers** bisexual or rarely unisexual, actinomorphic. **Perianth** spirally arranged; outer whorl occasionally sepaloid; inner whorl petaloid, white to pinkish. **Stamens** spirally arranged, anthers and filaments poorly differentiated; anthers introrse or extrorse (*Liriodendron*). **Ovary** superior; carpels free to basally fused, on a stalk, spirally arranged on an axis; ovules 2 to few per carpel; placentation marginal. **Fruit** a samaretum (*Liriodendron*) or an aggregate of follicles. **Seeds** with red to orange, fleshy coat (except *Liriodendron*).

乔木或**灌木**。**叶**具芳香味，常有腺点，常落叶性，单叶，互生（螺旋状）；全缘或分裂（4~10 裂，鹅掌楸属）；具叶柄；托叶盔状包围叶芽，早落，在小枝或叶柄上留下托叶环痕。**花序**通常顶生，单花；苞片佛焰苞状，落叶后具痕。**花**两性或极少为单性，辐射对称。**花被片**螺旋状排列；外轮偶为萼片状；内轮为花瓣状，白色至粉色。**雄蕊**螺旋状排列，花药与花丝分化不明显；花药内向或外向（鹅掌楸属）。**子房**上位；具子房柄，心皮离生至基部合生，在轴上螺旋状排列；每心皮含胚珠 2 枚至少数；边缘胎座。**果**为聚合翅果（鹅掌楸属）或聚合蓇葖果。**种子**具红色至橙色肉质种皮（除鹅掌楸属外）。

World 2 or 19/ca. 300, East and Southeast Asia, Central America, North America, Northern South America.
China 13/112.
This area 3/8.

全世界共 2 或 19 属约 300 种，分布于亚洲东部和东南部，中美洲，北美洲，南美洲北部。
中国产 13 属 112 种。
本地区有 3 属 8 种。

1. Leaf blade 4–10-lobed; samaretum/ 叶 4~10 裂；聚合翅果 .... **1. *Liriodendron*/ 鹅掌楸属**
1. Leaf blade unlobed; mature carpels not samaroid/ 叶不分裂；成熟心皮非翅果状
   2. Flowers terminal on axillary brachyblasts/ 花生于腋生短枝顶端 .... **2. *Michelia*/ 含笑属**
   2. Flowers terminal/ 花顶生 .... **3. *Yulania*/ 玉兰属**

MAGNOLIALES. Magnoliaceae. *Yulania liliiflora* (Desr.) D.L. Fu/ 木兰目 木兰科 紫玉兰

A. Flower segregated/ 花离析

B. Stamens and cross section/ 雄蕊及其横切面

C. Free carpels spirally arranged on an axis/ 离生心皮在轴上螺旋状排列

D. Flowering branch, both perianth and stamens removed/ 花枝，去除花被片和雄蕊

E. Flowering branch, perianth removed/ 花枝，去除花被片

F. Flowering branch, perianth partly removed/ 花枝，去除部分花被片

(Scale/ 标尺：A 5 cm；　B–C 5 mm；　D–F 1 cm)

1-1. *Liriodendron chinense* (Hemsl.) Sarg./ 鹅掌楸（Leaves lobed/ 叶分裂）；

1-2. *L.* × *sinoamericanum* P.C. Yieh ex C.B. Shang & Zhang R. Wang/ 杂种鹅掌楸（Samaretum/ 聚合翅果）；

2. *Michelia figo* (Lour.) Spreng./ 含笑花；

3. *Yulania biondii* (Pamp.) D.L. Fu/ 望春玉兰（Widely cultivated/ 广泛栽培）

019 Calycanthaceae 蜡梅科

Leaves simple, opposite / 单叶，对生
Flower bisexual, actinomorphic / 花两性，辐射对称
Cup-shaped or urceolate receptacle / 具杯状或坛状花托
Enlarged hypanthium bearing achenes inside / 增大的托杯内有瘦果

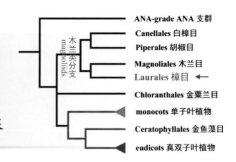

Shrubs or small trees. **Branchlets** dichotomous, quadrangular to subterete; with oil cells; buds covered with scales or infra-petiole buds. **Leaves** simple, opposite, margin entire or subentire, venation pinnate, petiolate; stipules absent. **Inflorescence** usually solitary, axillary or terminal on branchlets. **Flowers** bisexual, actinomorphic, usually fragrant, light-colored; pedicel short. **Perianth** parts distinct, spirally arranged on a cup-shaped or urceolate receptacle; perianth numerous, poorly differentiated. **Stamens** ribbon-like, filaments short to absent; anthers basifixed, extrorse, dehiscing by longitudinal slits; often staminodes 10–25. **Ovary** superior; carpels free, 5 to many or 1–3(–5); 1-loculed; ovules 1(–2); placentation marginal. **Fruit** an **aggregation** of achenes in enlarged hypanthium (= pseudocarp). **Seed** 1.

灌木或小乔木。**小枝**二歧分枝，四方形至近圆柱形；有油细胞；鳞芽或叶柄下芽。**叶**为单叶，对生，全缘或近全缘；羽状脉；有叶柄；托叶缺。**花序**为单花，腋生或生于小枝顶端。**花**两性，辐射对称，通常具芳香，色淡；花梗短。**花被片**离生，螺旋状着生于杯状或坛状花托上；多数分化不明显。**雄蕊**带状，花丝短或缺；花药基着，外向，纵裂；常具10~25枚退化雄蕊。**子房**上位；心皮分离，有5枚至多枚或1~3(~5)枚；1室；每室胚珠1(~2)枚；边缘胎座。**聚合瘦果**着生于增大的被丝托内（即为假果）。**种子**1粒。

World 3/10, North America, East Asia and Australia.
China 2/7.
This area 1/1.

全世界共3属10种，分布于北美、东亚和澳大利亚。
中国产2属7种。
本地区有1属1种。

***Chimonanthus* Lindl. 蜡梅属**

Chimonanthus praecox (L.) Link / 蜡梅

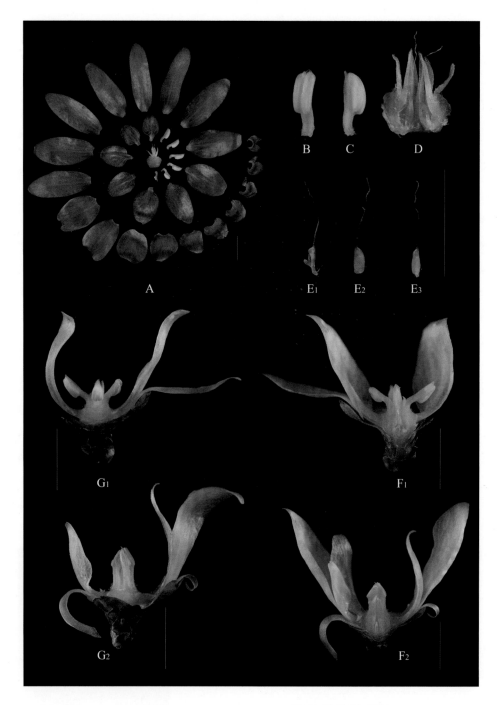

LAURALES. Calycanthaceae. *Chimonanthus praecox* (L.) Link/ 樟目 蜡梅科 蜡梅

A. Flower segregated/ 花离析

B. Frontal view of stamen/ 雄蕊正面观

C. Lateral view of stamen/ 雄蕊侧面观

D. Vertical section of urceolate receptacle/ 坛状花托纵切面

E_1. Vertical section of ovary/ 子房纵剖面

E_2. Lateral view of ovary/ 子房侧面观

E_3. Ventral view of ovary/ 子房腹面观

F_1. Vertical section of corolla, female phase/ 花冠纵切面，雌花期

F_2. Vertical section of corolla, male phase/ 花冠纵切面，雄花期

G_1. Vertical profile of corolla, female phase/ 花冠纵剖面，雌花期

G_2. Vertical profile of corolla, male phase/ 花冠纵剖面，雄花期

(Scale/ 标尺：5mm)

025　Lauraceae 樟科

Woody plant, cut parts aromatic / 木本植物，切开具香味
Leaves simple, alternate, entire / 单叶，互生，全缘
Anthers opening by flaps / 花药瓣裂
Ovary superior with a single ovule / 上位子房具单胚珠

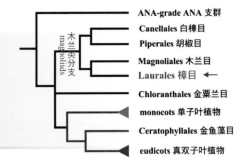

Trees or shrubs (*Cassytha*, a twining parasitic vine); cut parts aromatic. **Leaves** simple; usually alternate, or subopposite, or whorled; entire; triplinerved or trinerved; stipules absent. **Inflorescence** axillary, or pseudoterminal; usually branched, panicles, spikes, racemes; inflorescence often enveloped in bracts. **Flowers** bisexual and/or unisexual, very small; 3-merous, or 2-merous; 2 equal whorls of 3 tepals each; 3 whorls of fertile stamen, 1 inner whorl sterile; anthers 2-or 4-loculed opening by flaps; ovary superior with a single ovule. **Fruit** a drupe or berry with receptacle or pedicel enlarged surrounding the base, or entirely enclosing the fruit. **Seed** 1.

乔木或灌木（唯无根藤属为寄生性缠绕藤本）；切开具芳香气味。**叶**为单叶；常互生，或近对生、轮生；全缘；离基三出脉或三出脉托叶缺。**花序**腋生，或假顶生；常分枝，圆锥状，穗状，总状；花序常为苞片包被。**花**两性，或兼具单性，很小；3 基数，或 2 基数；花被片 2 轮等大，每轮 3 片；具 3 轮可育雄蕊，内轮不育；花药 2 室或 4 室，瓣裂；子房上位，具 1 枚胚珠。**果实**为核果或浆果，基部有膨大的花托或花梗包围，或果实全部陷于其中。**种子** 1 粒。

World 62/2, 000–2, 500, tropical and subtropical regions, especially Southeast Asia and tropical America.
China 25/445.
This area 7/12.

全世界共 62 属 2 000~2 500 种，分布于热带至亚热带，特别是东南亚和美洲热带地区。
中国产 25 属 445 属。
本地区有 7 属 12 种。

1. Twining parasitic vines, leafless/ 缠绕寄生藤本，无叶 .... **1. *Cassytha*/ 无根藤属**
1. Leafy trees or shrubs/ 具叶的乔木或灌木
    2. Inflorescence pseudoumbels or clusters/ 花序呈假伞形或簇生状
        3. Flowers 2-merous/ 花 2 基数 .... **2. *Neolitsea*/ 新木姜子属**
        3. Flowers 3-merous/ 花 3 基数
            4. Anthers 2-loculed/ 花药 2 室 .... **3. *Lindera*/ 山胡椒属**
            4. Anthers 4-loculed/ 花药 4 室 .... **4. *Litsea*/ 木姜子属**
    2. Inflorescence panicles, scattered/ 花序圆锥状，疏散
        5. Fruit base with a perianth cup/ 果下具果托 .... **5. *Cinnamomum*/ 樟属**

5. Fruit base without a perianth cup/ 果下无果托

　6. Persistent perianth lobes reflexed or patent, not tightly clasped at fruit base/ 宿存花被片反卷或展开，不紧包于果基部 ....**6. *Machilus*/ 润楠属**

　6. Persistent perianth erect or patent, tightly clasped at fruit base/ 宿存花被片直立或展开，紧包于果基部 ....**7. *Phoebe*/ 楠属**

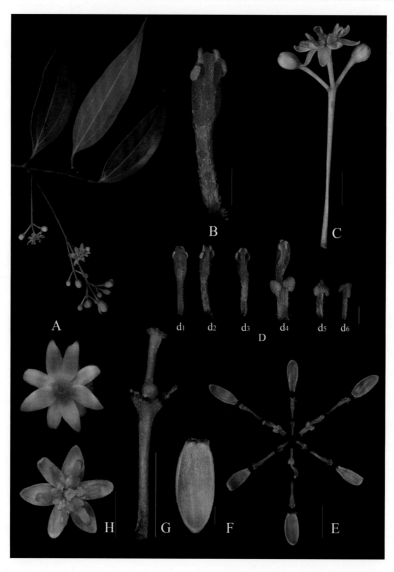

LAURALES. Lauraceae. *Cinnamomum burmannii* (Nees & T. Nees) Blume/ 樟目 樟科 阴香 (not distributed here/ 非本区域分布)

A. Flowering branch/ 花枝

B. Stamen/ 雄蕊

C. Cymes/ 聚伞花序

D_1. Dorsal view of stamen/ 雄蕊背面观

D_2. First whorl stamen/第一轮雄蕊

D_3. Second whorl stamen/第二轮雄蕊

D_4. Third whorl stamen/第三轮雄蕊

D_5. D_6. Fourth whorl stamen (sterile), ventral view (D_5), dorsal view(D_6)/ 第四轮雄蕊（不育），腹面观（D_5），背面观（D_6）

E. Flower segregated/ 花离析

F. Ventral view of perianth/ 花被片腹面观

G. Pistil/ 雌蕊

H. Dorsal view of flower (above); ventral view of flower (below)/ 花背面观（上方）；花腹面观（下方）

(Scale/ 标尺：A none/ 无；B,D,F 1 mm; C,E,G,H 5 mm)

1. *Cassytha filiformis* L./ 无根藤；
2. *Neolitsea aurata* (Hayata) Koidz var. *chekiangensis* (Nakai) Y.C. Yang & P.H. Huang/ 浙江新木姜子；
3. *Lindera erythrocarpa* Makino/ 红果山胡椒；
4. *Litsea cubeba* (Lour.) Pers./ 山鸡椒

5. *Cinnamomum japonicum* 'Chenii'/ 普陀樟；
6. *Machilus thunbergii* Siebold & Zucc./ 红楠；
7. *Phoebe sheareri* (Hemsl.) Gamble/ 紫楠

026 Chloranthaceae 金粟兰科

Leaves opposite, margin serrate / 叶对生，叶缘有锯齿
Node swollen and ocrea present / 茎节膨大，具托叶鞘
Flowers reduced / 花退化

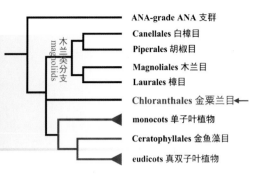

Trees, shrubs or **herbs** to **subshrubs**; stems with jointed nodes. **Leaves** aromatic, simple, opposite; margin serrate or crenate; petioles ± fused at base; stipules interpetiolar, usually small. **Inflorescence** capitula, spikes, panicles; sometimes bracteate. **Flowers** < 6 mm in diam., bisexual or unisexual, actinomorphic. **Perianth** reduced to absent. **Stamens** fused when 3(–5), filaments and anthers are poorly differentiated. **Ovary** inferior to superior; carpel 1; 1-loculed; ovule 1; placentation apical. **Fruit** a drupe.

乔木、灌木或**草本至亚灌木**；茎节膨大。**叶**具芳味香，单叶，对生；叶缘有锯齿或圆齿；叶柄在基部多少合生；托叶生于叶柄间，小。**花序**头状，穗状，圆锥状；有时具苞片。花直径小于 6 mm，两性或单性，辐射对称。**花被片**退化至缺。**雄蕊** 3（~5）枚时合生，花丝与花药分化不明显。**子房**下位至上位；心皮 1 枚；1 室；胚珠 1 枚；顶生胎座。**果实**为核果。

World 4/ca. 75, tropics and subtropics.
China 3/16.
This area 1/1.

全世界共 4 属约 75 种，分布于热带与亚热带。
中国产 3 属 16 种。
本地区有 1 属 1 种。

Chloranthus Sw. 金粟兰属

Chloranthus fortunei (A. Gray) Solms/ 水晶花（丝穗金粟兰）（a. Inflorescence/ 花序；b. Whole plants/ 植株）

027 Acoraceae 菖蒲科

Wetland herbs / 湿生草本
Leaves aromatic / 叶具芳香
Leaves narrow and overlapped at base / 叶窄，基部套叠
Leaf-like spathe long / 叶状佛焰苞长

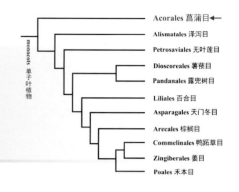

Evergreen, wetland **herbs**, glabrous. **Rhizome** creeping, much branched, lacunose, with aromatic oil cells. **Leaves** aromatic, basal, distichous, narrow ensiform, base sheathing, equitant, petiole absent and with a unifacial blade, venation parallel. **Inflorescence** dense spadixes, **axillary**, sessile, finger-like to slender and tail-like; long leaf-like spathe. **Flowers** small, bisexual, **perianths** 6, persistent, thin; **stamens** 6. **Ovary** 2- or 3-loculed. **Fruit** a red berry.

常绿**草本**，湿生，光滑。**根茎**匍匐，多分枝，多孔，细胞含芳香油。**叶**具芳香味，基生，2列，狭窄的剑形，基部鞘状套叠，无柄，等面叶，叶脉平行。肉穗**花序**，**腋生**，无总梗，指状至纤细鼠尾状；具较长的叶状佛焰苞。**花**小，两性，**花被片**6枚，宿存，薄；**雄蕊**6枚。**子房**2或3室。**果实**为红色浆果。

World 1/2–4, temperate Asia, subtropics and tropics.
China 1/2.
This area 1/2.

全世界共1属2~4种，分布于亚洲温带，亚热带及热带地区。
中国产1属2种。
本地区有1属2种。

***Acorus* L. 菖蒲属**

1. *Acorus calamus* L./ 菖蒲； 2. *A. gramineus* Sol. ex Aiton/ 金钱蒲（2a. Flowering plants/ 花期植株；2b. Fruiting plants/ 果期植株）

028　Araceae 天南星科

- Herbs / 草本
- Midrib usually pinnately branched / 中脉常羽状分枝
- Compound midrib / 复合中脉
- Spadix subtended by a spathe / 佛焰苞包被肉穗花序

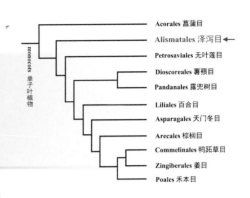

Herbs, climbing, terrestrial or epiphytic, rarely aquatic. **Leaves** usually divided into a blade, petiole and petiole sheath; often leathery; entire to deeply lobed or compound, often variegated, usually petiolate, very variable in shape; midrib compound, primary venation usually pinnately branched but sometimes pedate, arcuate or parallel, secondary venation reticulate or parallel-pinnate. **Inflorescence** consisting of a spadix subtended by a spathe. **Flowers** minute, bisexual or unisexual, actinomorphic, unisexual then female below. **Stamen** filaments free or fused; anthers basifixed, extrorse. **Ovary** superior; carpels fused or 1; style short to absent. **Fruit** a berry. **Seeds** very variable.

草本，攀缘、地生或附生，稀水生。叶通常分为叶片、叶柄和叶柄鞘；常革质；全缘至深裂或复叶状（假复叶），常有斑点，常具小叶柄，叶形多样；中脉（中肋）复合，主脉通常羽状分枝，但有时为鸟足状、弧形或平行脉，次级叶脉网状或羽状平行。花序为佛焰苞包被的肉穗花序。花小，两性或单性，辐射对称，单性时雌雄同序者花序上部为雄花，下部为雌花。雄蕊花丝离生或合生；花药基着，外向。子房上位；心皮多枚合生或仅1枚；花柱短或缺。果实为浆果。种子多样。

World 141/3,500–3,700, worldwide, especially tropical and subtropical regions.

China 30/190

This area 5/12.

全世界 141 属 3 500~3 700 种，全球广布，主要分布于热带和亚热带。

中国产 30 属 190 种。

本地区有 5 属 12 种。

1. Plant a free-floating or submersed aquatic/ 浮水或沉水植物
   2. Fronds rootless, without veins/ 叶状体无根，无叶脉 ....**1. *Wolffia*/ 无根萍属**
   2. Fronds with 1–21 roots, with 1–21 veins/ 叶状体 1~21 条根，具 1~21 条叶脉
      3. Fronds with 1 root and 1–5(–7) veins/ 叶状体 1 条根及 1~5（~7）条叶脉 ....**2. *Lemna*/ 浮萍属**
      3. Fronds with 2–21 roots and (3–)5–21 veins/ 叶状体有 2~21 条根及 (3~)5~21 条叶脉....**3. *Spirodela*/ 紫萍属**
1. Plant not an aquatic/ 非水生植物
   4. Female zone of spadix adnate to spathe/ 肉穗花序下部雌花序与佛焰苞贴生 ....**4. *Pinellia*/ 半夏属**
   4. Female zone of spadix free from spathe/ 肉穗花序下部雌花序与佛焰苞分离 ....**5. *Arum*/ 天南星属**

ALISMATALES. Araceae. *Pinellia ternata* (Thunb.) Ten. ex Breitenb./ 泽泻目 天南星科 半夏

A. Whole plant/ 植株

B. Vertical section of tuber/ 块茎纵切面

C_1. Pistil/ 雌蕊

C_2. Vertical profile of female flower/ 雌花纵剖面

D. Cross section of male spadix/ 雄花序横切面

E. Spathe throat (transverse septum)/ 佛焰苞喉部（横隔膜）

F. Leaves/ 叶

G. Tuber/ 块茎

H. Vertical profile of spathe/ 佛焰花序纵剖面

I. Vertical section of spathe, tube/ 佛焰花序纵切面, 管部

J. Vertical section of spathe/ 佛焰花序纵切面

(Scale/ 标尺：A–B 1 cm; C_1–E 1 mm; F–J 1 cm)

1. *Wolffia globosa* (Roxb.) Hartog & Plas/ 无根萍；
2. *Lemna aequinoctialis* Welw./ 稀脉浮萍；
3. *Spirodela polyrhiza* (L.) Schleid./ 紫萍；
4. *Pinellia pedatisecta* Schott/ 虎掌；
5-1. *Arisaema erubescens* (Wall.) Schott (Infructescence)/ 一把伞南星（果序）；
5-2. *A. heterophyllum* Blume/ 天南星

030 Alismataceae 泽泻科

Wetland herbs / 湿生草本
Leaves basal with arcuate veins / 叶基生，基出弧形脉
Sepals 3 persistent, petals 3 deciduous / 宿存花萼 3 枚，早落花瓣 3 枚
Embryo horseshoe-shaped / 胚马蹄形

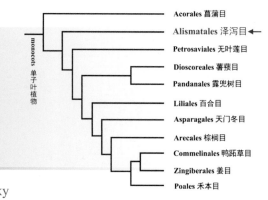

Perennial or rarely annual, **herbs**, marsh or aquatic, with milky sap or not, rhizomatous or cormous. **Leaves** basal, ±entire, veins somewhat arcuate and converging toward apex; shape of leaves variation; with sheathing petioles. **Inflorescence** often whorled at nodes of scape forming panicles, racemes, umbels, or solitary flower. **Flowers** bisexual or unisexual, actinomorphic; perianth free. **Sepals** 3, green, persistent. **Petals** 3, deciduous, ± crinkled, white or rarely yellow or pink, reduced to absent. **Stamen** filaments free; anthers usually basifixed, extrorse. **Ovary** superior; carpels free or sometimes basally fused, numerous, whorled on a flat receptacle or spirally arranged on a convex receptacle; placentation basal with 1(–2) ovules or sometimes many; style persistent. **Fruit** an achene or follicle. **Seeds** curved, with a horseshoe-shaped embryo; endosperm absent.

多年生或稀为一年生，**草本**，沼生或水生，具乳汁或无，具根状茎或球茎。**叶**基生，几全缘，叶脉多少弧形，在叶尖汇合；叶形多变；叶柄具鞘。**花序**轮生于花葶的节上，形成圆锥状、总状、伞形花序，或为单花。**花**两性或单性，辐射对称；花被片离生。**花萼** 3 枚，绿色，宿存。**花瓣** 3 枚，早落，多少皱褶，白色或稀为黄色或粉色，或退化至不存在。**雄蕊**花丝分离；花药通常基着，外向。**子房**上位；心皮离生或有时在基部合生，多数，轮生于扁平花托或螺旋状排列于凸起花托上；基底胎座具 1 (~2) 枚胚珠或有时多数；花柱宿存。**果实**为瘦果或蓇葖果。**种子**弯曲，胚马蹄形；无胚乳。

World 16/100, worldwide, especially in Northern Hemisphere tropical and temperate regions.
China 6/18.
This area 2/3.

全世界共 16 属 100 种，世界广布，北半球热带和温带地区分布较多。
中国产 6 属 18 种。
本地区有 2 属 3 种。

1. Flowers unisexual or polygamous, stamens 9 to many/ 花单性或一雄多雌杂性，雄蕊 9 至多数 .... **1. *Sagittaria*/ 慈姑属**
1. Flowers bisexual, stamens 6/ 花两性，雄蕊 6 枚 .... **2. *Alisma*/ 泽泻属**

1. *Sagittaria Lancifolia* 'Ruminoides'/ '红茎' 泽泻慈姑（Showing unisexual flowers, not distributed in this area/ 仅展示单性花，非本区域分布）；

2. *Alisma orentale* (Sam.) Juz./ 东方泽泻（Showing bisexual flowers, not distributed in this area/ 仅展示两性花，非本区域分布）

032 Hydrocharitaceae 水鳖科

Submerged herbs / 沉水草本
Leaves simple and linear / 单叶，线形
Fleshy capsule / 蒴果肉质

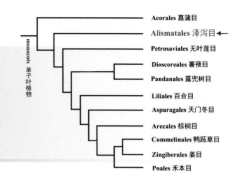

Herbs, submerged aquatic, in fresh or brackish water or marine. **Stems** elongated or stoloniferous. **Leaves** simple, linear, scale-like to leaf-like, aggregated at base or cauline whorled. **Inflorescence** solitary flower or few-aggregated; often 2 bracts fused, spathe-like. **Flowers** unisexual, or bisexual, actinomorphic or zygomorphic; floating on water surface, male flowers solitary or numerous, female flowers 1–2. **Perianth** 1–2 whorles, 3 per whorl, free; outer whorl sepaloid, inner whorl petaloid. **Stamens** up to six whorled. **Ovary** usually inferior; carpels fused or 1; 1-loculed; ovules 1 to many per locule; placentation marginal or basal. **Fruit** a fleshy capsule dehiscing by decay of pericarp. **Seeds** numerous and multi-shaped.

沉水**草本**，生于淡水或淡盐水或海水中。**茎**纤细或有匍匐茎。**叶**为单叶，线形，鳞片状至叶状，密集基生或轮叶茎生。**花序**为单花或少量簇生；2 枚苞片合生，佛焰苞状。**花**单性，或为两性，辐射对称或两侧对称；花开放时浮于水面，雄花单生或多数，雌花 1~2 朵。**花被片** 1~2 轮，每轮 3 枚，离生；外轮萼片状，内轮花瓣状。**雄蕊**可达 6 枚轮生。**子房**通常下位；心皮合生或仅 1 枚；1 室，每室胚珠 1 至多数；边缘胎座或基底胎座。**果实**为肉质蒴果，果皮腐烂开裂。**种子**多数，形状多样。

World 18/140, worldwide.
China 11/34. (*Egeria densa* Planchon introduced in China, excluded here)
This area 2/2.

全世界共 18 属 140 种，世界广布。
中国产 11 属 34 种（水蕴草为引入种，不计在内）。
本地区有 2 属 2 种。

1. Leaves cauline; stems elongated/ 叶茎生；茎细长 .... **1.** *Hydrilla*/ 黑藻属
1. Leaves all basal; stems short/ 叶基生；茎短缩 .... **2.** *Vallisneria*/ 苦草属

ALISMATALES. Hydrocharitaceae. *Vallisneria natans* (Lour.) H. Hara/ 泽泻目 水鳖科 苦草

A. Whole plant/ 植株

B. Male inflorescence and roots/ 雄花序与根系

C_1. Inside of male spathe/ 雄佛焰苞片内侧

C_2. Male inflorescence/ 雄花序

C_3. Vertical profile of male inflorescence/ 雄花序纵剖面

C_4. Male spathe/ 雄佛焰苞

D. Male flowers/ 雄花

E. Seeds/ 种子

(Scale/ 标尺：A–B 1 cm; C_1–E 1 mm)

1. *Hydrilla verticillata* (L. f.) Royle/ 黑藻；
2. *Vallisneria natans* (Lour.) H. Hara/ 苦草

038 Potamogetonaceae 眼子菜科

Perennial, aquatic herbs / 多年生水生草本
Leaves submerged or floating, often dimorphic / 叶沉水或浮水，多两型
Inflorescence dense spike / 密穗状花序
Ovary superior, carpels free / 子房上位，心皮离生

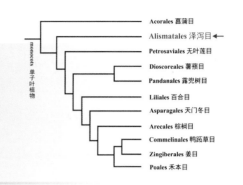

Perennial, aquatic **herbs**. **Leaves** submerged or floating, often dimorphic, simple, alternate, opposite or ±whorled, rarely scale-like, sheathing, sometimes a conspicuous basal ligule; margin entire to slightly toothed; petioles often present; sometimes stipule-like structures. **Inflorescence** dense spikes or 2-several flowers aggregated. **Flowers** bisexual or unisexual (plants monoecious or rarely dioecious), actinomorphic; sometimes spathe-like bracts. **Perianth** rounded, shortly clawed or absent. **Stamens** inserted on claws; anthers extrorse. **Ovary** superior; carpels free to partly fused; 1-loculed; ovule 1; placentation basal or apical. **Fruit** a drupe or berry.

多年生水生草本。**叶**沉水或浮水，多两型，单叶，互生、对生或多少轮生，稀鳞片状，具叶鞘，有时有明显叶舌；全缘或稍有锯齿；叶柄常存在；有时具托叶状结构。**花序**密穗状或2至数朵花集生。**花**两性或单性（雌雄同株或稀雌雄异株），辐射对称；苞片有时佛焰苞状。**花被片**圆形，稍具爪或无。**雄蕊**贴生于花被基部；花药外向。**子房**上位；心皮离生或部分合生；1室；胚珠1枚；基底胎座或顶生胎座。**果实**为核果或浆果。

World 4/102, worldwide, especially in temperate regions.
China 3/25.
This area 1/1.

全世界共4属102种，全球广布，尤其是温带地区。
中国产3属25种。
本地区有1属1种。

Potamogeton L. 眼子菜属

Potamogeton distinctus A. Benn./ 眼子菜

043　Natheciaceae 沼金花科

Leaves basal and tufted / 叶成簇基生
Inflorescence racemes or spikes, terminal / 总状或穗状花序，顶生
Perianth persistent in fruit / 花被果期宿存

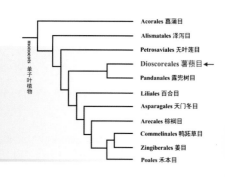

Perennial, usually rhizomatous **herbs**. **Leaves** basal, tufted, lanceolate to linear, with a conspicuous midvein; margin entire. **Scape** simple, erect, usually with a few small, bract-like leaves. **Inflorescence** terminal, racemes, or spikes, rarely corymbs. **Flowers** small, bisexual or rarely polygamous, actinomorphic; pedicellate or subsessile; pedicel bearing a bract and bracteole. **Perianth** tube campanulate or urceolate, 6-lobed, persistent in fruit. **Stamen** filaments basally fused, attached to base or tube of perianth. **Ovary** superior to inferior, 3-loculed; ovules many per locule; placentation basal to axile; style 1, 3-lobed and persistent. **Fruit** a loculicidal capsule enveloped by persistent perianth. **Seeds** small and numerous.

多年生**草本**，通常有根状茎。**叶**成簇基生，披针形至线形，中脉明显；全缘。花葶直立不分枝，通常具几枚苞片状叶。**花序**顶生，总状或穗状，稀为伞形花序。花小，两性或杂性，辐射对称；短柄或无柄；花梗具1枚苞片和1枚小苞片；花被筒钟形或坛状，6裂，果期宿存。雄蕊花丝常基部合生，贴生于花瓣或花冠筒基部。**子房**上位至下位，3室；每室有胚珠多数；基底胎座或中轴胎座；花柱1个，3裂，宿存。**果实**为蒴果，室背开裂，包藏于宿存的花被内。种子细小，多数。

World 5/41, interrupted with patchily in North temperate regions, Venezuela and Guyana, and scattered in west Malesia.
China 1/16.
This area 1/2.

全世界共5属41种，间断分布于北温带，委内瑞拉和圭亚那，马来西亚西部零星分布。
中国产1属16种。
本地区有1属2种。

Aletris L. 粉条儿菜属

Aletris scopulorum Dunn/ 短柄粉条儿菜（a. Inflorescence/ 花序；b. Infructescence/ 果序；c. Roots/ 根）

045 Dioscoreaceae 薯蓣科

Herbs twining / 缠绕藤本
Rhizomatous or tuberous / 具根状茎或块茎
Unisexual flowers small, dioecious / 单性花小，雌雄异株
Capsule 3-winged / 蒴果具 3 翅

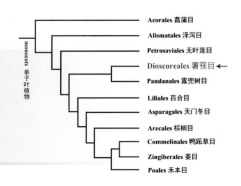

Perennial, **herbs** twining, rootstock rhizomatous or tuberous. **Leaves** simple or palmately compound, alternate or sometimes opposite, basal arcuate veins 3–9, main veins converging toward apex while lateral veins reticular; leaf bases cordate; margin entire to palmatifid; pulvinus always present; bulblets axillary or absent. **Inflorescence** usually axillary or terminal; spikes, panicles or umbel. **Flowers** small, unisexual (plants dioecious), actinomorphic. **Perianth** free, 6 in 2 whorls. **Stamens** 6 or 3. **Ovary** inferior, 3-loculed, ovules 2 per locule. **Fruit** 3-winged capsule dehiscing apically at maturity or a berry. **Seeds** with a membranous wing.

多年生缠绕**藤本**，具贮藏性的根状茎或块茎。**叶**为单叶或掌状复叶，互生，有时对生，基出 3~9 条弧形脉，主脉在叶尖汇合，侧脉网状；叶基心形；全缘至掌状分裂；常有叶枕；叶腋内有珠芽或无。**花序**常腋生或顶生；穗状、圆锥花序或伞形花序。**花**小，单性（雌雄异株），辐射对称。**花被片**离生，6 枚，2 轮。**雄蕊** 6 或 3 枚。**子房**下位，3 室，每室 2 枚胚珠。**果实**为蒴果，三棱形，每棱翅状，成熟后顶端开裂，或为浆果。**种子**有膜质翅。

World 4/ca. 870, largely in the tropics and subtropics.
China 2/58.
This area 1/6.

全世界 4 属约 870 种，主要分布于热带和亚热带。
中国产 2 属 58 种。
本地区有 1 属 6 种。

Dioscorea L. 薯蓣属

Dioscorea cirrhosa Lour./ 薯莨（a. Flowering branch/ 花枝；b. Fruiting branch / 果枝）

DIOSCOREALES. Dioscoreaceae. *Dioscorea nipponica* Makino/ 薯蓣目 薯蓣科 穿龙薯蓣

A. Male inflorescens/ 雄花序
B. Leaf on abaxial side/ 叶背面
C. Vertical section of male flower/ 雄花纵切
D. Dorsal view of male flower/ 雄花背面观
E. Frontal view of male flower/ 雄花正面观
F. Stamen oppsite the perianth/ 花药与花被片对生
(Scale/ 标尺：A-B 1 cm; C-F 1 mm)

Dioscorea tenuipes Franch. & Sav./ 细柄薯蓣
(a. Capsule/ 蒴果；b. Male inflorescence/ 雄花序；
c. Infructescence/ 果序）

053 Melanthiaceae 藜芦科

Stems erect, unbranched / 茎直立，不分枝
Inflorescences terminal / 花序顶生
Scape often bracteate / 花葶常具苞片
Perianth nectaries / 花被片具蜜腺

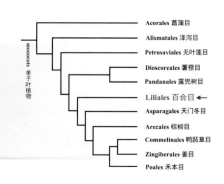

Perennial, **herbs** with rhizomatous or bulbous, rarely cormous. **Stems** erect, unbranched. **Leaves** simple, alternate (usually spiral), margin entire or toothed, sometimes sheathing; venation parallel to palmate. **Inflorescence** terminal, usually racemes or spikes, panicles or solitary flowers; scapose often with scaly leaves; often bracteate. **Flowers** bisexual, actinomorphic or zygomorphic. **Perianth** 2-whorled, free or fused, whorls similar or ±distinct; perianth often with nectaries, persistent in fruit. **Stamens** 1–3-whorled; anthers ±extrorse, or introrse. **Ovary** superior or inferior, carpels free or fused; 3-loculed, 2 to many ovules per locule; placentation usually axile, rarely parietal; stigma lobed. **Fruit** a capsule, follicle or a berry-like capsule. **Seeds** numerous.

多年生**草本**，具根状茎或鳞茎，稀具球茎。**茎**直立，不分枝。**叶**为单叶，互生（常螺旋状着生），全缘或具齿，有时具叶鞘；叶脉平行至掌状。**花序**顶生，常为总状、穗状或圆锥花序或为单花；花葶常具鳞片状叶；常具苞片。**花**两性，辐射对称或两侧对称。**花被片**2轮，离生或合生，两轮同形或多少有区别；花被常具蜜腺，果期宿存。**雄蕊**1~3轮；花药多少外向，或内向。**子房**上位或下位，心皮离生或合生；3室，每室胚珠2至多枚；常为中轴胎座，稀为侧膜胎座；柱头分裂。**果实**为蒴果、蓇葖果或浆果状蒴果。**种子**多数。

World 18/ca. 160, mainly in nortern temperate and cool temperate regions, few in South America.
China 7/49.
This area 1/1.

全世界共18属约160种，主要分布于北半球温带和寒温带地区，少数至南美洲。中国产7属49种。本地区有1属1种。

Paris L. (s.l.) 重楼属（广义）

1-1. *Paris polyphylla* Sm. var. *stenophylla* Franch./ 狭叶重楼；
1-2. *P. polyphylla* Sm. var. *chinensis* (Franch.) H. Hara / 华重楼

059 Smilacaceae 菝葜科

Climbing woody plants / 攀缘木本
Basal arcuate veins, mains veins converged / 基出弧形脉，主脉汇合
Petiole with sheath and tendril / 叶柄具鞘和卷须
Inflorescence an axillary umbel / 腋生伞形花序

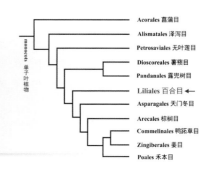

Woody to herbaceous **climbers**, rarely erect small shrubs, usually with short, thick rhizomes. **Stems** usually prickly. **Leaves** simple, alternate (2-ranked), margin entire, primary veins 3–7, basal arcuate, which converging toward apex; petiole usually winged sheath, with paired lateral tendrils. **Inflorescence** an axillary umbel, or compound umbels, often with a swollen receptacle. **Flowers** small, unisexual, dioecious. **Perianth** 6, free or slightly fused. **Stamen** usually 6 or absent. **Ovary** superior, 3-loculed, ovules 1–2 per locule; stigmas 3-lobed. **Fruit** a berry, red to black. **Seeds** 1–3.

木质至草质**攀缘**藤本，稀为直立小灌木，常具短粗的根状茎。**茎**常具刺。**叶**为单叶，互生（2列），全缘，具3~7基出弧形脉，在叶尖汇合；叶柄两侧常具翅状鞘，侧生1对卷须。**花序**为单个腋生的伞形花序，或为复伞形花序，花序托常膨大。**花**小，单性异株。**花被片**6枚，离生或稍合生。**雄蕊**通常6枚或缺。**子房**上位，3室，胚珠每室1~2枚；柱头3裂。**果实**为浆果，红色至黑色。**种子**1~3粒。

World 1/310+, pantropical to temperate regions.
China 1/92.
This area 1/4.

全世界共1属310余种，泛热带至温带分布。
中国产1属92种。
本地区有1属4种。

Smilax L. 菝葜属

1-1. *Smilax nipponica* Miq./ 白背牛尾菜；

1-2. *S. glabra* Roxb./ 土茯苓

060 Liliaceae 百合科

Bulbous herbs, leaves simple / 草本具鳞茎，单叶
Stems erect, unbranched / 直立茎，不分枝
Terminal few flowers / 顶生少数花
Perianth petaloid, thick / 花被片均为花瓣状，肥厚

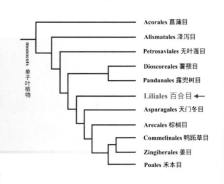

Perennial, bulbous **herbs** rarely rhizomatous. **Stems** erect, unbranched or branched above middle. **Leaves** simple, alternate, opposite, or whorled, amplexicaul. **Inflorescence** determinate, with few flowers, or reduced to a single flower, terminal. **Flowers** bisexual, usually actinomorphic, large and showy. **Perianth** 6 in 2 whorls, petaloid, thick, often with spots or striae, free. **Stamens** 6, filaments free or attached to perianth; anthers extrorse. **Ovary** superior, carpels 3, fused; 3-loculed; ovules (5–) many per locule; placentation axile or rarely parietal; stigma 3-lobed. **Fruit** a capsule, rarely a berry.

多年生**草本**，具鳞茎，稀有根状茎。**茎**直立，不分枝或中部以上分枝。**叶**为单叶，互生，对生或轮生，基部抱茎。**花序**为有限花序，具少数花，或退化为单花，顶生。**花**两性，常辐射对称，大而显著。**花被片**6枚，2轮，花瓣状，肥厚，常有斑点或线纹，离生。**雄蕊**6枚，花丝分离或贴生于花被片；花药外向。**子房**上位，心皮3枚，合生；3室，每室5至多枚胚珠；中轴胎座，稀为侧膜胎座；柱头3裂。**果实**为蒴果，稀为浆果。

World 16/635, worldwide, mainly in northern temperate regions.
China 13/148.
This area 4/7.

全世界共 16 属 635 种，世界广布，主要分布于北温带。
中国产 13 属 148 种。
本地区有 4 属 7 种。

1. Plants rhizomatous/ 植株具根状茎 .... **1. *Tricyrtis*/ 油点草属**
1. Plants bulbiferous/ 植株具鳞茎
　2. Bulbs with fleshy, farinaceous scales; flowers usually nodding; tepals each with a concave nectary near base adaxially/ 鳞茎具肉质，淀粉质鳞片；花常俯垂；花被片内侧基部具蜜腺窝 ...... **2. *Fritillaria*/ 贝母属**
　2. Bulbs without fleshy, farinaceous scales; flowers erect; tepals without a concave nectary/ 鳞茎不具肉质，淀粉质鳞片；花直立；花被片无蜜腺窝
　　3. Anthers dorsifixed and versatile; leaves cauline / 花药背着，丁字状；叶多茎生 .... **3. *Lilium*/ 百合属**
　　3. Anthers basifixed; flower usually solitary; leaves basal / 花药基部着生；花多单生；叶基生 ....
　　　 .... **4. *Amana*/ 老鸦瓣属**

LILIALES. Liliaceae. *Fritillaria thunbergii* Miq./ 百合目 百合科 浙贝母

A. Flowering branch/ 花枝

B. Cross section of ovary/ 子房横切面

C. Stigma/ 柱头

D. Flower segregated/ 花离析

E. Stamens/ 雄蕊

F. Perianth in dorsal and ventral view/ 花被片，背腹面观

G. Pistil/ 雌蕊

H. Young fruit/ 幼嫩果实

(Scale/ 标尺：A none/ 无；B–C 1 mm; D 1 cm; E 1 mm; F–H 1 cm)

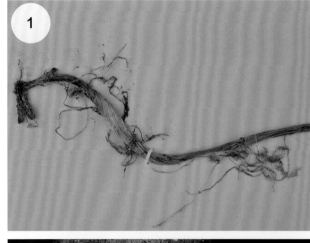

1. *Tricyrtis chinensis* Hir. Takah. bis (Showing creeping rhizome)/ 油点草（展示匍匐根状茎）；
2. *Fritillaria thunbergii* Miq./ 浙贝母；
3. *Lilium lankongense* Franch./ 匍匐百合（Showing dorsifixed anthers, not distributed in this area/ 仅展示花药背着结构，非分布于本区域）；
4. *Amana edulis* (Miq.) Honda/ 老鸦瓣

061 Orchidaceae 兰科

Flowers usually resupinate / 花常倒置
Labellum present / 具唇瓣
Gynandrium / 具合蕊柱
Ovary inferior / 子房下位

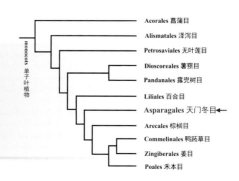

Herbs with protocorms; terrestrial, lithophytic or epiphytic, often with pseudobulbs (stem internodes swollen), or climbing; root often with a velamen. **Leaves** alternate or occasionally opposite, usually distichous, margin entire, usually basally sheathing, often fleshy or leathery. **Inflorescence** spikes, racemes or panicles, rarely solitary flowers. **Flowers** bisexual or very rarely unisexual, usually resupinate (twisted through 180°), zygomorphic. **Perianth** 6, 2 whorls; median tepal of inner whorl highly modified (lip or labellum). **Stamens** united with style to form a column (gynostemium); pollen usually aggregated into pollinia, often 2, 4, 6 or 8. **Ovary** inferior; carpels fused; 1-loculed with parietal placentation or rarely 3-loculed with axile placentation; ovules many per locule; stigma 3-lobed, 1 sterile and modified into a rostellum. **Fruit** usually a capsule. **Seeds** very numerous, dust-like, lacking endosperm.

草本，具有原球茎；地生、腐生或附生，常有假鳞茎（茎节间膨大），或为攀缘藤本；常具根被。**叶**互生或偶对生，常2列，全缘，多具叶鞘，常肉质或革质。**花序**穗状、总状或圆锥状，偶单花。**花**两性或偶为单性，通常倒置（扭转180°），两侧对称。**花被片**6枚，2轮，内轮中央1枚高度特化（特称唇瓣）。**雄蕊**与花柱完全融合成合蕊柱；花粉通常聚合成花粉块，一般为2枚、4枚、6枚或8枚。**子房**下位；心皮合生；1室侧膜胎座或稀为3室的中轴胎座；胚珠每室多枚；柱头3裂，1枚不育变态为蕊喙。**蒴果**。**种子**极小而多，粉尘状，无胚乳。

World 750/28,500, worldwide.
China 181/1 663 (Zhou et al., 2016).
This area 8/8.

全世界共750属28 500种，世界广布。
中国产181属1 663种 (Zhou et al., 2016)。
本地区有8属8种。

1. Plants growing from pseudobulbs / 植株具假鳞茎
　2. Pollinia 8, terrestrial / 花粉团8个，地生植物 .... **1. *Bletilla*** / 白及属
　2. Pollinia 4, epiphytic / 花粉团4个，附生植物 .... **2. *Bulbophyllum*** / 石豆兰属
1. Plants never growing from pseudobulbs / 植株不具假鳞茎
　3. Leaves terete / 叶圆柱形
　　4. Leaf 1, terrestrial / 叶1枚，地生兰 .... **3. *Microtis*** / 葱叶兰属

4. Leaves more than 1, epiphytic/ 叶多于 1 枚，附生植物 ....**4. *Luisia*/ 钗子股属**

3. Leaves flat or ensiform / 叶扁平或剑形
   5. Plant with tuber at stem base / 茎基部具块茎
      6. Lip often spurless/ 唇瓣通常无距 ....**5. *Herminium*/ 角盘兰属**
      6. Lip often spurred/ 唇瓣常具距 ....**6. *Shizhenia*/ 时珍兰属**
   5. Plant without tuber at stem base / 茎基部无块茎
      7. Spur slender, 3–5 cm; pollinia 2/ 距细长，3~5 cm；花粉团 2 个 ....**7. *Neofinetia*/ 风兰属**
      7. Spur saccate; pollinia 4/ 距囊状；花粉团 4 个 ....**8. *Cleisostoma*/ 隔距兰属**

ASPARAGALES. Orchidaceae. *Goodyera nankoensis* Fukuy./ 天门冬目 兰科 南湖斑叶兰 (not distributed here/ 非本区域分布)

A. Inflorescence/ 花序
B. Ovary/ 子房
C. Cross section of ovary/ 子房横切面
D. Lateral view of flower/ 花侧面观
E. Frontal view of corolla/ 花冠正面观
F. Perianth segregated/ 花被片离析
G. Pollinia/ 花粉块
H. Frontal view of gynostemium/ 蕊柱正面观
(Scale/ 标尺： A 1 cm; B–H 1 mm)

1. *Bletilla striata* (Thunb.) Rchb. f./ 白及；
2. *Bulbophyllum chondriophorum* (Gagnep.) Seidenf./ 城口卷瓣兰；
3. *Microtis unifolia* (G. Forst.) Rchb. f./ 葱叶兰；
4. *Luisia hancockii* Rolfe/ 纤叶钗子股（4a. Habitat/ 生境；4b. Flower/ 花）

5. *Herminium lanceum* (Thunb. ex Sw.) Vuijk/ 叉唇角盘兰；

6. *Shizhenia pinguicula* (Rchb. f. & S. Moore) X.H. Jin/ 时珍兰（6a. Flower/ 花；6b. Tuber/ 块茎）；

7. *Neofinetia falcata* (Thunb.) H.H. Hu/ 风兰；

8. *Cleisostoma arietium* Garay/ 牛角隔距兰（Showing saccate spur, not distributed in this area, photo by Xin Zhong/ 仅展示囊状距结构，非本区域分布，钟鑫拍摄）

066　Hypoxidaceae 仙茅科

Tufted herbs / 丛生草本
"V"shaped in leaf cross section / 叶横切面为 "V" 形
Leaves basal aggregated, sessile / 叶基簇生，无柄
Leaves with conspicuous hairs / 叶被毛明显

Perennial herbs, tufted; rhizomes subglobose, bulbed or tuberous. **Leaves** basally aggregated, "V" shaped in cross section, margin entire, sheathing, sessile, often with conspicuous hairs. **Inflorescence** racemes, spikes, heads or umbels, or solitary flower, scape leafless. **Flowers** bisexual, actinomorphic. **Perianth** 6, 2-whorled, persistent. **Stamens** 6, rarely 3, inserted at base of perianth. **Ovary** inferior; carpels fused, 3-loculed with axile placentation, locules 3–6 or 1 with parietal placentation; ovules 3 to many per locule. **Fruit** a capsule.

多年生**草本**，丛生；具近球形的根状茎、鳞茎或块茎。**叶**簇生于基部，横切面为"V"形，全缘，具鞘，无柄，常明显被毛。**花序**为总状、穗状、头状或伞形，或为单花，花葶无叶。花两性，辐射对称。**花被片** 6 枚，2 轮，宿存。**雄蕊** 6 枚，稀为 3 枚，嵌生于花被片基部。**子房**下位；心皮合生，3 室中轴胎座，3~6 室或 1 室的侧膜胎座；每室胚珠 3 枚至多数。**果实**为蒴果。

World 7–9/100–200, mainly in South Hemisphere to tropical Africa mountains and North American, few in China and Australia.
China 2/8.
This area 1/1.

全世界共 7~9 属 100~200 种，主要分布于南半球至非洲热带山区和北美，少数见于中国和澳大利亚。
中国产 2 属 8 种。
本地区有 1 属 1 种。

Hypoxis L. 小金梅草属

Hypoxis aurea Lour. / 小金梅草（a. Capsule/ 蒴果；b. Whole plant/ 全株）

070 Iridaceae 鸢尾科

Leaves alternate, distichous / 叶互生，2 列
Equitant / 叶基套叠
Inflorescence a scorpioid cyme / 蝎尾状聚伞花序

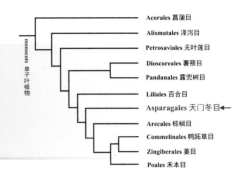

Perennial or rarely annual **herbs**, underground parts a rhizome, cormous or bulb. **Leaves** alternate, often distichous, ensiform to linear, base sheathing, equitant, and with a unifacial blade, parallel veined. **Inflorescence** determinate, a scorpioid cyme, often highly modified, variable, sometimes reduced to a solitary flower. **Flowers** usually large and showy, actinomorphic to zygomorphic. **Perianth** 6, petaloid, 2-whorled (often differentiated). **Stamens** 3, rarely 2, anthers extrorse. **Ovary** inferior, 3-loculed with axile placentation. **Fruit** usually a capsule, rarely a berry. **Seeds** with aril and wings or not.

多年生或稀为一年生**草本**，地下部分通常具根状茎、球茎或鳞茎。**叶**互生，常 2 列，剑形至线形，基部成鞘状，套叠，具单面叶，平行脉。**花序**为有限花序，蝎尾状聚伞，常高度特化，多变，有时退化为单花。**花**多大而显著，辐射对称至两侧对称。**花被片** 6 枚，花瓣状，2 轮（多不同型）。**雄蕊** 3 枚，稀为 2 枚，花药外向。**子房**下位，3 室中轴胎座。**果实**常为蒴果，极少为浆果。**种子**具假种皮和翅或无。

World 66/2, 035–2, 085, widely in tropics, subtropics and temperate regions, especially in south Africa and America tropics.
China 2/61.
This area 2/5.

全世界共 66 属 2 035~2 085 种，广布于热带、亚热带和温带地区，尤其是非洲南部及美洲热带地区。
中国产 2 属 61 种。
本地区有 2 属 5 种。

1. Flower >2.5 cm in diam.; style branches petaloid/ 花直径大于 2.5 cm；花柱分枝呈花瓣状 .... **1. *Iris*/ 鸢尾属**
1. Flower 0.8~1 cm in diam.; style branches not petaloid/ 花直径 0.8~1 cm；花柱分枝非花瓣状 .... **2. *Sisyrinchium*/ 庭菖蒲属**

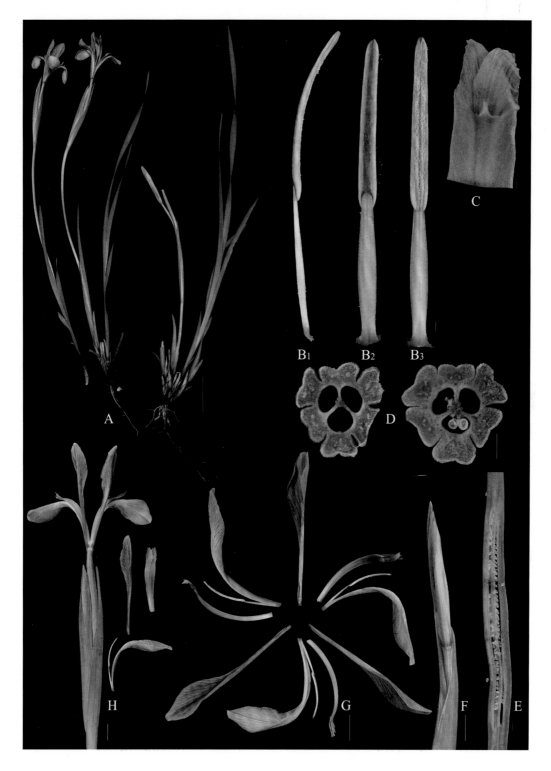

ASPARAGALES. Iridaceae. *Iris lactea* Pall./ 天门冬目 鸢尾科 马蔺

A. Whole plant/ 植株
B_1. Lateral view of stamen/ 雄蕊侧面观
B_2. Dorsal view of stamen/ 雄蕊背面观
B_3. Ventral view of stamen/ 雄蕊腹面观
C. Stigma/ 柱头
D. Cross section of ovary/ 子房横切面
E. Vertical section of ovary/ 子房纵切面
F. Flower bud/ 花蕾
G. Flower segregated/ 花离析
H. Vertical profile of corolla/ 花冠纵剖面

(Scale/ 标尺：A 5 cm; B_1–E 1 mm; F–H 1 cm)

1. *Iris proantha* Diels var. *valida* (S.S. Chien) Y.T. Zhao / 粗壮小鸢尾;

2. *Sisyrinchium palmifolium* L. / 簇花庭菖蒲（Showing flower structure, not distributed in this area/ 仅展示花部结构，非分布于本区域）

072 Asphodelaceae 阿福花科

Herbs with rhizomatous / 草本，具根状茎
Leaves basal, sheath closed / 叶基生，叶鞘闭合
Flowering stems shoot up from axil of leaves /
花葶从叶丛侧方抽出

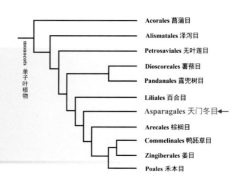

Perennial or rarely annual **herbs**, rhizomatous, sometimes shrubs to trees. **Leaves** basal or nearly basal, simple, distichous or spiral, long and narrow, fibrous not obvious, sometimes succulent, sheath closed. **Inflorescence** racemes, panicles or scorpioid, erect, monosymmetry, bracts present; scape often leafless, shoot up from axil of leaves; pedicels often articulated. **Flowers** bisexual, actinomorphic or zygomorphic. **Perianth** 6, petaloid, in 1 or 2 whorls, free or connate into tube. **Stamens** 6, filaments free. **Ovary** superior, 1 or 3-loculed, usually septal nectaries; plancentation axile; style slender. **Fruit** capsules, nut, schizocarp or berry.

多年生或稀为一年生**草本**，具根状茎，有时为灌木至乔木。**叶**基生或近基生，单叶，2列或螺旋状，狭长，不甚纤维化，有时肉质，具闭合叶鞘。**花序**为总状、圆锥花序或蝎尾状花序，直立，单面对称，具苞片；花葶少叶，从叶丛侧方抽出；花梗常具关节。**花**两性，辐射对称或两侧对称。**花被片**6枚，花瓣状，1轮或2轮，离生或合生成管状。**雄蕊**6枚，花丝分离。**子房**上位，1室或3室，常具子房壁间蜜腺；中轴胎座；花柱细长。**果实**为蒴果、坚果、分果或浆果。

World 41/900, mainly in Australia, Eurasia, Africa and western South America.
China 4/17.
This area 2/3.

全世界共 41 属 900 种，主要分布于澳大利亚、欧亚大陆、非洲和南美洲西部。
中国产 4 属 17 种。
本地区有 2 属 3 种。

1. Flower > 5 cm in diam., perianth more than 5 cm long; fruit a capsule / 花直径大于 5 cm，花被片长 5 cm 以上；果为蒴果 .... **1. *Hemerocallis*/ 萱草属**
1. Flower < 4 cm in diam., perianth no more than 1 cm long; fruit a berry/ 花直径小于 4 cm，花被片长不过 1 cm；果为浆果 .... **2. *Dianella*/ 山菅属**

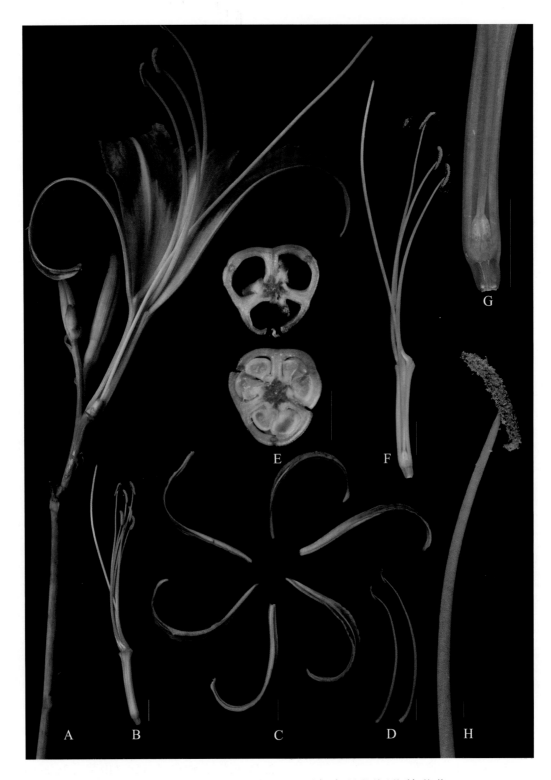

ASPARAGALES. Asphodelaceae. *Hemerocallis fulva* (L.) L./ 天门冬目 阿福花科 萱草

A. Scape/ 花葶

B. Pistil and stamen/ 雌雄蕊

C. Flower segregated/ 花离析

D. Stamen/ 雄蕊

E. Cross section of ovary/ 子房横切面

F. Vertical section of flower, perianth removed/ 花纵切面，除花被片

G. Vertical section of perianth tube/ 花被筒纵切面

H. Filament and anther/ 花丝与花药

(Scale/ 标尺： A none/ 无； B–D 1 cm; E 1 mm; F–G 1 cm; H 1 mm)

1. *Hemerocallis fulva* (L.) L. / 萱草;

2. *Dianella ensifolia* (L.) DC./ 山菅 （2a. Berry/ 浆果; 2b. Flower/ 花）

073　Amaryllidaceae 石蒜科

Leaves usually distichous, nearly basal / 叶常2列，近基生
Inflorescence often an umbel, flower showy / 常单伞形花序，花显著
Sometimes a corona present / 有时具副花冠

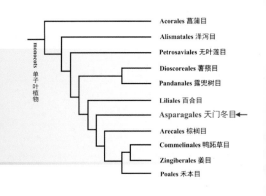

Perennial or biennial **herbs** from a bulb with contractile roots; stems reduced. **Leaves** aromatic, often deciduous, alternate, usually distichous, nearly basal, with parallel venation, sheathing at base. **Inflorescence** determinate, often umbel, solitary flowers, cymes, spikes, or panicles; scapose. **Flowers** actinomorphic or zygomorphic, showy, each associated with a filiform bract. **Perianth** 6, petaloid; sometimes present a corona. **Stamens** 6, usually attached to perianth, introrse. **Ovary** superior or inferior; carpels 3, septal nectaries; 3-loculed, placentation axile, basal or apical. **Fruit** a capsule or occasionally a berry.

多年生或二年生**草本**，有鳞茎，具收缩根；茎退化。**叶**具芳香味，常落叶性，互生，常2列，近基生，具平行脉，基部有鞘。**花序**为有限花序，常为伞形、单花、聚伞形、穗状或圆锥花序；具花葶。**花**辐射对称或两侧对称，显著，花具1枚线状苞片。**花被片**6枚，花瓣状；有时具副花冠。**雄蕊**6枚，常贴生于花被片上，内向。**子房**上位或下位；心皮3枚，具子房壁间蜜腺；3室，中轴胎座，基底胎座或顶生胎座。**果实**为蒴果，或偶为浆果。

World 68/1, 616, temperate regions.
China 6/ca. 161.
This area 3/9.

Narcissus tazetta L. var. *chinensis* M. Roem. / 水仙

全世界共68属1 616种，分布于温带地区。
中国产6属约161种。
本地区有3属9种。

1. Corona present/ 具副花冠 ....**1. *Narcissus*/ 水仙属**
1. Corona absent/ 无副花冠
　　2. Leaves absent in flowering; ovary inferior/ 植株花期无叶；子房下位 ....**2. *Lycoris*/ 石蒜属**
　　2. Leaves present in flowering; ovary superior/ 植株花期有叶；子房上位 ....**3. *Allium*/ 葱属**

ASPARAGALES. Amaryllidaceae. *Narcissus tazetta* L. var. *chinensis* M. Roem./ 天门冬目 石蒜科 水仙

A. Whole plant/ 植株
B. Corolla/ 花冠
C. Stigma/ 柱头
D. Flower segregated/ 花离析
E. Corona/ 副花冠
F. Vertical section of flower, perianth lobes removed/ 花纵切面，除花被裂片
G. Cross section of ovary/ 子房横切面
H. Stamen and vertical section of ovary/ 雄蕊与子房纵切面
I. Anther/ 花药

(Scale/ 标尺：A–B none/ 无；C 1 mm; D–F 1 cm; G–I 1 mm)

1. *Narcissus tazetta* L. var. *chinensis* M. Roem. / 水仙；
2. *Lycoris sprengeri* Comes ex Baker/ 换锦花（2a. Habitat/ 生境；2b. Inflorescence/ 花序）；3. *Allium tuberosum* Rottler ex Spreng./ 韭（3a. Inflorescense umbel/ 伞形花序；3b. Whole plant/ 植株）

074 Asparagaceae 天门冬科

Scape leaned one side / 花葶常偏向一侧斜出
Perianth 6, without spots / 花被片 6 枚，通常无斑点
Seeds usually black / 种子常为黑色

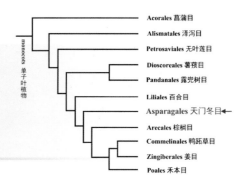

Shrubby, **tree-like** or perennial **herb** with rhizomes or bulbs, rarely climber, caespitose, epiphytic or succulent. **Leaves** basally aggregated when stemless or along stems, simple, alternate (distichous or spiral), sometimes scale-like; margin entire or toothed; venation usually parallel, rarely reticulate; often sheathing. **Inflorescence** racemes, panicles, spikes, solitary flower or umbel; often scapose with bracteate leaned one side, without spathe-like bracts or with ≥ 3. **Flowers** bisexual or rarely unisexual, actinomorphic or zygomorphic. **Perianth** 6, free or fused, without spots. **Stamens** 6, filaments usually free or basally fused, sometimes attached to perianth; anthers introrse. **Ovary** superior or inferior; usually septal nectaries; 3-loculed with placentation axile or apical. **Fruit** a capsule, berry, drupe or nut. **Seeds** usually black or pale brown.

灌木状、乔木状或多年生**草本**，具根状茎或鳞茎，稀攀缘藤本，丛生、附生或为多肉植物。**叶**丛生于无茎的植株基部或茎生，单叶，互生（二裂或螺旋状排列），有时为鳞片状；全缘或有锯齿；叶脉常平行，稀为网脉；常有叶鞘。**花序**为总状、圆锥状、穗状，或为单花、伞形花序；花葶具苞片，常偏向一侧斜出，无佛焰苞或具 3 枚以上。**花**两性，稀单性，辐射对称或两侧对称。**花被片** 6 枚，离生或合生，无斑点。**雄蕊** 6 枚，花丝常分离或基部联合，有时贴生于花被片上；花药内向。**子房**上位或下位；常具子房壁间蜜腺；3 室的中轴胎座或顶生胎座。**果实**为蒴果、浆果、核果或坚果。**种子**常为黑色或灰棕色。

World 153/ca. 2,500, worldwide, except Arctic.
China 25/ca. 258.
This area 8/13.

全世界共 153 属约 2 500 种，世界广布，除北极外。
中国产 25 属约 258 种。
本地区有 8 属 13 种。

ASPARAGALES. Asparagaceae. *Hosta ventricosa* (Salisb.) Stearn/ 天门冬目 天门冬科 紫萼

A. Inflorescence/ 花序

B. Pistil/ 雌蕊

C. Vertical profile of corolla/ 花冠纵剖面

D. Bract/ 苞片

E. Stamen/ 雄蕊

F. Stigma/ 柱头

G. Cross section of ovary/ 子房横切面

（Scale/ 标尺：A none/ 无；B–D 1 cm；E–G 5 mm）

1. Plants bulbiferous/ 植株具鳞茎 .... **1. *Barnardia*/ 绵枣儿属**
1. Plants not bulbiferous/ 植株不具鳞茎
 2. Leaves reduced to scales/ 叶退化为鳞片 .... **2. *Asparagus*/ 天门冬属**
 2. Leaves not reduced to scales/ 叶不为鳞片状
  3. Fruit bursting irregularly at an early stage and exposing seeds/ 果未成熟前已不整齐开裂并显露种子
   4. Flowers erect or suberect; filaments longer than or as long as anthers;seeds blackish/ 花直立或近直立；花丝长于或等长于花药；种子黑色 .... **3. *Liriope*/ 山麦冬属**
   4. Flowers ± nodding; filaments much shorter than anthers; seeds blue/ 花多少下垂；花丝远短于花药；种子蓝色 .... **4. *Ophiopogon*/ 沿阶草属**
  3. Fruit never bursting before seeds maturity/ 果未成熟前不开裂
   5. Leaves cauline/ 叶茎生 .... **5. *Polygonatum*/ 黄精属**
   5. Leaves basal/ 叶基生
    6. Leaves in a rosette/ 叶莲座状着生 .... **6. *Agave*/ 龙舌兰属 ***
    6. Leaves not in a rosette/ 叶非莲座状着生
     7. Leaf blade linear or narrowly oblanceolate; scape arising from a leaf axil, erect, shorter than leaves, fruit a berry/ 叶片线形或狭倒披针形；花葶自叶腋伸出，直立，短于叶片，浆果 .... **7. *Reineckea*/ 吉祥草属**
     7. Leaf blade ovate; scape terminal, usually with a few bract-like cauline leaves; fruit a capsule/ 叶片卵形；花葶顶生，常具数枚苞片状茎生叶；蒴果 .... **8. *Hosta*/ 玉簪属**

1. *Barnardia japonica* (Thunb.) Schult. & Schult. f./ 绵枣儿
（1a. Bulb/ 鳞茎；1b. Inflorescense/ 花序）；

2. *Asparagus cochinchinensis* (Lour.) Merr. / 天门冬；

3. *Liriope spicata* (Thunb.) Lour. / 山麦冬（3a. Seeds/ 种子；3b. Stamens/ 雄蕊）；

4. *Ophiopogon japonicus* (L. f.) Ker Gawl. / 麦冬

5. *Polygonatum cyrtonema* Hua / 多花黄精；

6. *Agave americana* L. / 龙舌兰（Natrulized/ 归化）；

7. *Reineckea carnea* (Andrews) Kunth/ 吉祥草（7a. Inflorescence/ 花序；7b. Infructescence/ 果序）；

8. *Hosta ventricosa* (Salisb.) Stearn/ 紫萼（8a. Leaves/ 叶片；8b. Inflorescence/ 花序）

076 Arecaceae 棕榈科

Plicate leaves / 叶片褶扇状
Base petitole sheathing and fibered / 叶柄基部鞘状具纤维
Inflorescence spikes surrounded by spathe / 穗状花序具佛焰苞

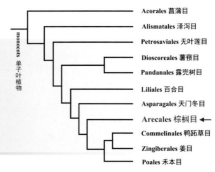

Massive to minute woody plants; **trees**, **shrubs** with unbranched or rarely branched trunks, or climbing. **Leaves** usually spirally arranged, crowded in a terminal crown, splitting to palmate fashion as the leaf expands; leave blade plicate; pinnate, palmate, or costapalmate, with parallel to divergent venation, the petiole often with a flap (hastula), sheathing, various fibered; stipule absent. **Inflorescence** usually axillary panicle or spike surrounded by spathe, rarely solitary flower. **Flowers** usually sessile; unisexual or rarely bisexual, actinomorphic. **Perianth** often inconspicuous, 2 whorles. **Stamens** 6 or 3 or many. **Ovary** superior, 3-loculed, ovule 1 per locule, placentation usually apical or basal. **Fruit** a drupe, often fibrous, or rarely a berry; usually **1-seeded**, endosperm sometimes ruminate.

大型至小型木本植物；**乔木**或**灌木**，不分枝或少分枝，或为藤本。**叶**通常螺旋状排列，集生顶端成树冠，叶片开展时为棕榈状，叶片褶扇状、羽状、掌状或肋掌状分裂，叶脉平行至分叉，叶柄常具副翼，基部鞘状，具各式纤维；无托叶。**花序**常为腋生圆锥花序或具佛焰苞的穗状花序，极少为单花。**花**常无柄；单性或稀为两性，辐射对称。**花被片**不明显，2 轮。**雄蕊** 6 枚或 3 枚或多数。**子房**上位，3 室，每室 1 胚珠，常为顶生胎座或基底胎座。**果实**为核果，常纤维质，或稀为浆果；**种子常为 1 粒**，胚乳有时嚼烂状。

World 183/2, 450, humid tropics and subtropics, warm temperate regions.
China 18/77.
This area 2/2.

全世界共 183 属 2 450 种，分布于热带与亚热带湿润地区和暖温带。
中国产 18 属 77 种。
本地区有 2 属 2 种。

1. Leaves palmate or costapalmate/ 叶掌状或肋掌状分裂 .... **1. *Trachycarpus*/ 棕榈属**
1. Leaves pinnate or bipinnate/ 叶羽状或 2 回羽状 .... **2. *Phoenix*/ 刺葵属**

1. *Trachycarpus fortunei* (Hook.) H. Wendl. / 棕榈；
2. *Phoenix roebelenii* O' Brien / 软叶刺葵（Natrulized/归化）

078　Commelinaceae 鸭跖草科

Leaf sheath closed / 叶鞘基部闭合
Leaf sheath lacking ligule / 叶鞘无叶舌
Tepal deliquescent / 花被片易分解
Filaments often hairy / 花丝多毛

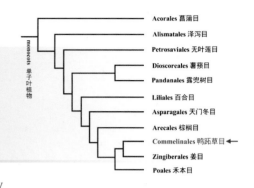

Perennial **herbs**, rarely annual, sometimes rhizomatous, rarely bulbous, sometimes climbers or epiphytes, with well-developed stems that are swollen at the nodes, or sometimes short, hairs often present. **Leaves** usually succulent, simple, entire, often with the opposite halves rolled separately against the midrib in bud; basal sheath closed; lacking a ligule, often pseudopetiolate. **Inflorescence** cymes or thyrses, rarely solitary flower; often subtended by a folded, leafy bracts. **Flowers** usually bisexual, actinomorphic or zygomorphic. **Perianth** 2 whorles, heterochlamydeous. **Sepals** 3, fused, sometimes free; green or sometimes petaloid. **Petals** 3, free or less often basally fused; quickly deliquesce; 1 petal sometimes differently colored or reduced. **Stamen** filaments often hairy. **Ovary** superior, 3-loculed, placentation axile. **Fruit** usually a loculicidal capsule, rarely a berry. **Seeds** with a conspicuous conical cap.

多年生**草本**，稀为一年生，有时具根状茎，稀为鳞茎，有时为攀缘或附生植物，发达的茎在节部略膨大，或有时缩短，植株通常被毛。**叶片**多肉质，单叶，全缘，常在芽中沿中脉对折；叶鞘基部闭合，无叶舌，常具假叶柄。**花序**为聚伞或聚伞圆锥花序，稀为单花；通常被折叠的叶状苞片所包被。**花**常两性，辐射对称或两侧对称。**花被片**2轮，异被花。**花萼**3枚，合生，有时离生；绿色或有时为花瓣状。**花瓣**3枚，离生或偶为基部合生；快速自体分解；1枚花瓣有时具不同颜色或退化。**雄蕊**花丝多毛。**子房**上位，3室，中轴胎座。**果实**常为室背开裂的蒴果，稀为浆果。**种子**具明显的圆锥状突起。

World 40/650, mainly in tropics, few in subtropics, rarely in temperate regions.
China 15/59.
This area 3/6.

全世界共40属650种，主要分布在热带地区，少量在亚热带地区，温带地区罕见。
中国产15属59种。
本地区有3属6种。

1. Inflorescence penetrated leaf sheath, sessile, capitate/ 花序穿透叶鞘而出，无总梗，头状 .... **1. *Amischotolype*/ 穿鞘花属**
1. Inflorescence neither penetrated leaf sheath nor sessile nor capitate/ 花序不穿透叶鞘，亦不成无总梗的头状花序
  2. Involucral bracts spathe-like/ 总苞片佛焰苞状 ....**2. *Commelina*/ 鸭跖草属**
  2. Involucral bracts present or absent, never spathe-like/ 总苞片有或无非佛焰苞状 ....**3. *Murdannia*/ 水竹叶属**

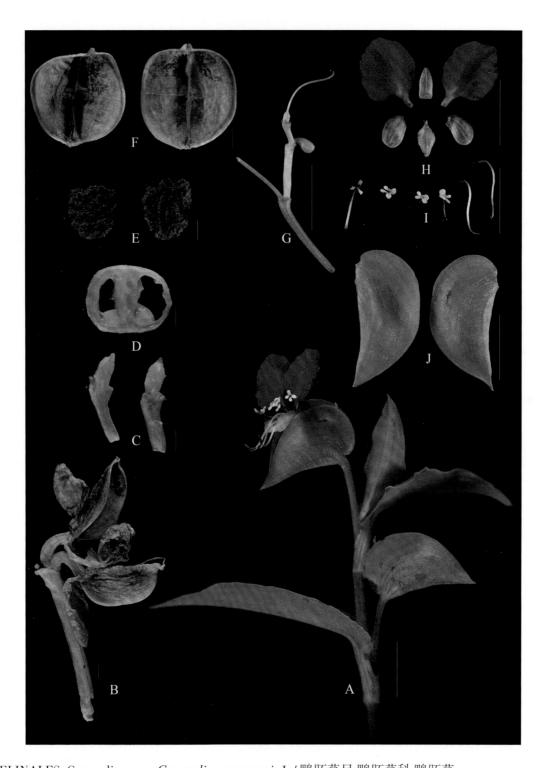

COMMELINALES. Commelinaceae. *Commelina communis* L./ 鸭跖草目 鸭跖草科 鸭跖草

A. Flowering branch/ 花枝
B. Infructescence/ 果序
C. Young fruit/ 幼果
D. Cross section of ovary/ 子房横切面
E. Seeds/ 种子
F. Fruit valves/ 果瓣
G. Cincinnus, perianth removed/ 蝎尾状聚伞花序，除花被片
H. Perianth/ 花被片
I. Fertile stamens 3, staminodes 3 (from the left, the 2nd to 4th)/ 可育雄蕊 3 枚，退化雄蕊 3 枚（左起，第 2~4 枚）
J. Involucral bracts/ 总苞片

(Scale/ 标尺： A 1 cm; B-F 1 mm; G-J 1 cm)

1. *Amischotolype hispida* (Less. & A. Rich.) D.Y. Hong/ 穿鞘花；
2. *Commelina communis* L./ 鸭跖草（2a. Involucral/ 总苞；2b. Flower/ 花；2c. White flower population/ 白花种群）；
3. *Murdannia spirata* (L.) G. Brückn./ 矮水竹叶（3a. Lateral view of flower/ 花侧面观；3b. Frontal view of flower/ 花正面观）

079 Philydraceae 田葱科

Leaves sheaths equitant / 叶鞘套叠
Bracts spathe-like / 苞片佛焰状
Flower bisexual, stamen 1 / 两性花，雄蕊1枚
Ovary superior; seeds tiny / 子房上位；种子细小
Inflorescence branched, bracts spathe-like / 花序分枝，苞片佛焰苞状
Flowers bisexual, sessile, zygomorphic / 花两性，无柄，两侧对称
Perianth 2-merous; stamen 1 / 花被片2基数，雄蕊1枚

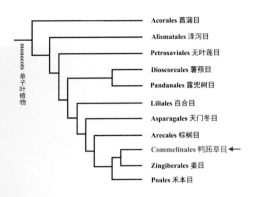

Perennial **herbs** with rhizomatous or cormous, erect. **Leaves** basally aggregated, simple, cauline ones spirally arranged; leaf blade linear or ensiform, veins parallel; leaf sheaths equitant. **Inflorescence** spike or racemes; often branched; bracts spathe-like. **Flowers** bisexual, sessile, zygomorphic. **Perianth** free or basally fused, yellow or whitish, segments 4, in 2 whorls, petaloid. **Stamen** 1. **Ovary** superior, 1-loculed, placentation parietal or 3-loculed with axile placentation; ovules many per locule; style simple. **Fruit** a loculicidal, 3-valved capsule or berry-like.

多年生直立**草本**，具根状茎或球茎。**叶**基部簇生，单叶，茎生叶螺旋状排列；叶线形或剑形，平行脉；叶鞘套叠。**花序**穗状或总状；常分枝；苞片佛焰苞状。**花**两性，无柄，两侧对称。**花被片**离生或基部合生，黄色或白色，4枚，排成2轮，花瓣状。**雄蕊**1枚。**子房**上位，1室，侧膜胎座或3室的中轴胎座；胚珠每室多枚；花柱单一。**果实**为蒴果，室背开裂成3瓣或浆果状。

World 3/6, mainly in Australia, few in West Pacific Islands and mainland Southeast Asia.
China 1/1.
This area 1/1.

全世界共3属6种，主要分布于澳大利亚，少数到西太平洋群岛及东南亚大陆。
中国产1属1种。
本地区有1属1种。

Philydrum Banks & Sol. ex Gaertn. 田葱属

Philydrum lanuginosum Banks & Sol. ex Gaertn. / 田葱（a. Habitat / 生境；b. Flower / 花；c. Capsule / 蒴果）

080 Pontederiaceae 雨久花科

Aquatic herbs / 水生草本
Spongy stem / 茎海绵质
Sheath spathe-like / 叶鞘佛焰苞状

Perennial or annual, rhizomatous or cormous, aquatic or halophytic **herbs**, usually grown in freshwater or marshes. **Leaves** submerged or emerged, simple, usually alternate (usually distichous), sheath open; margin entire. **Inflorescence** terminal spikes, racemes, panicles, umbel-like or solitary flower; bracts 2, spathe-like. **Flowers** bisexual, ±zygomorphic. **Perianth** basally fused or ±free; usually purple, blue or yellow. **Stamens** usually 1, 3 or 6, sometimes staminodes 2. **Style** 1. **Ovary** superior; sometimes septal nectaries; locules 1 or 3; ovules 1 to many per locule; placentation usually axile. **Fruit** a capsule or nut-like. **Seeds** small, longitudinally ribbed or smooth; endosperm copious, mealy; embryo central in seed, straight, terete.

多年生或一年生水生或盐生草本，具根状茎或球茎，常生于淡水或沼泽。叶沉水或浮水，单叶，常互生（多2列），叶鞘开放；全缘。花序为顶生的穗状、总状、圆锥状、伞形花序或为单花；苞片2枚，佛焰苞状。花两性，多少两侧对称。花被片基部合生或多少离生；常为紫色、蓝色或黄色。雄蕊多为1枚、3枚或6枚，有时具2枚退化雄蕊。花柱1个。子房上位；有时具子房壁间蜜腺；1或3室；胚珠每室1至多枚；常为中轴胎座。果实为蒴果或坚果状。种子小，有纵肋或光滑；胚乳丰富、粉状；胚位于种子中央，劲直，圆柱形。

World 9/33, widely in tropics and subtropics.
China 2 (one introduced) /5.
This area 2/2.

全世界共9属33种，主要分布于热带和亚热带。
中国产1属，另有1属引入，共5种。
本地区有2属2种。

1. Flowers distinctly pedicellate; actinomorphic; petiole not inflated/ 花明显具梗；辐射对称；叶柄不膨大 ....**1. *Monochoria*/ 雨久花属**
1. Flowers sessile; zygomorphic; petiole inflated/ 花无梗；两侧对称；叶柄膨大 ....**2. *Eichhornia*/ 凤眼莲属**

1. *Monochoria vaginalis* (Burm. f.) C. Presl ex Kunth / 鸭舌草；
2. *Eichhornia crassipes* (Mart.) Solms / 凤眼莲

085 Musaceae 芭蕉科

Prominent midrib with parallel lateral veins / 具明显的中脉和横出平行脉
Pseudostems composed of leaf sheaths / 叶鞘形成假茎
Bracts spathe-like, large, bright / 佛焰苞大而鲜艳
Inflorescence terminal, pendulous / 顶生花序，下垂

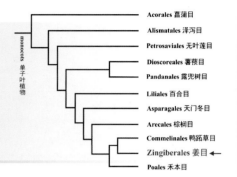

Herbs, growing from sympodial rhizomes or a massive, sympodial corm. **Pseudostems** composed of closely packed leaf sheaths, unbranched. **Leaves** large, spirally arranged, petiolate; leaf blade entire, prominent midrib with parallel lateral veins. **Inflorescence** terminal or rarely axillary, pendulous cymes; bracts spirally arranged, often brilliantly colored, spathe-like, large. **Flowers** unisexual or rarely bisexual, zygomorphic. **Perianth** in 2 whorls, petaloid; outer whorl fused; inner whorl 2 fused and 1 free. **Stamens** 5, filaments free; anthers 2-loculed, basifixed; sometimes staminode 1. **Ovary** inferior, septal nectaries; 3-loculed with axile placentation; ovules numerous per locule, anatropous. **Fruit** a berry, fleshy or leathery and dry, indehiscent. **Seeds** hard, without arillate; embryo straight, surrounded by a ± well–developed endosperm and a mealy perisperm.

草本，由合轴生长的根状茎或大型合轴生长的球茎发育。常有由叶鞘套叠而成的**假茎**，不分枝。**叶**通常较大，螺旋形排列，具叶柄；全缘，具明显的中脉和横出平行脉。**花序**多为顶生或稀腋生，聚伞花序下垂；苞片螺旋状排列，常呈大型鲜艳的佛焰苞状。花单性稀两性，两侧对称。**花被片** 2 轮，花瓣状；外轮合生，内轮 2 枚合生，1 枚分离。**雄蕊** 5 枚，花丝分离；花药 2 室，基着；有时具 1 枚退化雄蕊。**子房**下位，具子房壁间蜜腺；3 室的中轴胎座；每室胚珠多数，倒生胚珠。**果**为浆果，肉质或革质，干燥，不裂。**种子**坚硬，不具假种皮；胚直伸，多少包被有胚乳和粉质外胚乳。

World 3/ca. 40, Asia and tropical Africa.
China 3/14.
This area 1/1.

全世界共 3 属约 40 种，分布于亚洲和非洲热带地区。
中国产 3 属 14 种。
本地区有 1 属 1 种。

Musa L. 芭蕉属

Musa balbisiana Colla/ 野蕉（a. Habitat/ 生境； b. Infructescence/ 果序， photo by XiangXiu SU/ 苏享修拍摄）

089 Zingiberaceae 姜科

Rhizome often spicy aromatic / 根状茎常有辛辣气味
Stamen 1, staminodes petaloid / 发育雄蕊 1 枚，退化雄蕊花瓣状
Style often in groove of fertile stamen /
花柱常生于可育雄蕊凹槽内

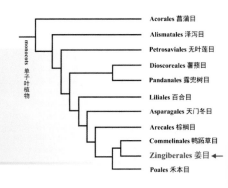

Perennial **herbs**, usually rhizomatous, often spicy-aromatic; stems erect, often short, sometimes leafy. **Leaves** distichous, simple, sometimes bladeless and reduced to sheaths; leaf sheaths open; ligule usually present; venation with prominent midrib and parallel lateral veins; petiole present or not. **Inflorescence** panicles, rarely cymes, spikes or solitary flower; bracts petaloid. **Flowers** bisexual, zygomorphic. **Perianth** 2-whorled; sepals usually fused into tubular, thin, split along side, sometimes spathe-like, apex 3-toothed or 3-lobed; petals basally fused, distally 3-lobed; lobes varying in size and shape. **Stamen** 1, filament short, anther 2-loculed, introrse, dehiscing by slits or occasionally pores; staminodes 2–4, petaloid, inner whorl usually fused (= labellum), outer whorl petaloid or reduced to absent. **Ovary** inferior, 3-loculed initially, 1– or 3-loculed when mature; ovules numerous per locule; placentation parietal, basal, or axile; style often in groove of fertile stamen; stylodia 2, reduced to nectaries at apex of ovary. **Fruit** a capsule, fleshy or dry, dehiscent or indehiscent, sometimes berry-like. **Seeds** few to many, aril often lobed or lacerate.

多年生**草本**，常具根状茎，具辛辣气味；茎直立，常短缩，有时多叶。**叶**二列，单叶，有时无叶片并退化成叶鞘；叶鞘开放；常具叶舌；具明显中脉和平行的横出脉；叶柄有或无。**花序**为圆锥花序，稀为聚伞、穗状花序或单花；苞片花瓣状。**花**两性，两侧对称。**花被片** 2 轮；萼片常合生成细管状，沿一侧开裂，有时呈佛焰苞状，先端 3 齿或 3 裂；花瓣基部合生，下唇先端 3 裂，裂片大小与形状各异。**雄蕊** 1 枚，花丝短，花药 2 室，内向，纵裂或偶尔孔裂；退化雄蕊 2~4 枚，花瓣状，内轮常合生（唇瓣），外轮花瓣状或退化消失。**子房**下位，初为 3 室，成熟后为 1 或 3 室；胚珠每室多数；侧膜胎座、基底胎座或中轴胎座；花柱常生于可育雄蕊凹槽内；分枝花柱 2 个，在子房顶端退化为蜜腺。**果实**为蒴果，肉质或干燥，开裂或不裂，有时呈浆果状。**种子**少数至多数，常具分裂或撕裂的假种皮。

World 50/1, 300, tropical and subtropical regions, especially in Southeast Asia.
China 20/216.
This area 1/2 (Naturalized).

全世界共 50 属 1 300 种，主要分布于热带和亚热带地区，特别是东南亚。
中国产 20 属 216 种。
本地区有 1 属 2 种（归化）。

Alpinia **Roxb.** 山姜属

1-1. *Alpinia japonica* (Thunb.) Miq. / 山姜（1-1a. Infrutescence/ 果序；1-1b. Rhizomatous/ 根状茎）；

1-2. *A. zerumbet* (Pers.) B.L. Burtt & R.M. Sm./ 艳山姜（1-2a. Flowers/ 花；1-2b. Infrutescence/ 果序；1-2c. Seeds/ 种子）

090 Typhaceae 香蒲科

Inflorescence dense, terminal spikes / 花序密集，顶生穗状
Perianth chaffy elongate scales / 花被片为被膜片的长形鳞片
Female flowers with fine hairs at base / 雌花基部具丝状毛
Ovary on a long capillary stalk / 子房具发状子房柄

Perennial **herbs** with creeping rhizomes, aquatic or in marshes. **Leaves** alternate, distichous, erect, emersed or floating, sheathed at base. **Inflorescence** dense, terminal spikes or heads. **Flowers** minute, numerous, unisexual, actinomorphic; sometimes female flowers with fine hairs at base. **Perianth** of 3–6 chaffy elongate scales or hairy bract-like perianth or absent. **Stamens** 1-whorled; anthers basifixed, dehiscing longitudinally. **Ovary** superior, 1-loculed, rarely 2-loculed, narrow at base or on a long capillary stalk; ovule 1; placentation apical. **Fruit** minute, drupe or nut-like, indehiscent. **Seeds** with thin testa.

多年生水生或湿生**草本**，具匍匐根状茎。**叶**互生，二列，**直立**，挺水或浮水，叶基有鞘。**花序**密集，顶生穗状或头状。花细小，多数，单性，辐射对称；雌花有时基部具丝状毛。**花被片**为3~6枚被膜片的长形鳞片或被毛的苞片状花被片或缺。**雄蕊**1轮；花药基着，纵向开裂。**子房**上位，1室，稀2室，基部狭窄或生于发状子房柄上；胚珠1枚；顶生胎座。**果实**小，核果或坚果状，不裂。**种子**种皮薄。

World 2/35, worldwide.

China 2/23.

This area 1/2.

全世界共2属35种，世界广布。

中国产2属23种。

本地区有1属2种。

Typha L. 香蒲属

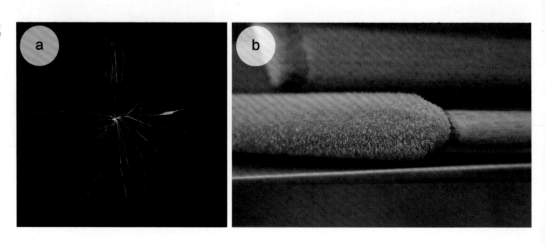

Typha angustifolia L./ 水烛（a. Fruit/ 果实；b. Spikes/ 穗状花序）

094 Eriocaulaceae 谷精草科

Leaves rosette, grass-like, fenestrate / 叶莲座状，似禾草，具窗孔
Scapes twisted, base surrounded by spathe-like sheath / 花葶扭转，基部被佛焰状鞘
Bracts involucral, small, dry, scaly / 总苞片小，干，鳞片状
Flowers small, unisexual / 花小，单性

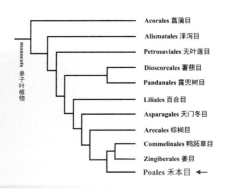

Perennial or annual **herbs** with rhizomes; xerophytic or aquatics. **Leaves** rosette, grass-like, linear, fenestrate, simple, alternate (usually spiral), margin entire, sheathing. **Inflorescence** capitate, less often spikes; scapes thin, twisted, angled, base surrounded by spathe-like sheath; involucral, small, dry, scaly. **Flowers** 2- or 3-merous, small (2–4 mm long), unisexual, with both sexes usually in same head; actinomorphic or sometimes zygomorphic. **Perianth** 2-whorled, in male flowers often inconspicuous; petal sometimes with nectary glands on outer surface. **Stamens** 1–2 whorls; anthers introrse; staminodes common in female flowers. **Ovary** superior; (1–)3-loculed; ovule 1 per locule, placentation apical; style 1; stigmas 2–3. **Fruit** a capsule, thin, loculicidal, or rarely an achene. **Seeds** small; testa usually reticulate and prickly; endosperm with abundant starch grains.

多年生或一年生**草本**，具根状茎；旱生或湿生。**叶**莲座状，似禾草，线形，具窗孔，单叶，互生（常螺旋状），全缘，具叶鞘。**花序**头状，少为穗状；花葶细，扭转，具棱角，基部被佛焰苞状鞘；总苞片小，干，鳞片状。**花** 2 或 3 基数，小（长 2~4 mm），单性，头状花序常具雌雄花；辐射对称或有时两侧对称。**花被片** 2 轮，在雄花中常不显；花瓣背面有时具蜜腺。**雄蕊** 1~2 轮；花药内向；雌花常具不育雄蕊。**子房**上位；(1~)3 室；胚珠每室 1 枚，顶生胎座；花柱 1 个，柱头 2~3 个。**果实**为蒴果，薄，室背开裂，或稀为瘦果。**种子**小；种皮常具网格状和刺突；胚乳具丰富淀粉粒。

World 11/ca.1,400, widely in tropics and subtropics, especially in tropical America.
China 1/ca. 35.
This area 1/1.

全世界共 11 属约 1 400 种，广泛分布于热带和亚热带，尤其热带美洲。
中国产 1 属约 35 种。
本地区有 1 属 1 种。

Eriocaulon L. 谷精草属

Eriocaulon buergerianum Körn/ 谷精草
（a. Habitat/ 生境；b. Roots/ 根系）

097 Juncaceae 灯芯草科

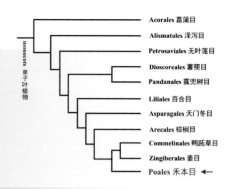

Stems erect, terete / 茎直立，圆柱形
Leaves 3-ranked, or reduced to a bladeless sheath / 叶 3 列茎生或仅存叶鞘
Perianth glume-like, dull-coloured / 花被片颖片状，颜色不显
Stigmas 3, feathery / 柱头 3 个，羽毛状

Perennial or annual **herbs**, rarely shrub-like, tufted or with erect or creeping rhizome; stems erect, terete or laterally flattened. **Leaves** linear, simple, alternate, 3-ranked or 2-ranked; occasionally reduced to a bladeless or nearly bladeless sheath at base of stem (cataphyll). **Inflorescence** a panicles, corymbs, or solitary flower. **Flowers** bisexual or unisexual (plants dioecious or monoecious), mostly wind pollinated, actinomorphic, usually small (<1 cm long). **Perianth** (3 or) 6, in (1 or) 2 whorls, free, glume-like, dull-coloured. **Stamens** 2-whorled, rarely 1-whorled, alternating with tepals; filaments thin; anthers basifixed, 2-loculed, dehiscing longitudinally. **Ovary** superior, 1-loculed, or divided by 3 septa and 3-loculed; ovules 3 and inserted at base of ovary, or numerous and biseriate on 3 parietal placentation; style 1; stigmas 3, feathery. **Fruit** a capsule, 1–3-valved, loculicidal. **Seeds** small, sometimes with tail-like appendage; embryo straight, enclosed by fleshy endosperm.

多年生或一年生**草本**，稀为灌木状，簇生或具直立或匍匐根状茎；茎直立，圆柱形或两侧扁平。**叶**线形，单叶，互生，3 列或 2 列；有时退化或在茎基部仅存叶鞘（低出叶）。**花序**圆锥状，伞房状或为单花。**花**两性或单性（植物雌雄异株或同株），多为风媒花，辐射对称，常小（短于 1cm）。**花被片**（3 或）6 枚，（1 或）2 轮，离生，颖片状，颜色不显。**雄蕊** 2 轮，稀 1 轮，与花被片互生；花丝薄；花药基着，2 室，纵裂。**子房**上位，1 室或 3 瓣裂 3 室；胚珠 3 枚，嵌生于子房基部，或为多数，于 3 个侧膜胎座上排为 2 列；花柱 1 个，柱头 3 个，羽毛状。**果实**为蒴果，1~3 瓣，室背开裂。**种子**小，有时具尾状附属物；胚直伸，为肉质胚乳所包。

World 7/ca. 450, mainly in temperate and cold regions, also in tropical montane habitats.
China 2/92.
This area 2/9.

全世界共 7 属约 450 种，主要分布于温带和寒带地区，热带山地也有。
中国产 2 属 92 种。
本地区有 2 属 9 种。

1. Leaves glabrous, leaf sheaths open; capsule with in many seeds/叶片光滑，叶鞘开放；蒴果种子多数 ....**1. *Juncus*/ 灯芯草属**

1. Leaves long white ciliate at margin, leaf sheaths closed; capsule 3-seeded/叶片边缘具白色长纤毛，叶鞘闭合；蒴果种子 3 枚 ....**2. *Luzula*/ 地杨梅属**

1. *Juncus setchuensis* Buchenau/ 野灯芯草（1a. Stems erect, terete/ 茎直立，圆柱形; 1b. Infrutescence/ 果序; 1c. Creeping rhizome/ 匍匐根状茎）;

2-1. *Luzula multiflora* (Ehrh.) Lej./ 多花地杨梅;
2-2. *L. campestris* (L.) DC./ 地杨梅（2-2a. Leaves/ 叶; 2-2b. Infrutescence/ 果序）

098 Cyperaceae 莎草科

Culms (=stems) solid, triangular / 秆（茎）坚硬，三棱形
Inflorescence terminal, composed of spikelets / 花序顶生，由小穗组成
Perianth absent or reduced to bristles or scales / 花被片缺失或退化成刚毛或鳞片
Fruit partially or completely enclosed by an enlarged basal prophyll / 果实部分或全部被增大的先出叶所包裹

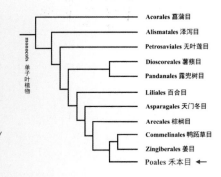

Annual or perennial **herbs**, rhizomatous to stoloniferous, rarely shrubs or woody climbers; culms (=stems) solid, triangular. **Leaves** basal and/or cauline, usually linear or reduced, simple, alternate (usually 3-ranked, rarely spiral or distichous); sheaths open or closed; ligules often present. **Inflorescence** terminal, composed of spikelets arranged in spikes, panicles, corymbs, heads or rarely solitary flower; bracts involucral, 1-many, leafy or scale-like; spikelets with 1-many. **Flowers** bisexual or unisexual with plants monoecious or rarely dioecious, actinomorphic; usually within distichous or imbricate and spiral bracts (= glumes). **Perianth** absent or reduced to bristles or scales. **Stamens** 1–3 or 6, or rarely many; anthers basifixed. **Ovary** superior, 2– or 3-carpellate, or 4; 1-loculed, ovule 1; style divided or rarely undivided, base sometimes persistent and variously shaped in fruit; stigmas 2 or 3; placentation basal or rarely apical. **Fruit** usually a hard biconvex or trigonous nutlet, rarely with a succulent or corky exocarp, surface smooth or variously minutely patterned, sometimes partially or completely enclosed by an enlarged basal prophyll (utricle).

一年生或多年生**草本**，具根状茎至匍匐根茎，稀为灌木或木质藤本；秆（茎）坚硬，三棱形。**叶**基生和/或茎生，常线形或退化，单叶，互生（常三列，稀螺旋状或二列）；叶鞘开放或闭合；常具叶舌。**花序**顶生，由小穗组成的穗状、圆锥状、伞房状和头状花序或稀为单花；总苞片1至多枚，叶状或鳞片状；小穗1至多数。**花**两性或单性，若单性则多为雌雄同株，稀雌雄异株，辐射对称；苞片（颖片）常为二列、覆瓦状或螺旋状。**花被片缺失或退化成刚毛或鳞片**。**雄蕊**1~3枚或6枚，或稀多数；花药基着。**子房**上位，心皮2枚或3枚，或4枚；1室，胚珠1枚；花柱分裂或稀不裂，基部有时宿存，在果期形状各异；柱头2个或3个；基底胎座或稀为顶生胎座。**果实**常为双凸状或三棱形坚果，稀有肉质或壳质外果皮，表面光滑或有各种精细纹饰，有时部分或全部被增大的先出叶（果囊）所包裹。

World 106/5, 400, worldwide, especially in North Temperate zone.
China 33/865.
This area 14/49.

全世界共 106 属 5 400 种，世界广布，尤其是北温带。
中国产 33 属 865 种。
本地区有 14 属 49 种。

1. All flowers unisexual/ 花全部为单性
   2. Female flowers and nutlets not enclosed by a utricle/ 雌花和小坚果不被果囊所包裹 ....**1. *Scleria*/ 珍珠茅属**
   2. Female flowers and nutlets enclosed by a utricle/ 雌花和小坚果被果囊所包裹 ....**2. *Carex*/ 薹草属**
1. At least some flowers bisexual/ 花多少为两性
   3. Spikelets much reduced, with 0–2 glumes, spikelets densely clustered into spikes or in a capitate/ 小穗简化，具 0~2 枚鳞片，小穗密集簇生于穗状或头状花序 ....**3. *Lipocarpha*/ 湖瓜草属**
   3. Spikelets not as above, elongated, with spirally or distichously arranged glumes/ 小穗与上不同，伸长，鳞片螺旋状或二列
     4. Spikelets usually with bisexual and male flowers/ 小穗常具两性花和雄花
       5. Nutlets biconvex, with persistent style base; stigmas 2/ 小坚果双凸状，花柱基宿存；柱头 2 枚 ....**4. *Rhynchospora*/ 刺子莞属**
       5. Nutlets trigonous, without persistent style base; stigmas 3/ 小坚果三棱形，无宿存花柱基；柱头 3 枚 ....**5. *Cladium*/ 一本芒属**
     4. Spikelets usually with bisexual flowers only/ 小穗只有两性花
       6. Style jointed with ovary and clearly demarcated from it/ 花柱与子房合生且有明显界限
         7. Leaf blades absent; perianth bristles 3–8/ 无叶片，下位刚毛 3~8 ....**6. *Eleocharis*/ 荸荠属**
         7. Leaf blades usually present; perianth bristles absent/ 通常具叶片；无下位刚毛 ....**7. *Fimbristylis*/ 飘拂草属**
       6. Style continuous with ovary and not demarcated from it/ 花柱与子房无明显界限
         8. Glumes distichous/ 鳞片排成 2 列
           9. Stigmas 3; nutlets trigonous/ 柱头 3 个；小坚果三棱形 ....**8. *Cyperus*/ 莎草属**
           9. Stigmas 2; nutlets biconvex/ 柱头 2 个；小坚果双凸状
              10. Spikelets with more than 2 glumes; spikelet axis and glumes persistent/ 小穗鳞片超过 2 枚；小穗轴和鳞片宿存 ....**9. *Pycreus*/ 扁莎属**
              10. Spikelets with 1 or 2 glumes; spikelet axis deciduous, spikelets falling whole/ 小穗鳞片 1 或 2 枚；小穗轴脱落，小穗整个脱落 ....**10. *Kyllinga*/ 水蜈蚣属**
         8. Glumes spirally arranged/ 鳞片螺旋状排列
           11. Perianth bristles present, 3 outer ones needle-like, 3 inner one suqamellate/ 下位刚毛存在，外轮 3 枚针状，内轮 3 枚具小鳞片 ....**11. *Fuirena*/ 芙兰草属**
           11. Perianth bristles present but not as above, or absent/ 花被不具刚毛或具刚毛但非上面所述
              12. Inflorescence paniculate/ 圆锥花序 ....**12. *Scirpus*/ 藨草属**
              12. Inflorescence a single spikelet, or up to 3 spikelets, or capitate/ 花序仅单个小穗，或 3 个，或头状
                 13. At least 2 involucral bracts over 1.5 cm, longest involucral bract leaf-like, erect to spreading/ 至少有 2 枚总苞片长于 1.5 cm，最长总苞片叶状，直立或开展 ....**13. *Bolboschoenus*/ 三棱草属**
                 13. One involucral bract over 1.5 cm, longest involucral bract erect, culm-like and apparently continuous with culm, inflorescence appearing to be lateral/ 仅 1 枚总苞片长于 1.5cm，最长总苞片直立，似秆，与秆无明显界限，花序犹如侧生 ....**14. *Schoenoplectus*/ 水葱属**

1. *Scleria* sp./ 珍珠茅属；
2. *Carex pumila* Thunb./ 矮生薹草；
3. *Lipocarpha microcephala* (R. Br) Kunth/ 湖瓜草
   （photo by XinXin ZHU 朱鑫鑫）；
4. *Rhynchospora rubra* (Lour.) Makino/ 刺子莞
   （4a. Nutlets and spikelets/ 小坚果与小穗；4b. Infructescence/ 果序）；
5. *Cladium jamaicense* Crantz subsp. *chinense* (Nees) T. Koyama/ 一本芒
6. *Eleocharis geniculata* (L.) Roem. & Schult./ 黑籽荸荠

7. *Fimbristylis sericea* R. Br./ 绢毛飘拂草；
8. *Cyperus cyperoides* (L.) Kuntze/ 砖子苗；
9. *Pycreus flavidus* (Retz.) T. Koyama/ 球穗扁莎；
10. *Kyllinga brevifolia* Rottb. var. *leiolepis* (Franch. & Sav.) H. Hara/ 无刺鳞水蜈蚣；
11. *Fuirena ciliaris* (L.) Roxb./ 毛芙兰草；
12. *Scirpus karuisawensis* Makino/ 华东藨草（photo by XinXin ZHU/ 朱鑫鑫拍摄）；
14. *Schoenoplectus tabernaemontani* (C. C. Gmel.) Palla/ 水葱

103 Poaceae 禾本科

Culms hollow, round, and noded / 秆中空，圆柱形，有节
Leaves distichous, sheath free / 叶二列，叶鞘开放
Inflorescence consisted by spikelets / 花序由小穗组成
Caryopsis / 颖果

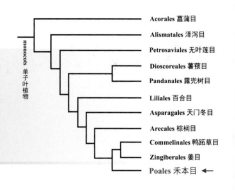

Herbs, rhizomatous, stoloniferous or annual. **Culms** (=stems) rounded or flattened, noded, often hollow. **Leaves** simple, alternate and usually distichous; sheaths usually free and overlapping, or fused; ligule at junction of blade and sheath, membranous, or a row of hairs, rarely absent; blade usually linear, rarely broad and short, flat or rolled, parallel-veined. **Inflorescence** a spikes, panicles, cymes, or racemes, consisted by spikelets. **Spikelets** of closely overlapping basal bracts (glumes) and distichous florets. **Florets** composed of lemmas (= lower bracts), paleas (=upper bracts) and between them flowers. **Flowers** mostly bisexual, usually with ovary, stamens, and 2 or 3 minute fleshy scales (lodicules). **Stamens** often 3, with versatile anthers. **Ovary** superior; 1-loculed, ovule 1 per locule; styles (1–) 2 (–3), stigmas usually 2, feathery. **Fruit** a caryopsis, with the pericarp adhering to the seed, or a capsule, nut, utricle or berry.

草本，常具地下茎或匍匐茎或为一年生。**秆**（茎）圆形或稍扁，节显著，节间常中空。**叶**为单叶，互生，通常二列；叶鞘通常开放，边缘重叠或愈合；叶舌位于叶鞘和叶片连接处，膜质或为一圈毛，稀无叶舌；叶片常线形，稀宽短，卷成管状或开展，平行脉。**花序**为由小穗组成的穗状、圆锥状、聚伞圆锥状或总状花序。**小穗**由二列紧密重叠的颖片和二列小花组成。**小花**由外、内稃和其中的花组成。花大多两性，通常由子房、雄蕊和 2 或 3 片肉质浆片组成。**雄蕊**常 3 枚，花药丁字着生。**子房**上位；1 室，每室胚珠 1 枚；花柱 (1~)2(~3)，柱头常 2 个，羽毛状。**果实**为颖果，果皮紧密包被种子，或为蒴果、坚果、胞果或浆果。

World 700/11, 000, worldwide.
China 227/1, 797.
This area 70/ca.130 (18 tribes).

全世界共 700 属 11 000 种，世界广布。
中国产 227 属 1 797 种。
本地区有 70 属约 130 种（隶属 18 族）。

1. Bamboos, culms woody/ 竹类，秆木质 .... **I. Bambuseae/ 簕竹族**
1. Herbs, occasionally reeds or culms cane-like/ 草本，有时芦苇状或秆似甘蔗状
    2. Spikelets arranged in pairs (rarely 3) on racemes/ 总状花序各节的小穗成对（稀 3）
        3. Spikelets disarticulating above glumes/ 小穗成熟时脱节于颖之上方
            4. Lemmas awnless/ 小花的外稃均无芒 .... **II. Isachneae/ 柳叶箬族**
            4. Upper lemma often awned/ 上方小花的外稃常具芒 .... **III. Arundinelleae/ 野古草族**

3. Spikelets falling entire at maturity/ 小穗成熟时整体脱落
    5. Fertile floret awnless/ 两性小花无芒 ....**IV. Paniceae**/ 黍族
    5. Fertile floret often with geniculate awn/ 两性小花常具膝状芒 ....**V. Andropogoneae**/ 高粱族
2. Spikelets arranged singly in panicles or racemes/ 小穗在圆锥花序或总状花序轴上单生
  6. Spikelets with strictly 2 florets; lower floret staminate or barren, upper floret fertile/ 小穗具 2 朵花；下方的小花为雄性或为不育花，上方的小花可育
    7. Spikelets disarticulating above glumes/ 小穗成熟时脱节于颖之上方
      8. Lemmas awnless/ 小花的外稃均无芒 ....**II. Isachneae**/ 柳叶箬族
      8. Upper lemma often awned/ 上方小花的外稃常具芒 ....**III. Arundinelleae**/ 野古草族
    7. Spikelets falling entire at maturity/ 小穗成熟时整体脱落
      9. Fertile floret awnless/ 两性小花无芒 ....**IV. Paniceae**/ 黍族
      9. Fertile floret often with geniculate awn/ 两性小花常具膝状芒 ....**V. Andropogoneae**/ 高粱族
  6. Spikelets with 1 to many florets (if 2 florets, then both fertile, or the lower fertile)/ 小穗具 1 至多朵小花 (如为 2 朵花，则均为两性花，或下方的小花两性)
    10. Spikelets with 2 or more fertile florets/ 小穗具 2 至多朵两性花
      11. Inflorescence of one or more racemes/ 花序由 1 或多个总状花序构成
        12. Ligule a row of hairs; lemmas 1–3-veined/ 叶舌为一圈毛；外稃具 1 ~ 3 条脉 ....**VI. Eragrostideae**/ 画眉草族
        12. Ligule membranous; lemmas 5 or more veined/ 叶舌膜质；外稃 5 至多脉
          13. Leaf sheaths tubular/ 叶鞘管状 ....**VII. Meliceae**/ 臭草族
          13. Leaf sheaths not tubular/ 叶鞘不为管状
            14. Ovary with hairy apical appendage/ 子房顶端具被毛的附属物
              15. Spikelets shortly pedicellate/ 小穗具短梗 ....**VIII. Brachypodieae**/ 短柄草族
              15. Spikelets sessile/ 小穗无梗 ....**IX. Triticeae**/ 小麦族
            14. Ovary glabrous/ 子房无毛 ....**X. Poeae**/ 早熟禾族
      11. Inflorescence a panicles/ 圆锥花序
        16. Spikelets with 2 florets, rachilla extension absent/ 小穗具 2 朵小花, 小穗轴不延伸 ....**II. Isachneae**/ 柳叶箬族
        16. Spikelets with several florets, or if 2 then rachilla extend/ 小穗具多花, 如仅具 2 朵花, 则小穗轴延伸
          17. Leaf sheaths tubular, margin joined for most or all of length/ 叶鞘管状，边缘大部分或全部连合 ....**VII. Meliceae**/ 臭草族
          17. Leaf sheaths not tubular, margin free/ 叶鞘不为管状，边缘分离
            18. Culms reed-like, usually tall/ 主秆芦苇状，通常高大 ....**XI. Arundineae**/ 芦竹族
            18. Culms mostly slender, if tall then not reed-like/ 主秆通常瘦弱，如果高大则不为芦苇状
              19. Ligule a row of hairs/ 叶舌为一圈毛 ....**VI. Eragrostideae**/ 画眉草族
              19. Ligule membranous/ 叶舌膜质
                20. Leaf blades with obvious cross veins/ 叶片具明显的横脉 ....**XII. Centotheceae**/ 假淡竹叶族

20. Leaf blades without cross veins/ 叶片无横脉
 21. Glumes usually as long as spikelet, always longer than lowest lemma/ 颖片通常与小穗等长，并总是长于第一外稃 .... **XIII. Aveneae**/ 燕麦族
 21. Glumes shorter than spikelet, usually shorter than lowest lemma/ 颖片短于小穗，通常比第一外稃短
  22. Ovary glabrous or hairy, styles arising from its apex/ 子房无毛或被毛，花柱自顶端伸出 .... **X. Poeae**/ 早熟禾族
  22. Ovary with a hairy apical appendage, styles arising beneath it/ 子房顶端具被毛的附属物，花柱生于子房的前下方 .... **XIV. Bromeae**/ 雀麦族
10. Spikelets with 1 fertile floret, sometimes with additional staminate or barren florets/ 小穗具 1 朵两性花，有时有雄花或不育花
 23. Glumes absent or both very short/ 颖片缺或内外颖均很短小 .... **XV. Oryzeae**/ 稻族
 23. Glumes well developed, at least the upper/ 颖片正常发育，至少内颖如此
  24. Leaf blades with cross veins/ 叶片具横脉
   25. Leaf blades with twisted pseudopetiole; caryopsis globose/ 叶片具扭转的假叶柄；果球形 .... **XVI. Phaenospermateae**/ 显子草族
   25. Leaf blades not pseudopetiolate or twisted; caryopsis ovoid to trigonous/ 叶片无假叶柄，不扭转；颖果卵形至三棱形
    26. Glumes persistent/ 颖片宿存 .... **XII. Centotheceae**/ 假淡竹叶族
    26. Glumes deciduous/ 颖片脱落 .... **II. Isachneae**/ 柳叶箬族
  24. Leaf blades without cross veins/ 叶片无横脉
   27. Inflorescence composed of one or more racemes/ 花序由 1 或多个总状花序构成
    28. Glumes shorter than floret; lemma awnless or with terminal straight awn/ 颖片短于小花；外稃无芒或先端具直芒 .... **X. Poeae**/ 早熟禾族
    28. Glumes longer than floret, or lemma with dorsal or geniculate awn/ 颖片长于小花，如短，则外稃背部具芒或具膝状芒
     29. Lemma 5-veined; spikelets orbicular/ 外稃 5 脉，小穗圆形 .... **XIII. Aveneae**/ 燕麦族
     29. Lemma 1–3-veined; spikelets not as above/ 外稃 1~3 条脉，小穗不为上述形态 .... **XVII. Cynodonteae**/ 虎尾草族
   27. Inflorescence a panicles/ 圆锥花序
    30. Ligule a row of hairs/ 叶舌为一圈毛 .... **VI. Eragrostideae**/ 画眉草族
    30. Ligule membranous/ 叶舌膜质 .... **XVIII. Stipeae**/ 针茅族

POALES. Poaceae. *Poa annua* L./ 禾本目 禾本科 早熟禾

A. Whole plant/ 植株
B. Spikelet segregated/ 小穗离析
C. Inflorescence/ 花序
D. Ligule/ 叶舌
E. Spikelets/ 小穗
F. Pistil and stamen/ 雌雄蕊
G. Pistil and stamen, amplified/ 雌雄蕊，放大
(Scale/ 标尺： A 1 cm; B 1mm; C 1 cm; D-G 1 mm)

一、Bambuseae/ 簕竹族
I. Bambuseae/ 簕竹族

1. Rhizome pachymorph/ 地下茎为合轴型 .... **1. *Bambusa*/ 簕竹属**
1. Rhizome leptomorph/ 地下茎为单轴型
    2. Spikelets sessile/ 小穗无柄 .... **2. *Phyllostachys*/ 刚竹属**
    2. Spikelets pedicellate/ 小穗具柄 .... **3. *Pleioblastus*/ 苦竹属**

1. *Bambusa ventricosa* McClure/ 佛肚竹（Natrulized/ 归化）；
2. *Phyllostachys nidularia* Munro / 篌竹；
3. *Pleioblastus amarus* (Keng) Keng f./ 苦竹（photo by Hai HE/ 何海拍摄）

II. Isachneae/ 柳叶箬族

1. Upper lemma indurate/ 上方的花外稃变硬 .... **1. *Isachne*/ 柳叶箬属**
1. Upper lemma membranous/ 上方的花外稃膜质 .... **2. *Sphaerocaryum*/ 稃荩属**

1. *Isachne albens* Trin./ 白花柳叶箬；
2. *Sphaerocaryum malaccense* (Trin.) Pilg./ 稃荩

III. Arundinelleae/ 野古草族

Only *Arundinella* Raddi

仅野古草属 1 属

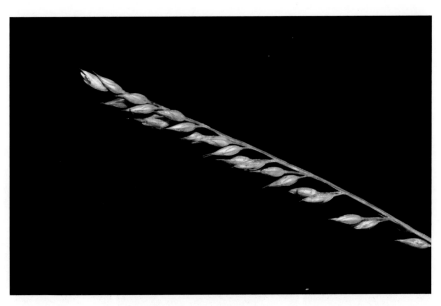

Arundinella hirta (Thunb.) Tanaka/ 毛秆野古草（photo by XinXin ZHU/ 朱鑫鑫拍摄）

IV. Paniceae/ 黍族

1. Plants dioecious/ 植物体雌雄异株 ....**1. *Spinifex*/ 鬣刺属**
1. Plants bisexual/ 植物体两性
  2. Spikelets subtended by bristles or a spiny involucre/ 小穗基部具有刚毛或具刺的总苞
    3. Bristles persisting on the axis after the spikelets have fallen/ 小穗脱落后刚毛仍宿存于穗轴上 ....**2. *Setaria*/ 狗尾草属**
    3. Bristles or spines falling as involucres around the spikelet/ 刚毛或刺连同小穗一起脱落 ....**3. *Pennisetum*/ 狼尾草属**
  2. Spikelets not subtended by bristles/ 小穗基部无刚毛
    4. Inflorescence a spike-like panicles/ 花序为成穗状的圆锥花序
      5. Spikelets laterally compressed/ 小穗两侧压扁
        6. Lower lemma entire, awnless/ 下方的花外稃全缘，无芒 ....**4. *Cyrtococcum*/ 弓果黍属**
        6. Lower lemma bilobed or shortly awned/ 下方的花外稃2裂或具短芒 ....**5. *Melinis*/ 糖蜜草属**
      5. Spikelets dorsally compressed/ 小穗背腹压扁
        7. Panicles spike-like/ 圆锥花序穗状 ....**6. *Sacciolepis*/ 囊颖草属**
        7. Panicles open/ 圆锥花序开展
          8. Upper glume as long as or only slightly shorter than the spikelet/ 内颖与小穗等长或略短 ....**7. *Panicum*/ 黍属**
          8. Upper glume 2/3 spikelet length or less/ 内颖长不超过小穗长的2/3 ....**8. *Digitaria*/ 马唐属**
    4. Inflorescence of unilateral racemes/ 花序为单侧的总状花序
      9. Spikelets laterally compressed; lower glumes awned/ 小穗两侧压扁或外颖具芒 ....**9. *Oplismenus*/ 求米草属**
      9. Spikelets dorsally compressed; lower glume awnless/ 小穗背腹压扁，外颖无芒
        10. Spikelets densely packed in 4 rows or congested into clusters/ 小穗密集成4列或成簇 ....

.... **10.** *Echinochloa*/ 稗属
  10. Spikelets mostly in 1 or 2 rows/ 小穗通常排成 1 或 2 列
    11. Spikelets supported on a basal bead-like swelling/ 小穗具珠状的基盘 ....
.... **11.** *Eriochloa*/ 野黍属
    11. Spikelets without a basal bead-like swelling/ 小穗不具珠状的基盘 .... **12.** *Paspalum*/ 雀稗属

1. *Spinifex littoreus* (Burm. f.) Merr./ 老鼠芳（1a. Infructescence/ 果序；1b. Male inflorescence/ 雄花序）；
2. *Setaria faberi* R.A.W. Herrm./ 大狗尾草；
3. *Pennisetum alopecuroides* Spreng./ 狼尾草；
4. *Cyrtococcum patens* (L.) A. Camus/ 弓果黍；
5. *Melinis repens* (Willd.) Zizka/ 红毛草；
6. *Sacciolepis indica* (L.) Chase/ 囊颖草；
7-1. *Panicum sumatrense* Roth/ 细柄黍；
7-2. *P. repens* L./ 铺地黍；
8. *Digitaria ciliaris* (Retz.) Koeler/ 纤毛马唐；
9. *Oplismenus undulatifolius* (Ard.) P. Beauv. var. *binatus* S.L. Chen & Y.X. Jin/ 双穗求米草

10. *Echinochloa colona* (L.) Link/ 光头稗；
11. *Eriochloa villosa* (Thunb.) Kunth/ 野黍；
12. *Paspalum scrobiculatum* L. var. *orbiculare* (G. Forst.) Hack./ 圆果雀稗（12a. Inflorescence/ 花序；12b. Caryopsis/ 颖果）

V. Andropogoneae/ 高粱族

1. Spikelets all unisexual/ 小穗单性 .... **1. *Coix*/ 薏苡属**
1. Spikelets all bisexual/ 小穗均为两性
  2. Spikelets of a pair similar in shape, usually both fertile/ 成对小穗均可育且同形
    3. Spikelets apex truncate/ 小穗先端平截 .... **2. *Pogonatherum*/ 金发草属**
    3. Spikelets apex narrow/ 小穗先端狭窄
      4. Rachis joint absent/ 穗轴无关节
        5. Spikelets awnless; panicles contracted or spike-like/ 小穗无芒；圆锥花序紧缩狭窄而呈穗状 .... **3. *Imperata*/ 白茅属**
        5. Spikelets awned; panicles loose/ 小穗常有芒；圆锥花序开展
          6. Racemes rachis fragile/ 花序的分枝细弱 .... **4. *Spodiopogon*/ 大油芒属**
          6. Racemes rachis tough/ 花序分枝强壮 .... **5. *Miscanthus*/ 芒属**
      4. Rachis joint present/ 穗轴有关节
        7. Inflorescence on a shortened central axis/ 总状花序生于短缩的主轴 .... **6. *Eulalia*/ 黄金茅属**
        7. Inflorescence on an elongated central axis/ 总状花序生于延长的主轴
          8. Inflorescence nearly sessile/ 花序近无梗 .... **7. *Saccharum*/ 甘蔗属**
          8. Inflorescence pedicellate/ 花序有梗 .... **4. *Spodiopogon*/ 大油芒属**
  2. Spikelets of a pair different in shape and sex/ 成对小穗不同型也不同性
    9. Rachis internodes and pedicels stout/ 穗轴节间及小穗柄粗短
      10. The second lemma awned/ 第 2 外稃具芒
        11. Racemes solitary/ 总状花序单独 1 枚 .... **8. *Apluda*/ 水蔗属**
        11. Racemes 2 or more/ 总状花序 2 枚或更多 .... **9. *Ischaemum*/ 鸭嘴草属**
      10. The second lemma awnless/ 第 2 外稃无芒

12. Racemes usually digitate/ 总状花序近指状排列 .... **10.** *Phacelurus*/ 束尾草属
12. Racemes all solitary/ 总状花序均为单生
  13. Inflorescence terminal/ 花序顶生 .... **11.** *Eremochloa*/ 假俭草属
  13. Inflorescence from leaf axil/ 花序腋生 .... **12.** *Rottboellia*/ 筒轴茅属
9. Rachis internodes and pedicels slender/ 穗轴节间及小穗柄细长
  14. Awn arising from low down on lemma/ 芒近稃体基部处着生 .... **13.** *Arthraxon*/ 荩草属
  14. Awn not arising from low down on lemma/ 芒并非着生于稃体基部
    15. Racemes paniculate arranged/ 总状花序呈圆锥状排列
      16. Racemes of 2–7 spikelet pairs/ 总状花序含 2~7 对小穗 .... **14.** *Sorghum*/ 高粱属
      16. Racemes of more than 8 spikelet pairs/ 总状花序含 8 对以上小穗 .... **15.** *Bothriochloa*/ 孔颖草属
    15. Racemes solitary or paired/ 总状花序单独 1 个或成对
      17. Lower glume of sessile spikelet without keels/ 无柄小穗第一颖无脊 .... **16.** *Themeda*/ 菅属
      17. Lower glume of sessile spikelet 2-keeled/ 无柄小穗第一颖内折成 2 脊 .... **17.** *Cymbopogon*/ 香茅属

1. *Coix lacryma-jobi* L./ 薏苡；
2. *Pogonatherum crinitum* (Thunb.) Kunth/ 金丝草（2a. Whole plant/ 植株；2b. Spikelets/ 小穗）；
3. *Imperata cylindrica* (L.) Raeusch. var. *major* (Nees) C.E. Hubb./ 大白茅（3a. Spikelets/ 小穗；3b. Culm node/ 节）；
4. *Spodiopogon sibiricus* Trin./ 大油芒（photo by XinXin ZHU 朱鑫鑫拍摄）

5. *Miscanthus floridulus* (Labill.) Warb. ex K. Schum. & Lauterb./ 五节芒；

6. *Eulalia wightii* (Hook. f.) Bor/ 魏氏金茅（Showing shortened central axis, not distributed in this area, photo by XinXin ZHU/ 朱鑫鑫拍摄）；

7. *Saccharum spontaneum* L./ 甜根子草；

8. *Apluda mutica* L./ 水蔗草；

9. *Ischaemum* sp./ 鸭嘴草属；

10. *Phacelurus* sp./ 束尾草属；

11. *Eremochloa ophiuroides* (Munro) Hack./ 假俭草；

12. *Rottboellia cochinchinensis* (Lour.) Clayton/ 筒轴茅（photo by XinXin ZHU/ 朱鑫鑫拍摄）；

13. *Arthraxou hispidus* (Thunb.) Makino/ 荩草；

14-1. *Sorghum propinquum* (Kunth) Hitchc./ 拟高粱；

14-2. *S. nitidum* (Vahl) Pers./ 光高粱；

15. *Bothriochloa ischaemum* (L.) Keng/ 白羊草（photo by Yuan WANG/ 汪远拍摄）

16. *Themeda triandra* Forssk./ 黄背草； 17. *Cymbopogon goeringii* (Steud.) A. Camus/ 橘草

VI. Eragrostideae/ 画眉草族

1. Spikelets with 1 floret/ 小穗具 1 朵花 .... **1.** *Sporobolus*/ 鼠尾粟属
1. Spikelets with 2 or more florets/ 小穗具 2 至多朵花
    2. Lemmas emarginate or 2-toothed at apex/ 外稃顶端微凹或具 2 齿
        3. Inflorescence a large panicles/ 花序为大型圆锥花序 .... **2.** *Neyraudia*/ 类芦属
        3. Inflorescence composed of racemes/ 花序由总状花序构成 .... **3.** *Leptochloa*/ 千金子属
    2. Lemmas usually entire at apex, glabrous/ 外稃顶端通常全缘，无毛
        4. Spikelets sessile/ 小穗无柄
            5. Racemes terminating in a spikelet/ 花序有顶生小穗 .... **4.** *Eleusine*/ 穇属
            5. Racemes terminating without spikelet/ 花序无顶生小穗 .... **5.** *Dactyloctenium*/ 龙爪茅属
        4. Spikelets pedicellate/ 小穗具柄 .... **6.** *Eragrostis*/ 画眉草属

1. *Sporobolus fertilis* (Steud.) Clayton/ 鼠尾粟（1a. Infructescence/ 果序； 1b. Caryopsis/ 颖果）；
2. *Neyraudia reynaudiana* (Kunth) Keng ex Hitchc./ 类芦（2a. Large panicles/ 大型圆锥花序； 2b. Spikelets/ 小穗）；
3. *Leptochloa chinensis* (L.) Nees/ 千金子

4. *Eleusine indica* (L.) Gaertn./ 牛筋草（4a. Infructescence/ 果序；4b. Spikelets/ 小穗）；

5. *Dactyloctenium aegyptium* (L.) Willd./ 龙爪茅（5a. Infructescence/ 果序；5b. Inflorescence/ 花序）；

6-1. *Eragrostis ferruginea* (Thunb.) P. Beauv./ 知风草；

6-2. *E. cumingii* Steud./ 珠芽画眉草

VII. Meliceae/ 臭草族

Only *Schizachne* Hack.
仅裂稃茅属 1 属

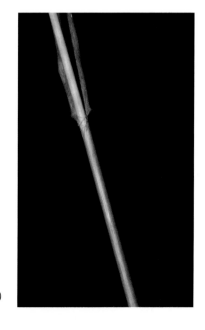

Schizachyrium delavayi (Hack.) Bor/ 旱茅（Showing leaf sheath, not distributed in this area, photo by XinXin ZHU/ 仅展示叶鞘，非本区域分布，朱鑫鑫拍摄）

VIII. Brachypodieae/ 短柄草族

Only *Brachypodium* P. Beauv.

仅短柄草属 1 属

IX. Triticeae/ 小麦族

Only *Elymus* L.

仅披碱草属 1 属

1. *Brachypodium sylvaticum* (Huds.) P. Beauv./ 短柄草（photo by XinXin ZHU/ 朱鑫鑫拍摄）；
2. *Elymus kamoji* (Ohwi) S.L. Chen/ 柯孟披碱草

X. Poeae/ 早熟禾族

1. Spikelets pedicellate/ 小穗具柄
    2. Plants annual/ 一年生植物
        3. Lemmas awned/ 外稃具芒 ....**1. *Vulpia*/ 鼠茅属**
        3. Lemmas awnless/ 外稃无芒
            4. Spikelets pedicels filiform; lemmas base cordate/ 小穗柄丝状；外稃基部心形 ....**2. *Briza*/ 凌风草属**
            4. Spikelets pedicels slender; lemmas base not cordate/ 小穗柄细瘦；外稃基部不呈心形 ....
            ....**3. *Poa*/ 早熟禾属**
    2. Plants perennial/ 多年生植物
        5. Lemmas rounded on back/ 外稃背面圆 ....**4. *Festuca*/ 羊茅属**
        5. Lemmas keeled throughout/ 外稃从基部到顶部具龙骨突
            6. Florets horizontally spreading/ 小花水平排列 ....**2. *Briza*/ 凌风草属**
            6. Florets spreading upward/ 小花向上伸展 ....**3. *Poa*/ 早熟禾属**
1. Spikelets sessile/ 小穗无柄
    7. Spikelets with several florets/ 小穗具数朵小花 ....**5. *Lolium*/ 黑麦草属**
    7. Spikelets with one floret/ 小穗具 1 朵花 ....**6. *Parapholis*/ 假牛鞭草属**

1. *Vulpia myuros* (L.) C.C. Gmel./ 鼠茅；
2. *Briza minor* L./ 银鳞茅（2a. Inflorescence/ 花序；2b. Plant/ 植株；naturalized/ 归化）；
3. *Poa annua* L./ 早熟禾（3a. Plant/ 植株；3b. Spikelets/ 小穗，photo by WeiLiang MA/ 由马炜梁拍摄）；
4. *Festuca parvigluma* Steud./ 小颖羊茅；
5. *Lolium* sp./ 黑麦草属；
6. *Parapholis incurva* (L.) C.E. Hubb./ 假牛鞭草（6a. Inflorescence/ 花序；6b. Habitat/ 生境）

XI. Arundineae/ 芦竹族

1. Floret callus with long spreading hairs; lemma glabrous; ligule ciliate/ 小穗基盘被开展长柔毛；外稃无毛；叶舌被纤毛 .... **1. *Phragmites*/ 芦苇属**
1. Floret callus pilose; lemmas back with spreading long silky-white hairs; ligule membranous / 小穗基盘被短柔毛；外稃背部有白色丝质开展长毛；叶舌膜质 .... **2. *Arundo*/ 芦竹属**

1. *Phragmites australis* (Cav.) Trin. ex Steud./ 芦苇； 2. *Arundo donax* 'Versicolor'/ 芦竹'花叶'（Show whole plant, photo by Yuan WANG/ 仅展示植株，汪远拍摄）

XII. Centotheceae/ 假淡竹叶族

Only *Lophatherum* Brongn.

仅淡竹叶属 1 属

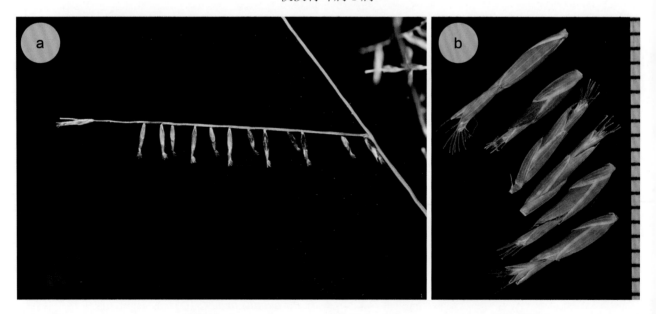

Lophatherum gracile Brongn./ 淡竹叶（a. Infructescence/ 果序； b. Matured spikelets/ 成熟小穗）

XIII. Aveneae/ 燕麦族

1. Spikelets with 2 or more fertile florets/ 小穗具 2 或多朵可育小花
    2. Ovary hairy/ 子房被毛 ....**1. *Avena*/ 燕麦属**
    2. Ovary glabrous or almost so/ 子房无毛或近无毛 ....**2. *Deyeuxia*/ 野青茅属**
1. Spikelets with 1 fertile floret/ 小穗具 1 朵可育小花
    3. Inflorescence of several racemes along a central axis/ 总状花序排列于穗轴的一侧 ....
    ....**3. *Beckmannia*/ 茵草属**
    3. Inflorescence a panicles, sometimes spike-like/ 圆锥花序，有时呈穗状
        4. Fertile floret accompanied by staminate or sterile florets/ 可育小花与雄性的或不育的小花聚生 ....
        ....**4. *Phalaris*/ 虉草属**
        4. Fertile floret solitary/ 可育小花单生
            5. Spikelets falling entirely/ 小穗完整地脱落
                6. Spikelets shed with a basal stipe/ 小穗连同基部的柄一起脱落 ....**5. *Polypogon*/ 棒头草属**
                6. Spikelets shed without a basal stipe/ 脱落的小穗基部不具柄 ....**6. *Alopecurus*/ 看麦娘属**
            5. Spikelets disarticulating above glumes/ 小穗脱节于颖之上
                7. Glumes slightly shorter than floret/ 颖片略短于小花 ....**2. *Deyeuxia*/ 野青茅属**
                7. Glumes equaling or longer than floret/ 颖片等于或长于小花
                    8. Callus glabrous or shortly hairy; lemmas hyaline/ 基盘无毛或具短毛；外稃透明 ....
                    ....**7. *Agrostis*/ 剪股颖属**
                    8. Callus bearded; lemmas membranous to firm/ 基盘具髯毛；外稃膜质至坚硬
                        9. Lemma callus hairs almost as long as or clearly shorter than floret/
                        外稃基盘的毛等于或明显短于小花 ....**2. *Deyeuxia*/ 野青茅属**
                        9. Lemma callus hairs often much exceeding floret/ 外稃基盘的毛通常显著长于小花 ....
                        ....**8. *Calamagrostis*/ 拂子茅属**

1. *Avena fatua* L./ 野燕麦；   2. *Deyeuxia effusiflora* Rendle/ 疏穗野青茅

3. *Beckmannia syzigachne* (Steud.) Fernald/ 菵草（3a. Spikelets/ 小穗；3b. Inflorescence/ 花序）；

4. *Phalaris arundinacea* L./ 虉草（photo by XinXin ZHU/ 朱鑫鑫拍摄）；

5. *Polypogon fugax* Nees ex Steud./ 棒头草（5a. Infructescence/ 果序；5b. Caryopsis/ 颖果）；

6. *Alopecurus aequalis* Sobol./ 看麦娘（6a.Inflorescence/ 花序；6b. Caryopsis/ 颖果）；

7. *Agrostis* sp./ 剪股颖属；

8. *Calamagrostis epigeios* (L.) Roth/ 拂子茅（8a. Infructescence/ 果序；8b. Caryopsis/ 颖果）

XIV. Bromeae/ 雀麦族

Only *Bromus* L.

仅雀麦属 1 属

1-1. *Bromus catharticus* Vahl/ 扁穗雀麦；

1-2. *B. remotiflorus* (Steud.) Ohwi/ 疏花雀麦

XV. Oryzeae/ 稻族

1. Florets sessile/ 小花无柄 ....**1. *Leersia*/ 假稻属**
1. Florets pedicellate/ 小花有柄 ....**2. *Zizania*/ 菰属**

1. *Leersia sayanuka* Ohwi/ 秕壳草；

2. *Zizania latifolia* (Griseb.) Turcz. ex Stapf/ 菰（photo by XinXin ZHU/ 朱鑫鑫拍摄）

XVI. Phaenospermateae/ 显子草族

Only *Phaenosperma* Munro ex benth

仅显子草属 1 属

Phaenosperma globosum Munro ex Benth./ 显子草

（a. Infructescence/ 果序；b. Inflorescence/ 花序；c. Whole Plants/ 植株）

XVII. Cynodonteae/ 虎尾草族

1. Inflorescence cylindrical, deciduous/ 花序圆柱形，脱落 ....**1. *Zoysia*/ 结缕草属**
1. Inflorescence not cylindrical, persistent or rarely deciduous/ 花序不为圆柱状，宿存，稀脱落
   2. Racemes borne along an axis/ 总状花序沿轴着生 ....**2. *Spartina*/ 米草属**
   2. Racemes digitate, subdigitate or solitary/ 总状花序指状、近指状排列，或仅 1 个
     3. Fertile floret solitary/ 可育小穗单生 ....**3. *Cynodon*/ 狗牙根属**
     3. Fertile floret accompanied by male or sterile florets/ 可育小穗伴生雄性或不育小穗
....**4. *Chloris*/ 虎尾草属**

1. *Zoysia japonica* Steud./ 结缕草；
2. *Spartina alterniflora* Loisel./ 互花米草；
3. *Cynodon dactylon* (L.) Pers./ 狗牙根（3a. Racemes digitate/ 总状花序指状；3b. Spikelets/ 小穗）

4. *Chloris formosana* (Honda) Keng ex B.S. Sun & Z.H. Hu/ 台湾虎尾草

（4a. Inflorescence/ 花序 ； 4b. Whole plant/ 植株）

XVIII. Stipeae/ 针茅族

Only *Achnatherum* P. Beauv.

仅芨芨草属 1 属

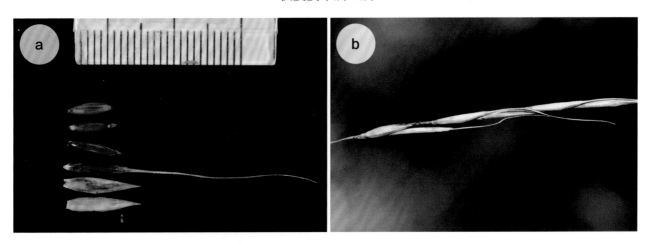

Achnatherum coreanum (Honda) Ohwi/ 大叶直芒草

（a. Caryopsis/ 颖果；b. Infructescence/ 果序）

104 Ceratophyllaceae 金鱼藻科

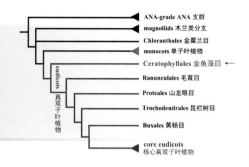

Perennial aquatic herbs, rootless / 多年生水生草本，无根
Leaves often dichotomously dissected / 叶常二叉状细裂
Flowers unisexual, monoecious / 花单性，雌雄同株
Seed 1, unitegmic; endosperm absent / 种子1粒，单珠被；无胚乳

Perennial aquatic **herbs**, rootless, but often anchored by colorless root-like branches. **Leaves** submerged or floating, simple, opposite to 3–12 whorled; often dichotomously dissected, filiform, entire to serrate, stipules absent. **Inflorescence** axillary solitary flower, ±sessile, male flowers and female flowers at different nodes. **Flowers** unisexual, plants monoecious, actinomorphic, inconspicuous, with a whorl of 7 to numerous bracts. **Perianth** bract-like, basally fused, apex spiny. **Stamens** 10 to numerous, free, anthers fused with appendages; anthers 2-loculed, dehiscing longitudinally. **Ovary** superior, carpel 1; 1-loculed; ovule 1; placentation apical; style persistent. **Fruit** an achene, often with 2 or more spines. **Seed** 1, unitegmic; endosperm and perisperm absent.

多年生水生**草本**，无根，但具无色根状分枝使植株附着。**叶**沉水或浮水，单叶，对生至3~12枚轮生；常二叉状细裂，丝状，全缘至有锯齿，托叶缺。**花序**为腋生单花，近无柄，雄花序与雌花序着生于不同节上。**花**单性，雌雄同株，辐射对称，不显著，苞片1轮，7至多枚。**花被片**苞片状，基部合生，先端刺状。**雄蕊**10枚至多数，分离，花药与附属物合生；花药2室，纵裂。**子房**上位，心皮1枚；1室；胚珠1枚；顶生胎座；花柱宿存。**果实**为瘦果，常有2个或多个刺状突起。**种子**1粒，单珠被；无胚乳和外胚乳。

World 1/6, worldwide.
China 1/3.
This area 1/1.

全世界共1属6种，世界广布。
中国产1属3种。
本地区有1属1种。

Ceratophyllum L. 金鱼藻属

Ceratophyllum demersum L./ 金鱼藻
（photo by WeiLiang MA/ 由马炜梁拍摄）

106 Papaveraceae 罂粟科

White or colored sap / 具白色或有色乳汁
Flowers usually 2-merous / 花常 2 基数
Stigma radial / 柱头放射状
Capsule dehiscing by valves or pores / 蒴果瓣裂或孔裂

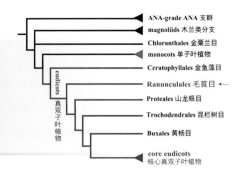

Annual, biennial or perennial **herbs** to soft-wooded shrubs, usually latex present, clear, white or less often yellow or orange. **Leaves** simple or compound; alternate or in a basal rosette, rarely opposite; leaf blade often lobed or dissected, entire to serrate, stipules absent. **Inflorescence** racemes, panicles, umbel or solitary flower. **Flowers** bisexual; usually 2-merous, rarely 3- or 4-merous; actinomorphic to zygomorphic. **Perianth** 2-whorled; sepals green or petaloid, deciduous; petals usually free, imbricate, sometimes crumpled in bud; sometimes spurred, rarely 8–12 or absent. **Stamens** free or in bundles; anthers basifixed, extrorse. **Ovary** superior or inferior; carpels 2 to several, fused; locules 1 to many; ovules 1 to many; placentation parietal or basal; stigma radial. **Fruit** a capsule, opening by apical pores, valves or longitudinal slits.

一年生、二年生或多年生**草本**至软木质灌木，常具乳汁，透明、白色或少有黄色或橙色。**叶**为单叶或复叶；互生或基生莲座状，稀对生；常浅裂或深裂，全缘至有锯齿，托叶缺。**花序**总状、圆锥状、伞形或为单花。**花**两性；常 2 基数，稀 3 或 4 基数；辐射对称至两侧对称。**花被片** 2 轮；花萼绿色或花瓣状，早落；花瓣常离生，覆瓦状，有时芽中皱褶；有时有距，稀 8~12 或缺。**雄蕊**离生或成束；花药基着，外向。**子房**上位或下位；心皮 2 至多枚，合生；1 至多室；胚珠 1 至多枚；侧膜胎座或基底胎座；柱头辐射状。**果实**为蒴果，顶孔开裂、瓣裂或纵裂。

World ca. 38/700+, Northern temperate, especially in Mediterranean, West Asia, Central Asia, East Asia and Southwest North America.
China 19/443.
This area 1/9.

全世界共约 38 属 700 余种，分布于北温带，尤以地中海地区、西亚、中亚、东亚和北美洲西南部居多。
中国产 19 属 443 种。
本地区有 1 属 9 种。

Corydalis DC. 紫堇属

1-1. *Corydalis heterocarpa* Siebold & Zucc./ 异果黄堇；
1-2. *C. incisa* (Thunb.) Pers./ 刻叶紫堇

108　Lardizabalaceae 木通科

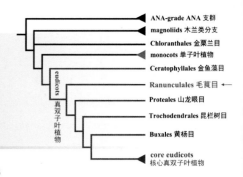

Woody climbers / 木质藤本
Leaves compound, usually palmate / 常掌状复叶
Flowers unisexual, 3-merous / 花单性，3 基数
Stigma conspicuous, sessile or subsessile / 柱头明显，花柱几无

Woody **climbers**, rarely erect shrubs; winter buds large. **Leaves** compound, palmate to 3-foliate, rarely pinnate; alternate (spiral); petioles and petiolules swollen at both ends; stipules absent or rarely present. **Inflorescence** racemes (often pendulous), corymbs, or spikes, rarely panicles or solitary flower. **Flowers** unisexual, plants dioecious, rarely monoecious, or polygamomonoecious, actinomorphic, 3–merous; bracteolate. **Perianth** (2–) 4-whorled; sepals 6, 2-whorled, free, petaloid, usually 3 in *Akebia*; petals free and reduced or absent, smaller than sepals, sometimes modified into nectaries. **Stamen** filaments basally fused or free, opposite the perianth; anthers basifixed; often staminodes 3–6 in female flowers. **Ovary** superior; carpels free, 3(–9) or numerous, many in *Sargentodoxa*; 1-loculed; ovules few to many per locule; placentation usually marginal; stigma conspicuous, sessile or subsessile. **Fruit** a fleshy follicles or berry. **Seeds** numerous, rarely solitary; seed coat crustaceous; endosperm copious.

木质**藤本**，稀为直立灌木；冬芽大。**叶**为复叶，掌状至 3 枚小叶，稀为羽状；互生（螺旋状）；叶柄和小叶柄两端膨大；托叶缺或稀存在。**花序**总状（常下垂），伞房状，或穗状，稀圆锥状或单花。**花**单性，雌雄异株，稀雌雄同株，或杂性同株，辐射对称，3 基数；具小苞片。**花被片**（2~）4 轮；**花萼** 6 枚，2 轮，离生，花瓣状，木通属常 3 枚；花瓣离生或退化至缺失，小于花萼，有时变态为蜜腺。**雄蕊**花丝基部合生或离生，与花被片对生；花药基着；雌花中常有 3~6 枚退化雄蕊。**子房**上位；心皮离生，3（~9）枚或多数，大血藤属心皮多数；1 室，胚珠每心皮少数至多数；常为边缘胎座；柱头明显，花柱几无。**果实**为肉质蓇葖果或浆果。**种子**多数，稀 1 粒；种皮脆壳质；胚乳丰富。

World 7/ca. 40, disjunct distribution between East Asian and South American.
China 5/ca. 34.
This area 2/4.

全世界共 7 属约 40 种，东亚与南美间断分布。
中国产 5 属约 34 种。
本地区有 2 属 4 种。

1. Leaflet apex rounded to obtuse or emarginate; filaments very short; fruit dehiscent/ 小叶先端圆钝或微凹；花丝极短；果实开裂 ....**1. *Akebia*/ 木通属**
1. Leaflet apex usually acuminate or caudate; filaments conspicuous; fruit indehiscent/ 小叶先端常锐尖或尾尖；花丝明显；果实不开裂 ....**2. *Stauntonia*/ 野木瓜属**

1. *Akebia quinate* (Houtt.) Decne./ 木通（Flowering branch/ 花枝）;
2. *Stauntonia leucantha* Y.C. Wu/ 钝药野木瓜（Fruiting branch/ 果枝）

109 Menispermaceae 防己科

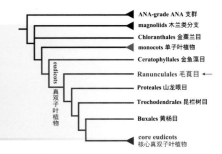

Woody climbers or twining vines / 木质攀缘或缠绕藤本
Petiole swollen at both ends / 叶柄两端肿胀
Carpels free, often on a gynophore / 心皮离生，常生于雌蕊柄上
Seeds often curved / 种子常弯曲

Woody climbers or twining vines, rarely erect shrubs or small trees, very rarely erect herbs or epiphytes. Leaves alternate, spiral; stipules absent; petiole swollen at both ends; leaf blade simple, sometimes palmately lobed, rarely trifoliolate, venation often palmate, less often pinnate. Inflorescence cymes, panicles, fascicles or racemes; sometimes cauliflorous; bracts usually small, rarely leafy. Flowers unisexual (plants dioecious), actinomorphic or rarely zygomorphic, usually 3-merous, inconspicuous. Sepals free, imbricate. Petals free, rarely fused or absent. Stamen filaments free or fused, sometimes stamens completely fused into synandrium; anthers introrse, 1- or 2-loculed or apparently 4-loculed, dehiscing longitudinally or transversely; sometimes staminodes in female flowers. Ovary superior; carpels free or 1, often on a stalk (= gynophore); ovules 2 reducing to 1 per carpel by abortion; placentation marginal; style very short to absent. Fruit a fleshy or dry drupe, sometimes aggregated. Seeds often curved, embryo mostly curved.

木质攀缘或缠绕藤本，稀直立灌木或小乔木，极稀为直立草本或附生植物。叶互生，螺旋状排列；托叶缺；叶柄两端膨大；单叶，有时掌状分裂，稀为 3 枚小叶，叶脉常为掌状，稀羽状。花序聚伞状、圆锥状、簇生或总状；有时茎生；苞片小，稀叶状。花单性（雌雄异株），辐射对称或稀为两侧对称，常 3 基数，不显著。花萼离生，覆瓦状。花瓣离生，稀合生或缺。雄蕊花丝离生或合生，有时雄蕊全部合生形成聚药雄蕊；花药内向，1 或 2 室或假 4 室，纵裂或横裂；有时雌花中具退化雄蕊。子房上位；心皮离生或仅 1 枚，常生于柄上（雌蕊柄）；胚珠每室 2 枚，1 枚被吸收而仅存 1 枚；边缘胎座；花柱极短或缺。果实为肉质或干燥核果，有时为聚合果。种子常弯曲，胚大多弯曲。

World 70/442, pantropics.

China 19/77.

This area 4/7.

全世界共 70 属 442 种，泛热带分布。

中国产 19 属 77 种。

本地区有 4 属 7 种。

1. Leaves peltate; carpel 1/ 叶片盾状着生；心皮 1 枚 .... **1. *Stephania*/ 千金藤属**
1. Leaves not peltate; carpels 3/ 叶片非盾状；心皮 3 枚
    2. Leaf blade lobed, if not, with main basal veins dividing or fusing before reaching margin/ 叶片分裂，如不裂，则基部主脉分叉或达到边缘前联结

3. Petal apex 2-lobed/ 花瓣先端 2 裂 ....**2. *Cocculus*/ 木防己属**
3. Petal apex not lobed/ 花瓣先端不裂 ....**3. *Sinomenium*/ 风龙属**
2. Leaf blade never lobed, with main basal veins and their outer branches leading directly to margin/ 叶片不裂，基部主脉及外侧脉直达叶缘 ....**4. *Diploclisia*/ 秤钩风属**

1. *Stephania japonica* (Thunb.) Miers/ 千金藤 ；
2. *Cocculus orbiculatus* (L.) DC./ 木防己 ；
3. *Sinomenium acutum* (Thunb.) Rehder & E.H. Wilson/ 风龙 ；
4. *Diploclisia affinis* (Oliv.) Diels/ 秤钩风 [Herbarium specimens (US) Henry 1887/ 美国国家历史博物馆馆藏标本 Henry1887 号]

110 Berberidaceae 小檗科

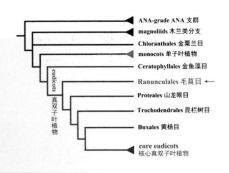

Stems sometimes spiny / 茎有时具刺
Petals hooded, pouched, or spurred / 花瓣盔状、袋状或有距
Anthers 2-loculed, dehiscing by valves / 花药 2 室，瓣裂
1-carpellate, style persistent in fruit as a beak / 单心皮，花柱在果实上宿存为喙

Biennial to perennial **herbs**, or **shrubs**, rarely small **trees**, evergreen or deciduous, sometimes rhizomatous or tuberous; stems sometimes spiny. **Leaves** alternate or opposite, compound or simple; margin toothed or rarely entire; stipules present or absent; venation pinnate or palmate. **Inflorescence** terminal or axillary, racemes, spikes, umbels, cymes, or panicles, or flowers fascicled or solitary; bracteate. **Flowers** pedicellate or sessile, bisexual, actinomorphic. **Perianth** free, usually 2- or 3-merous, rarely absent; sepals 6–9, often petaloid, free, in 2 or 3 whorls; petals 6, free, flat, hooded, pouched, or spurred; nectary present or absent. **Stamens** 6, opposite petals; anthers 2-loculed, dehiscing by valves or longitudinal silts. **Ovary** superior, 1-carpellate; ovules (1–) many; placentation marginal or appearing basal; style present or absent, sometimes persistent in fruit as a beak. **Fruit** a berry, capsule, follicle, nut, or utricle. **Seeds** 1 to numerous, sometimes arillate; endosperm copious.

　　二年生至多年生**草本**或灌木，稀小乔木，常绿或落叶，有时具根状茎或块茎；茎有时具刺。**叶**互生或对生，复叶或单叶；边缘有锯齿或稀全缘；托叶有或无；叶脉羽状或掌状。**花序**顶生或腋生，总状、穗状、伞形、聚伞或圆锥状，或簇生或为单花；具苞片。**花**有柄或无，两性，辐射对称。**花被片**离生，常 2 或 3 基数，稀缺失；花萼 6~9 枚，常花瓣状，离生，2 或 3 轮；花瓣 6 枚，离生，扁平、盔状、袋状或有距；蜜腺有或无。**雄蕊** 6 枚，与花瓣对生；花药 2 室，瓣裂或纵裂。**子房**上位，单心皮；胚珠 1 枚至多数；边缘胎座或基底胎座；花柱有或无，有时在果实上宿存成喙状。**果实**为浆果、蒴果、蓇葖果、坚果或胞果。**种子** 1 粒至多数，有时具假种皮；胚乳丰富。

World 15/ca. 650, mainly in North temperate and subtropical mountains.
China 11/ca. 303.
This area 2/3.

全世界共 15 属约 650 种，主要分布于北温带和亚热带山区。
中国产 11 属约 303 种。
本地区有 2 属 3 种。

1. Leaves 2 or 3 pinnately compound; margin of leaflets entire; anthers dehiscing by longitudinal slits/ 叶为二回或三回羽状复叶；小叶全缘；花药纵裂 ………………………………………………… **1.** *Nandina*/ 南天竹属
1. Leaves simple or pinnately compound; margin of leaflets toothed; anthers dehiscing by 2 valves/ 单叶或一回羽状复叶；小叶有锯齿；花药 2 瓣裂 ………………………………………… **2.** *Mahonia*/ 十大功劳属

1. *Nandina domestica* Thunb./ 南天竹（1a. Anthers dehiscing by longitudinal slits/ 花药纵裂，photo by XinXin ZHU/ 朱鑫鑫拍摄；1b. Three Pinnately compound/ 三回羽状复叶）；

2. *Mahonia bealei* (Fortune) Carrière/ 阔叶十大功劳 (2a. Infrcutescence/ 果序；2b. Pinnately compound/ 一回羽状复叶)

111 Ranunculaceae 毛茛科

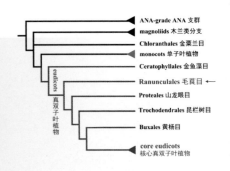

Stamens numerous, free / 雄蕊多数，离生
Flowers actinomorphic, sepals free / 花辐射对称，花萼分离
Ovary superior, carpels numerous, free / 子房上位，心皮多数，离生
Fruit a group of follicles or achenes / 果实为聚合的蓇葖果或瘦果

Perennial or annual **herbs**, sometimes subshrubs or herbaceous or woody vines. **Leaves** simple to compound, alternate (usually spiral) or opposite; margin pinnatisect to pinnatifid, toothed or rarely entire. **Inflorescence** terminal, simple or compound monochasium, dichasium, simple or compound racemes, or solitary flower. **Flowers** bisexual or unisexual, actinomorphic, rarely zygomorphic. **Sepals** 3–6 or more, free, petaloid or sepaloid, imbricate or sometimes valvate in bud. **Petals** present or absent, 2–8 or more, free, usually with nectaries; sometimes spurred or absent. **Stamens** numerous, rarely few, free; filaments linear or filiform; anthers latrorse, introrse, or extrorse; often staminodes present. **Ovary** superior; carpels numerous or few, rarely 1, free, rarely connate to various degrees; ovules 1 to many per locule; placentation marginal, basal, apical (rarely) or axile. **Fruit** a group of follicles or achenes, rarely capsules or berries. **Seeds** small, with copious endosperm and minute embryo.

多年生或一年生**草本**，有时为亚灌木状或草质或木质藤本。**叶**为单叶至复叶，互生（常螺旋状排列）或对生；羽状全裂至羽状半裂，有锯齿或稀全缘。**花序**顶生，简单或复合的单歧聚伞花序、二歧聚伞花序，简单或复合的总状花序，或为单花。**花**两性或单性，辐射对称，稀两侧对称。**花萼**3~6枚或更多，离生，花瓣状或萼片状，在芽中排列为覆瓦状或有时镊合状。**花瓣**有或无，2~8枚或更多，离生，常具蜜腺；有时有距或无。**雄蕊**多数，稀少量，离生；花丝线形或丝状；花药侧向纵裂，内向或外向开裂；常具退化雄蕊。**子房**上位；心皮多数或较少，稀1枚，离生，罕有不同程度合生；每室胚珠1至多枚；边缘胎座、基底胎座、顶生胎座（稀）或中轴胎座。**果实**为聚合蓇葖果或聚合瘦果，稀为蒴果或浆果。**种子**小，胚乳丰富，胚细小。

World ca. 55/2, 525, worldwide, especially in northern temperate and cold temperate regions.
China 35/ca. 923.
This area 5/18.

全世界共约55属2 525种，全球广布，主要分布于北半球温带和寒温带。
中国产35属约923种。
本地区有5属18种。

1. Ovary with 1 ovule; fruit an achene/ 子房有1枚胚珠；果为瘦果
   2. Perianth 1-whorled, petal absent/ 花被1轮；花瓣不存在
      3. Cauline leaves opposite/茎生叶对生 ....**1. *Clematis*/ 铁线莲属**

3. Cauline leaves alternate/ 茎生叶互生 ....**2. *Thalictrum*/ 唐松草属**
2. Perianth 2-whorled or more; petals present/ 花被 2 轮或多轮；花瓣存在 ....**3. *Ranunculus*/ 毛茛属**
1. Ovary with several or many ovules; fruit a follicle/ 子房有数枚或多数胚珠；果为蓇葖果
4. Flowers zygomorphic/ 花两侧对称 ....**4. *Delphinium*/ 翠雀属**
4. Flowers actinomorphic/ 花辐射对称 ....**5. *Semiaquilegia*/ 天葵属**

RANUNCULALES. Ranunculaceae. *Ranunculus muricatus* L./ 毛茛目 毛茛科 刺果毛茛

A. Whole plant/ 植株
B. Pistil/ 雌蕊
C. Stamen/ 雄蕊
D. Pistil and stamen segregated/ 雌雄蕊离析
E. Perianth segregated/ 花被片离析
F. Petal in dorsal and ventral view/ 花瓣背腹面观
(Scale/ 标尺： A 1 cm; B–F 1 mm)

1-1. *Clematis apiifolia* DC./ 女萎；

1-2. *C. lanuginosa* Lindl./ 毛叶铁线莲；

2. *Thalictrum fortunei* S. Moore/ 华东唐松草；

3. *Ranunculus ternatus* Thunb./ 猫爪草（3a. Perianth 2-whorled/ 花被 2 轮；3b. Petals 5/ 花瓣 5 枚）；

4. *Delphinium anthriscifolium* var. *calleryi* (Franch.) Finet & Gagnep./ 卵瓣还亮草；

5. *Semiaquilegia adoxoides* (DC.) Makino/ 天葵

112 Sabiaceae 清风藤科

Leaves alternate, stipules absent / 叶互生，无托叶
Flowers bisexual, small / 花两性，小
Disk with nectar, entire to lobed / 具蜜腺的花盘，全缘或分裂
Perianth unequal, petals opposite the sepals / 花被片不等大，花瓣与萼片对生

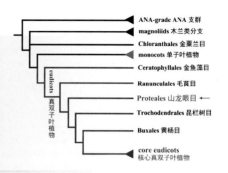

Trees, **shrubs**, or **woody climbers**, deciduous or evergreen. **Leaves** alternate, simple or odd pinnately compound; sometimes minute obscure reddish gland dots; stipules absent. **Inflorescence** axillary or terminal, usually in cymes or panicles, or sometimes solitary axillary flower; sometimes minutely bracteate. **Flowers** bisexual or polygamous-dioecious, actinomorphic or zygomorphic, minute; disk with nectar, entire to lobed. **Sepals** free or fused at base, imbricate, equal or unequal. **Petals** imbricate, equal or inner 2 much smaller than outer 3; opposite the sepals. **Stamens** 2 or (4–) 5, attached to petals, all fertile or outer 3 infertile; anthers 2-loculed, with narrow connectives or with thick cupular connectives. **Ovary** superior, sessile, 2(or 3)-loculed, with 1 or 2 half-anatropous ovules per locule; placentation axile to apical. **Fruit** a drupe or schizocarp, indehiscent.

乔木、灌木或**木质藤本**，落叶或常绿。叶互生，单叶或奇数羽状复叶；有时具细小不明显的红色腺点；无托叶。花序腋生或顶生，常为聚伞或圆锥状，有时单花腋生；有时苞片细小。花两性或杂性异株，辐射对称或两侧对称，细小；花盘具蜜腺，全缘或分裂。花萼离生或基部合生，覆瓦状，等大或不等大。花瓣覆瓦状，等大或内侧2枚远小于外侧3枚；与花萼对生。雄蕊2枚或（4~）5枚，着生于花瓣上，全部可育或外侧3枚不育；花药2室，具狭窄的药隔或具宽厚的杯状药隔。子房上位，无柄，2（或3）室，每室具1或2枚半倒生胚珠；中轴至顶生胎座。果实为核果或分果，不裂。

World 3/ca. 100, disjunct distribution in pantropical pacific rim.
China 2/46.
This area 2/3.

全世界共3属约100种，泛热带太平洋边缘间断分布。
中国产2属46种。
本地区有2属3种。

1. Woody climbers or scandent shrubs; flowers actinomorphic/ 木质藤本或攀缘灌木；花辐射对称 .... **1. *Sabia*/ 清风藤属**
1. Trees or shrubs, erect; flowers zygomorphic/ 直立灌木或乔木；花两侧对称 .... **2. *Meliosma*/ 泡花树属**

1. *Sabia discolor* Dunn/ 灰背清风藤；
2. *Meliosma rigida* Siebold & Zucc./ 笔罗子

113 Nelumbonaceae 莲科

Aquatic herbs and rhizomatous / 水生草本，具根状茎
Receptacle obconic and spongy / 花托倒圆锥状，海绵质
Seeds without endosperm; embryo large, green / 种子无胚乳；胚大型，绿色

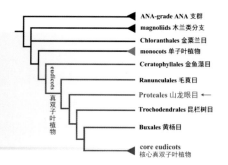

Perennial **herbs**, aquatic, rhizomes branched, repent, forming swollen terminal storage tubers late in growing season. **Leaves** arising from rhizome, alternate, emersed or floating, long petiolate; leaf blade centrally peltate, veins radially extended. **Inflorescence** solitary, axillary, long pedunculate. **Flowers** bisexual, actinomorphic. **Stamens** numerous; anthers dehising by longitudinal slits. **Perianth** numerous, free, outer ones most reduced, inner ones larger and petaloid. **Ovary** superior, styles very short; stigmas capitate; carpels numerous, free, separately and loosely embedded in cavities on flattened top of spongy receptacle; 1-loculed; ovule 1. **Fruit** nut-like, indehiscent. **Seeds** without endosperm and perisperm; embryo large, green.

多年生水生**草本**，根状茎分枝，匍匐，生长季末期先端形成膨大的储藏块茎。**叶**由根状茎发出，互生，挺水或浮水，叶柄长；叶片中心盾状，叶脉辐射延伸。**花序**为单花腋生，具长梗。**花**两性，辐射对称。**雄蕊**多数，花药纵裂。**花被片**多数，离生，最外轮退化，内轮较大，花瓣状。**子房**上位，花柱短，柱头头状；心皮多数，离生，疏松地嵌生在平顶海绵质花托内；1室；1胚珠。**果实**坚果状，不裂。**种子**无胚乳和外胚乳；胚大，绿色。

World 1/2, disjunct distribution in Southeast Asia, northern Australia, central America and North America.
China 1/1.
This area 1/1.

全世界共1属2种，间断分布于东南亚，澳大利亚北部，中美洲和北美洲。
中国产1属1种。
本地区有1属1种。

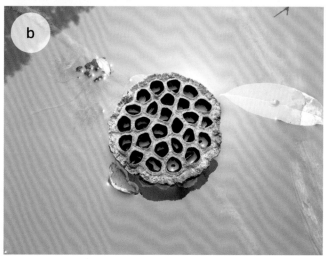

***Nelumbo* Adans. 莲属**

Nelumbo nucifera Gaertn./ 莲（a. Flower, photo by FengLuan LIU/ 花，刘凤栾拍摄；b. Receptacle/ 果托）

115　Proteaceae 山龙眼科

Inflorescence compound cone-like heads / 花序成球果状
Perianth 1-whorled, petaloid, often 2+2 / 花被片 1 轮，花瓣状，常 2+2
Stamens connective often prolonged / 雄蕊药隔常延伸

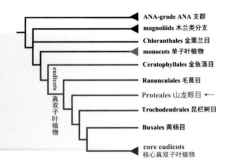

Trees or **shrubs**. **Leaves** alternate, rarely opposite or whorled, simple or compound; margin toothed, lobed or entire; stipules absent. **Inflorescence** simple or panicles with paired flowers, or simple to compound cone-like heads, spikes or racemes; often bracteate. **Flowers** bisexual or rarely unisexual and dioecious, actinomorphic or zygomorphic, often 2-merous. **Perianth** 1-whorled, valvate, usually petaloid, often 2+2, rarely 3 or 5; often nectary glands 1–4. **Stamens** 4, opposite the perianth; filaments usually adnate to perianth; anthers basifixed, usually 2-loculed, dehiscing longitudinally, connective often prolonged. **Ovary** superior, carpel 1, rarely 2; 1-loculed; ovules 1 or 2(or more), pendulous, placentation usually marginal or apical; style terminal, simple, often apically clavate. **Fruit** a follicle, achene, or drupe or drupaceous. **Seeds** 1 or 2(or few to many), sometimes winged; endosperm absent (or vestigial); embryo usually straight.

乔木或**灌木**。**叶**互生，稀对生或轮生，单叶或复叶；有锯齿、分裂或全缘；托叶缺。**花序**为简单或圆锥状，花双生，或简单至复合的球果状，排列成头状、穗状或总状；常具苞片。**花**两性或稀单性并雌雄异株，辐射对生或两侧对称，常 2 基数。**花被片** 1 轮，镊合状排列，常花瓣状，2+2，稀为 3 或 5；常有蜜腺 1~4。**雄蕊** 4 枚，与花被片对生；花丝常贴生于花被片上；花药基着，常 2 室，纵向开裂，药隔常延伸。**子房**上位，心皮 1 枚，稀 2 枚；1 室；胚珠 1 枚或 2 枚（或更多），下垂，边缘胎座或顶生胎座；花柱顶生，不分枝，顶端常棍棒状。**果实**为菁葵果、瘦果或核果，或核果状。**种子** 1 粒或 2 粒（或少数至多数），有时具翅；无胚乳（或仅有残迹）；胚常直伸。

World 77/1, 600, largely in South Hemisphere tropics and subtropics, especially in Australia and South Africa.
China 3/25.
This area 1/3.

全世界共 77 属 1 600 种，主要分布于南半球的热带和亚热带地区，特别是澳大利亚和南非。
中国产 3 属 25 种。
本地区有 1 属 3 种。

Helicia Lour. 山龙眼属

Helicia cochinchinensis Lour./ 小果山龙眼

（a. Infructescence/ 果序；b. Inflorescence/ 花序；
c. Fruits/ 果实；d. Leaves on young plants/ 幼苗叶片）

117 Buxaceae 黄杨科

Leaves simple, stipules absent / 单叶，无托叶
Flowers unisexual, monoecious / 花单性，雌雄同株
Female flower with bract-like perianth, ovary superior / 雌花花被苞片状，子房上位
Ovules 2 per locule, seeds black, shiny / 胚珠每室 2 枚，种子黑色具光泽

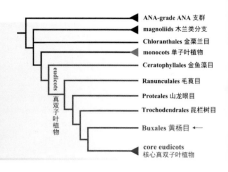

Trees, **shrubs** or rarely **subshrubs** to perennial **herbs**. **Leaves** simple, alternate or opposite, margin entire or dentate; stipules absent. **Inflorescence** racemes, spikes, panicles or solitary flower, bracteate. **Flowers** unisexual, usually monoecious, or rarely bisexual; actinomorphic. **Male flowers** with inconspicuous perianth or absent; stamens 2–6, or many, opposite the perianth; usually pistillodes present. **Female flowers** with bract-like perianths, 5–6. **Ovary** superior, carpels fused or 1; locules (2–)3(–4); usually ovules 2, rarely 1 per locule; placentation axile, rarely apical; styles free, persistent; stigma broadly recurved. **Fruit** a loculicidal capsule or drupe. **Seeds** black, shiny.

乔木、灌木或稀亚灌木至多年生草本。叶为单叶，互生或对生，全缘或具齿；托叶缺。花序总状、穗状、圆锥状或为单花，有苞片。花单性，常雌雄同株，稀两性；辐射对称。雄花花被不显或缺；雄蕊 2～6 枚或多数，与花被对生；常具退化雌蕊。雌花具苞片状花被，5~6 朵。子房上位，心皮合生或 1 枚；(2~)3(~4) 室；每室常 2 枚胚珠，稀仅 1 枚；中轴胎座，稀顶生胎座；花柱分离，宿存；柱头宽阔，弯曲。果实为室背开裂蒴果或核果。种子黑色，有光泽。

World 5/ca. 100, Northern Hemisphere regions.
China 3/ca. 28.
This area 2/4.

全世界共 5 属约 100 种，分布于北半球。
中国产 3 属约 28 种。
本地区有 2 属 4 种。

1. Leaves opposite; fruit a loculicidal capsule/叶对生；果实为室背开裂蒴果 ....**1. *Buxus*/** 黄杨属
1. Leaves alternate; fruit a berry, ± fleshy/叶互生；果实为浆果，多少肉质 ....**2. *Pachysandra*/** 板凳果属

1-1. *Buxus sinica* (Rehder & E.H. Wilson) M. Cheng/ 黄杨；
1-2. *B. rugulosa* Hatus./ 皱叶黄杨（Showing capsule, not distributed in this area, photo by XinXin ZHU/ 仅展示蒴果，非本区域分布，朱鑫鑫拍摄）；
2. *Pachysandra terminalis* Siebold & Zucc./ 顶花板凳果

123 Altingiaceae 蕈树科

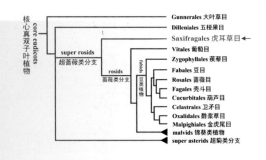

Trees, resinous / 乔木，具树脂
Leaves palmatifid, intrapetiolar / 叶掌裂，具叶柄内托叶
Perianth absent, female flowers in globular heads / 花被缺，雌花成球形头状花序
Fruit a capsule, many united forming multiple fruits / 果实为蒴果，多个聚合

Trees; resinous. **Leaves** palmatifid, simple, alternate (usually spiral); margin toothed; venation palmate to pinnate; petioles present; stipules intrapetiolar, deciduous. **Flowers** unisexual, monoecious, actinomorphic. **Male flowers** in terminal racemes, each flower with 1–4 bracts; perianth absent; stamens many in clusters, anthers basifixed. **Female flowers** in globular heads; perianth absent; staminodes absent or needle-like. **Ovary** inferior to less often part-inferior; carpels fused; 2-loculed; ovules many per locule; placentation axile. **Fruit** a capsule, many united forming multiple fruits.

乔木；具树脂。**叶**掌状分裂，单叶，互生（常螺旋状）；边缘具齿；叶脉掌状或羽状；具叶柄；叶柄内托叶，早落。**花**单性，雌雄同株，辐射对称。**雄花**为顶生总状，每朵花具 1~4 枚苞片；花被缺；雄蕊多数成簇，花药基着。**雌花**花序头状，呈球形；花被缺；退化雄蕊缺或针状。**子房**下位至稀半下位；心皮合生；2 室；每室胚珠多枚；中轴胎座。**果实**为蒴果，多个聚合。

World 1/15, Eastern Mediterranean, Asia, Central America, tropical and subtropical North America.
China 1/10.
This area 1/1.

全世界共 1 属 15 种，分布于地中海东部、亚洲、中美洲、北美洲热带和亚热带地区。
中国产 1 属 10 种。
本地区有 1 属 1 种。

Liquidambar L. 枫香树属

Liquidambar formosana Hance / 枫香树

124 Hamamelidaceae 金缕梅科

Shrubs or trees, usually of stellate hairs / 灌木或乔木，常具星状毛
Stipules minute to large, usually paired / 托叶细小至大型，常成对
Petals ribbon-like and circinate in bud / 花瓣常为带状并在芽中拳卷
Styles 2, stigmas recurved / 花柱2个，柱头外弯

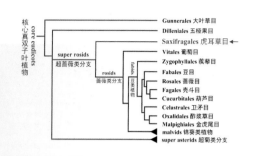

Shrubs or **trees**, evergreen or deciduous, bisexual, andromonoecious, or monoecious; usually with stellate hairs or stellate or peltate scales; buds perulate or naked. **Leaves** distichous or spiral, rarely subopposite or opposite, stipules minute to large, usually paired; leaf blade simple or palmately lobed, pinnately veined or palmately 3–5-veined. **Inflorescence** usually spikes or heads, rarely racemes or (condensed) cymes or panicles, axillary or terminal. **Flowers** small to medium-sized, bracteate and often bracteolate, bisexual or unisexual, actinomorphic or rarely zygomorphic, usually a hypanthium present. **Perianth** present or rarely absent; sepals 4 or 5(–10), sometimes absent, imbricate, usually persistent; petals absent or 4 or 5, yellow, white, greenish or red, often ribbon-like and circinate in bud, caducous. **Stamens** 4, 5, or many, free; anthers basifixed, introrse, usually dehiscing by 1–2-valves, rarely longitudinal slits, connective protruding; disk scales sometimes present. **Ovary** usually inferior or superior; 2-loculed, carpels free at apex; ovule mostly 1 per carpel, less often many; placentation axile; styles 2, stigmas recurved. **Fruit** a capsule, dehiscing septicidally, septifragally, or loculicidally and 4-valved.

灌木或乔木，常绿或落叶，两性花，雄全同株或雌雄同株；常被星状毛，或星状或盾状鳞片；芽有鳞片或裸露。叶二列或螺旋状，稀近对生或对生，托叶细小至大型，常成对；叶为单叶或掌状分裂，叶脉羽状或掌状3~5条。花序常穗状或头状，稀总状或（密集）聚伞或圆锥状，腋生或顶生。花小至中型，具苞片与小苞片；两性或单性，辐射对称或稀为两侧对称，常具被丝托。花被片存在或稀缺失；萼片4或5(~10)枚，有时缺，覆瓦状，常宿存；花瓣缺或4或5枚，黄色、白色、绿色或红色，常带状，芽中拳卷，早落。雄蕊4枚或5枚或多数，离生；花药基着，内向，常1~2瓣裂，稀纵裂，药隔伸出；花盘有时具鳞片。子房常下位或上位；2室，心皮先端分离；胚珠每室常1枚，稀多数；中轴胎座；花柱2个，柱头外弯。果实为蒴果，室间、室轴或室背裂开为4瓣。

World 27/106, tropics to temperate regions, especially from East Asia to Australia.

China 15/61.

This area 2/3.

全世界共27属106种，分布于热带至温带，尤其是东亚至澳大利亚。

中国产15属61种。

本地区有2属3种。

1. Petals absent/ 无花瓣 .... **1. *Distylium*/ 蚊母树属**
1. Petals present/ 有花瓣 .... **2. *Loropetalum*/ 檵木属**

SAXIFRAGALES. Hamamelidaceae. *Loropetalum chinense* (R. Br.) Oliv./ 虎耳草目 金缕梅科 檵木

A. Flowering branch/ 花枝

B. Cross section of ovary/ 子房横切面

C. Staminodes 4/ 退化雄蕊 4 枚

D. Calyx lobes/ 花萼裂片

E. Flowers clustered/ 花簇生

F–G. Corolla/ 花冠

H. Stamen with horn-like connective/ 雄蕊具角状药隔

I. Flower segregated/ 花离析

J. Dorsal view of leaf/ 叶背面观

K. Staminodes, amplified/ 退化雄蕊，放大

L. Stamens alternate with staminodes/ 雄蕊与退化雄蕊互生

M. Vertical profile of flower, petal removed/ 花纵剖面，除花瓣

(Scale/ 标尺： A 1 cm; B–M 1 mm)

1. *Distylium myricoides* Hemsl./ 杨梅蚊母树（Female inflorescence/ 雌花序）；
2. *Loropetalum chinense* (R. Br.) Oliv./ 檵木

126 Daphniphyllaceae 虎皮楠科

Woody, evergreen / 常绿木本
Leaves simple, fascicled, long petiolate / 单叶簇生，具长柄
Flowers unisexual, petals absent / 花单性，花瓣缺
Drupe, often glaucous / 核果，常被白霜

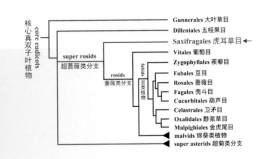

Trees or **shrubs**, evergreen; branchlets with leaf scars and lenticels. **Leaves** simple, alternate, usually clustered at branch ends; entire, long petiolate, stipules absent. **Inflorescence** racemes, axillary, bracteate at base. **Flowers** small, unisexual, dioecious, sometimes sterile, pedunculated; calyx 3–6-lobed; petals absent. **Male flowers**: stamens 5–12(–18), filaments shorter than anthers. **Female flowers**: ovary superior; carpels fused; 2-loculed; ovules (1–)2 per locule; placentation axile or apical; style branches 2, recurved, persistent. **Fruit** a drupe, often glaucous.

常绿**乔木**或**灌木**；小枝具叶痕和皮孔。**叶**为单叶，互生，常集生于小枝顶端，全缘，具长柄，托叶缺。**花序**总状腋生，基部具苞片。花小，单性，雌雄异株，有时不育，具花梗；花萼 3~6 裂，花瓣缺。**雄花**：雄蕊 5~12(~18) 枚，花丝短于花药。**雌花**：子房上位；心皮合生；2 室；胚珠每室（1~）2 枚；中轴或顶生胎座；柱头 2 裂，弯曲，宿存。**果实**为核果，常被白霜。

World 1/10, East Asia to Malesia.
China 1/10.
This area 1/1.

全世界共 1 属 10 种，分布于东亚至马来西亚。
中国产 1 属 10 种。
本地区有 1 属 1 种。

Daphniphyllum Blume 虎皮楠属

Daphniphyllum oldhamii (Hemsl.) K. Rosenthal/ 虎皮楠
（a. Habitat/ 生境 ； b. Leaves/ 叶片 ； c. Infructescence/ 果序）

128 Grossulariaceae 茶藨子科

Woody, often spiny / 木本，多刺
Leaves often palmately lobed / 叶多掌状分裂
Ovary inferior, hypanthium well-developed / 子房下位，花托发达
Juicy berry, calyx persistent / 浆果多汁，萼宿存

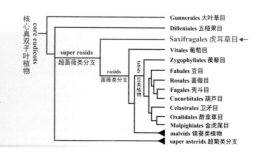

Shrubs, sometimes climbing, rarely small trees; often spiny; usually with glandular hairs. **Leaves** alternate, rarely clustered, petiolate, leaf blade palmately lobed; stipulate absent. **Inflorescence** racemes; bracts present, hairy. **Flowers** bisexual, or unisexual and dioecious, 4- or 5-merous; actinomorphic; hypanthium well-developed and petaloid. **Sepals** tube basally adnate to ovary, persistent. **Petals** alternate with and often smaller than sepal lobes, sometimes absent. **Ovary** inferior to part-inferior; carpels fused; 1-loculed; ovules 4-many; placentation parietal; style 2-lobed or divided to almost 1/2 its length. **Fruit** a juicy berry.

灌木，有时为攀缘藤本，稀小乔木；常具刺；常被腺毛。**叶**互生，稀簇生，有叶柄，常掌状分裂；托叶缺。**花序**为总状花序，具苞片，被毛。**花**两性，或单性，雌雄异株，4或5基数；辐射对称；托杯发达成花瓣状。**花萼**筒基部与子房合生，宿存。**花瓣**小于萼裂片，并与之互生，有时缺。**子房**下位至半下位；心皮合生；1室；胚珠4枚至多数；侧膜胎座；花柱2浅裂或深裂至1/2。**果实**为浆果，多汁。

World 1/150, northern temperate regions, also along the Andes.
China 1/59.
This area 1/2.

全世界共 1 属 150 种，分布于北温带，沿安第斯山脉也有分布。
中国产 1 属 59 种。
本地区有 1 属 2 种。

Ribes L. 茶藨子属

Ribes fasciculatum Siebold & Zucc. var. *chinense* Maxim./ 华蔓茶藨子（a. Flowering branch/ 花枝；b. Fruiting branch/ 果枝）

129 Saxifragaceae 虎耳草科

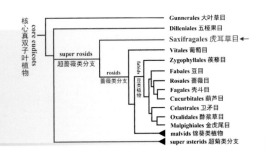

Leaves often basally aggregated / 叶常基部簇生
Hypanthium often cup-shaped / 托杯常成杯状
Inflorescence variety, scapose / 花序多样，具花葶
Septicidal capsule / 蒴果室间开裂

Perennial **herbs**, rarely annuals or biennials; sometimes succulent or xerophytes. **Leaves** often basally aggregated, simple or sometimes compound, alternate or less often opposite; venation pinnate or sometimes palmate; margin entire, lobed, or toothed; petioles present or absent; stipules absent or sometimes stipule-like structures. **Inflorescence** cymes, racemes, spikes, heads, fascicles, panicles or rarely solitary flower; scapose; bracteate. **Flowers** bisexual or rarely unisexual (androdioecious), actinomorphic or rarely zygomorphic, hypanthium free or fused to ovary base, often cup-shaped. **Sepals** fused or free, sometimes petaloid. **Petals** fused or free, sometimes clawed, lobed or entire. **Stamens** usually opposite the petals; anthers usually basifixed. **Ovary** superior to inferior; carpels usually fused (at least basally) or free; 1–3-loculed; ovules few-many; placentation axile to parietal; styles 2(–3). **Fruit** a septicidal capsule or sometimes a group of follicles.

多年生**草本**，稀一年或二年生；有时肉质或旱生植物。**叶**常基部簇生，单叶或有时复叶，互生或稀对生；叶脉羽状或偶掌状；全缘、浅裂或有锯齿；叶柄有或无；托叶缺或有时具托叶状结构。**花序**聚伞、总状、穗状、头状、簇生、圆锥状或稀为单花；有花葶；具苞片。**花**两性或稀单性（雄全异株），辐射对称或稀为两侧对称，托杯离生或与子房基部合生，成杯状。**花萼**合生或离生，有时花瓣状。**花瓣**合生或离生，有时具爪，分裂或全缘。**雄蕊**常与花瓣对生；花药常基着。**子房**上位至下位；心皮常合生（至少基部）或离生，1~3室；胚珠少数至多数；中轴胎座至侧膜胎座；花柱2（~3）个。**果实**为室间开裂蒴果或有时为聚合蓇葖果。

World 33–38/620+, mostly in northern temperate regions.
China 14/ca. 268.
This area 1/1.

全世界共33~38属620余种，主要分布于北温带。
中国产14属约268种。
本地区有1属1种。

Saxifraga L. 虎耳草属

Saxifraga stolonifera Curtis/ 虎耳草

(a. Inflorescence/ 花序; b. Flower/ 花)

130　Crassulaceae 景天科

Succulent perennial herbs / 肉质多年生草本
Flower actinomorphic, 3 or 5-merous / 花辐射对称，3 或 5 基数
Ovary superior, carpels free / 子房上位，心皮分离

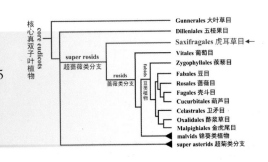

Succulent perennial or rarely annual **herbs**, subshrubs, or shrubs. **Leaves** alternate, opposite, or verticillate, usually simple; margin entire to toothed. **Inflorescence** terminal or axillary, cymes, corymbs, spikes, racemes, panicles, or solitary flower. **Flowers** usually bisexual, sometimes unisexual in *Rhodiola* (when plants dioecious or rarely gynodioecious), actinomorphic; 3 or 5-merous; hypanthium present. **Perianth** imbricate. **Sepals** almost free or basally connate, persistent. **Petals** free or connate. **Stamens** often twice as many as petals or equal to the number of petals. **Ovary** superior; carpels free, usually nectar scales at base of each carpel; placentation marginal. **Fruit** usually a follicle or capsule; erect or spreading, membranous or leathery, 1 to many seeded. **Seeds** small; endosperm scanty or not developed.

肉质多年生或稀一年生**草本**、亚灌木或灌木。**叶**互生、对生或轮生，常单叶；全缘至有锯齿。**花序**顶生或腋生，为聚伞、伞房状、穗状、总状、圆锥花序或为单花。**花**常两性，红景天属有时单性（雌雄异株或稀雌全异株），辐射对称；3 或 5 基数；具托杯。**花被片**覆瓦状。**萼片**几离生或基部合生，宿存。**花瓣**离生或合生。**雄蕊**数 2 倍于花瓣或与花瓣数相同。**子房**上位；心皮离生，心皮基部常具腺鳞；边缘胎座。**果实**常为菁葖果或蒴果；果瓣直立或开展，膜质或革质，具 1 至多粒种子。**种子**小，胚乳缺乏或不发育。

World 35/1, 500, widespread in tropical and temperate regions.

China 12/ca. 232.

This area 4/14.

全世界共 35 属 1 500 种，广布于热带与温带地区。

中国产 12 属约 232 种。

本地区有 4 属 14 种。

1. Carpels stipitate, free/ 心皮有柄，离生
  2. Basal leaves not in a conspicuous rosette/ 基部的叶不呈明显莲座状 ....**1. Hylotelephium/** 八宝属
  2. Basal leaves in a somewhat conspicuous rosette/ 基部的叶多少呈莲座状 ....**2. Orostachys/** 瓦松属
1. Carpels sessile, base usually connate/ 心皮无柄，基部常合生
  3. Leaves flattened, margin serrate or crenate/ 叶扁平，边缘具锯齿 ....**3. Phedimus/** 费菜属
  3. Leaves entire/ 叶全缘 ....**4. Sedum/** 景天属

1. *Hylotelephium erythrostictum* (Miq.) H. Ohba/ 八宝（ 1a. Leaves/ 叶；1b. Infructescence/ 果序）；
2. *Orostachys fimbriata* (Turcz.) A. Berger/ 瓦松；
3. *Phedimus aizoon* (L.)'t Hart/ 费菜（3a. Infructescence/ 果序；3b. Leaves/ 叶）；
4. *Sedum formosanum* N.E. Br./ 台湾佛甲草

134　Haloragaceae 小二仙草科

Flowers (2–) 4-merous, usually unisexual / 花 (2) 4 基数，常单性
Sepals valvate, petals imbricate / 花萼镊合状，花瓣覆瓦状
Ovary inferior, carpels fused / 子房下位，心皮合生
Stigma feathery / 柱头羽毛状

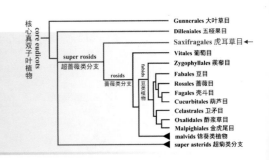

Small **trees**, **shrubs**, **subshrubs** or perennial or annual **herbs**; aquatic or terrestrial. **Leaves** simple or compound; alternate, opposite, or verticillate; margin entire, toothed or deeply pinnatisect; stipules absent. **Inflorescence** axillary or terminal, often spikes, corymbs, racemes or solitary flower. **Flowers** (2–) 4-merous, usually unisexual (plants monoecious, polygamomonoecious or less often dioecious) or bisexual, actinomorphic. **Sepals** free or absent, valvate. **Petals** free or absent, imbricate, sometimes hooded, deciduous. **Stamens** filaments short; anthers basifixed. **Ovary** inferior; carpels fused; 1–4-loculed, ovules 1–2 per locule; placentation apical; stigma 2–4-lobed, feathery. **Fruit** schizocarp, nut or drupe.

小**乔木**、**灌木**、**亚灌木**或多年生或一年生**草本**；水生或陆生。**叶**为单叶或复叶；互生、对生或轮生；全缘，有锯齿或羽状深裂；托叶缺。**花序**腋生或顶生，常成穗状、伞房状、总状或为单花。**花** (2–)4 基数，常单性（雌雄同株、杂性同株或稀为雌雄异株）或两性，辐射对称。**花萼**离生或缺，镊合状。**花瓣**离生或缺，覆瓦状，有时兜状，早落。**雄蕊**花丝短；花药基着。**子房**下位；心皮合生；1~4 室，胚珠每室 1~2 枚；顶生胎座；柱头 2~4 裂，羽毛状。**果实**为分果、坚果或核果。

World 8–10/120, worldwide, especially in Australia.
China 2/13.
This area 2/4.

全世界共 8~10 属 120 种，世界广布，特别是澳大利亚。
中国产 2 属 13 种。
本地区有 2 属 4 种。

1. Herbs terrestrial; leaves opposite or alternate/陆生草本；叶对生或互生 ....**1. *Gonocarpus*/** 小二仙草属
1. Herbs aquatic (in China); leaves whorled/ 水生草本（中国产）；叶轮生 ....**2. *Myriophyllum*/** 狐尾藻属

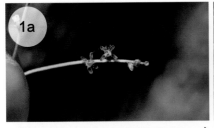

1. *Gonocarpus micranthus* Thunb./ 小二仙草
（1a. Inflorescence/ 花序；1b. Plant/ 植株）；

2. *Myriophyllum aquaticum* (Vell.) Verdc./ 粉绿狐尾藻（Natrulized/ 归化）

136 Vitaceae 葡萄科

Woody climbers often with leaf-opposed tendrils / 木质藤本常具与叶对生的卷须
Often swollen nodes on branches / 茎节膨大
Leaves gland-doted, with intrapetiolar stipules / 叶具腺点，常有叶柄内托叶
Petals deciduous, sometimes forming a calyptra / 花瓣早落，有时形成帽状体
Seeds endotestal, with 2 furrows / 种子具内种皮，具2道沟槽

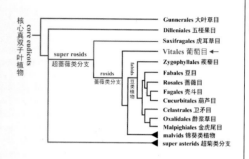

Woody climbers often with leaf-opposed tendrils, or sometimes **trees**, **shrubs** or **herbs**; stems sometimes with conspicuous lenticels, often swollen nodes on branches. **Leaves** gland-doted, deciduous, simple (often palmately lobed) or compound (3-foliate, palmate or pinnate), alternate or rarely opposite; venation often palmate; margin toothed, often with glandular teeth or rarely entire; intrapetiolar, usually deciduous. **Inflorescence** cymes, spikes, corymbs or panicles, often leaf-opposite, pseudo-terminal, or axillary. **Flowers** small, bisexual or polygamous, actinomorphic; 4 or 5(–7)-merous. **Sepals** fused or free, reduced to lobes or teeth. **Petals** free or basally fused, or distally connate forming a calyptra (*Vitis*), valvate, deciduous. **Stamens** opposite the petals; anthers introrse, dehiscing longitudinally, 4-loculed or rarely 2-loculed. **Floral disk** intrastaminal, ring-shaped, cupular, or gland-shaped. **Ovary** superior; carpels fused; 2-loculed; ovules (1–) 2 per locule; placentation axile or basal. **Fruit** a berry, 1–4(–6) seeded. **Seeds** endotestal, with an abaxial chalazal knot and an adaxial raphe with 2 furrows, one on each side; embryo straight, small; endosperm present.

木质藤本常具与叶对生的卷须，或有时为乔木、灌木或草本；茎有时具显著皮孔，茎节常膨大。**叶**有腺点，早落，单叶（常掌状分裂）或复叶（3出、掌状或羽状），互生或稀对生；叶脉常掌状，叶缘有锯齿，常具腺齿或稀全缘；有叶柄内托叶，常早落。**花序**为聚伞、穗状、伞房状或圆锥状，常与叶对生，假顶生或腋生。花小，两性或杂性，辐射对称；4或5(~7)基数。**花萼**合生或离生，退化为裂片或牙齿状。**花瓣**离生或基部合生，或先端合生成帽状体（葡萄属），镊合状，早落。**雄蕊**与花瓣对生；花药内向，纵裂，花药4室或稀2室。**花盘**位于雄蕊内，环状、杯状，或为腺体状。**子房**上位；心皮合生；2室；胚珠每室(1~)2枚；中轴胎座或基底胎座。**果实**为浆果，含1~4(~6)枚种子。**种子**具内种皮，合点远端具瘤突，近端有种脊，两侧各具一道沟槽；胚直伸，小；具胚乳。

World 14–15/ca. 800, tropical, subtropical and warm temperate regions.
China 9/156.
This area 5/18.

全世界共14~15属约800种，分布于热带、亚热带和暖温带地区。
中国产9属156种。
本地区有5属18种。

1. Petals distally connate forming a calyptra/ 花瓣顶端合生成帽状体 .... **1. *Vitis*/ 葡萄属**
1. Petals free/ 花瓣离生
   2. Flowers usually 4-merous/ 花常 4 基数
      3. Style inconspicuous, stigma usually 4-divided/ 花柱不明显，柱头 4 裂 .... **2. *Tetrastigma*/ 崖爬藤属**
      3. Style conspicuous, stigma undivided/ 花柱明显，不分裂 .... **3. *Cayratia*/ 乌蔹莓属**
   2. Flowers usually 5-merous/ 花常 5 基数
      4. Tendril usually with suckers/ 卷须具吸盘 .... **4. *Parthenocissus*/ 爬山虎属**
      4. Tendril without suckers/ 卷须不具吸盘 .... **5. *Ampelopsis*/ 蛇葡萄属**

1. *Vitis wilsoniae* H.J. Veitch/ 网脉葡萄；
2. *Tetrastigma hemsleyanum* Diels & Gilg/ 三叶崖爬藤；
3. *Cayratia japonica* (Thunb.) Gagnep./ 乌蔹莓；
4. *Parthenocissus tricuspidata* (Siebold & Zuss.) Planch./ 地锦（4a. Inflorescence/ 花序；4b. Suckers/ 吸盘）；
5. *Ampelopsis glandulosa* (Wall) Momiy./ 蛇葡萄

138 Zygophyllaceae 蒺藜科

Stems spiny, often swollen nodes / 茎具刺，茎节常膨大
Leaves usually even-pinnate / 偶数羽状复叶
Flowers actinomorphic, disk often present / 花辐射对称，常具花盘
Filaments often with nectary scales; anthers dorsifixed / 花丝常具腺鳞；花药背着

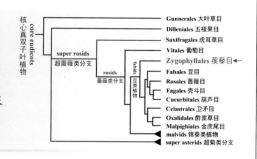

Trees, **shrubs**, **subshrubs** or perennial **herbs**, rarely annual; stems spiny, often swollen nodes. **Leaves** sometimes gland-doted, aromatic, sometimes deciduous, compound (usually even–pinnate or 2-foliate) or rarely simple, opposite or less often alternate, often asymmetric; stipules free or connate, spiny, scaly or leafy, rarely absent. **Inflorescence** axillary or terminal, solitary flower, racemes, or cymes. **Flowers** bisexual or rarely unisexual, actinomorphic or zygomorphic, disk often present. **Sepals** free or rarely basally fused, imbricate or valvate. **Petals** free or absent, often clawed. **Stamen** filaments often with nectary scales; anthers dorsifixed, introrse; sometimes staminodes 4–5. **Ovary** superior; carpels fused; locules (2–)4–5(–12); ovules 1 to many per locule; placentation axile. **Fruit** a loculicidal or septicidal capsule, with wings or not, or rarely a drupe.

乔木、**灌木**、**亚灌木**或多年生**草本**，稀一年生；茎具刺，茎节常膨大。**叶**有时具腺点，芳香，有时早落，复叶（常偶数羽状或仅2小叶）或稀为单叶，对生或少为互生，基部常偏斜；托叶分离或合生，刺状、鳞片状或叶状，稀缺。**花序**腋生或顶生，单花、总状或聚伞花序。**花**两性或稀单性，辐射对称或两侧对称，常具花盘。**萼片**离生或稀基部合生，覆瓦状或镊合状。**花瓣**离生或缺，常具爪。**雄蕊**花丝常具腺鳞；花药背着，内向；有时有4~5枚退化雄蕊。**子房**上位；心皮合生；(2~)4~5(~12)室；胚珠每室1至多枚；中轴胎座。**果实**为室背或室间开裂蒴果，有或无翅，或稀为核果。

World 22/ca. 280, widely distributed in arid and saline habitats across the world.
China 3/22.
This area 1/1.

全世界共22属约280种，广泛分布于全球的干旱和盐碱地区。
中国产3属22种。
本地区有1属1种。

***Tribulus* L. 蒺藜属**

ZYGOPHYLLALES. Zygophyllaceae. *Tribulus terrestris* L./ 蒺藜目 蒺藜科 蒺藜

A. Fruit branch/ 果枝
B. Fruit/ 果实
C. Even-pinnate/ 偶数羽状复叶
D. Seed/ 种子
E. Vertical section of fruit/ 果实纵切面
F. Mericarp/ 分果片
G. Lateral view of fruit/ 果实侧面观
H. Frontal view of fruit/ 果实正面观
(Scale/ 标尺：1 mm)

Tribulus terrestris L./ 蒺藜（a. Fruit/ 果实；b. Fruiting branch/ 果枝）

140 Fabaceae 豆科

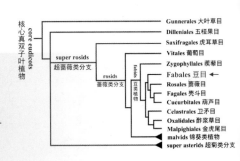

Leaves compound, often pulvinate / 复叶，常有叶枕
Petals free, often papilionaceous / 花瓣离生，常为蝶形花冠
Filaments often fused into a tube or 2 bundles / 花丝常为管状或成 2 束
Carpel 1, placentation marginal; legume / 单心皮，边缘胎座，荚果

Trees, **shrubs**, **herbs** or woody **climbers**; very often bearing root-nodules that harbor nitrogen-fixing bacteria. **Leaves** sometimes deciduous, usually compound (pinnate, bipinnate, palmate, 1–3-foliate) or simple, rarely phylloclades, alternate or rarely opposite to whorled; petioles present to absent, often pulvinate; stipules and stipels usually present, sometimes absent, intrapetiolar, sometimes leafy, spines or glands. **Inflorescence** racemes, corymbs, spikes, panicles, fascicles, heads or solitary flower; sometimes cauliflorous. **Flowers** bisexual or sometimes unisexual (plants monoecious in some Mimosoideae), actinomorphic (Mimosoideae), ± zygomorphic (Caesalpinioideae) to very zygomorphic (Papilionoideae). **Sepals** free or fused into a tube, rarely reduced. **Petals** free, imbricate or valvate, rarely reduced; often papilionaceous with an upper petal (= standard), 2 lateral petals (= wings), and lower 2 petals usually fused (= keel). **Stamens** 10 or many, filaments often fused into a tube or 2 bundles (often 9+1 or rarely 5+5); anthers 2-loculed, basifixed or dorsifixed. **Ovary** superior; carpel usually 1 or rarely a few free; 1-loculed; ovules 1 to many per locule; placentation marginal; style 1; stigma 1. **Fruit** usually a legume, dry or fleshy, dehiscent or indehiscent, sometimes jointed and breaking up into 1-seeded segments. **Seeds** without or with very scanty endosperm.

乔木、灌木、草本或**木质藤本**；常有根瘤与固氮菌共生。**叶**有时早落，常为复叶（羽状，二回羽状，掌状，1~3 枚小叶）或单叶，稀为叶状枝，互生或稀对生至轮生；具叶柄或缺，常有叶枕；具叶柄内托叶和小托叶，偶缺，为叶状、刺状或腺体状。**花序**总状、伞房状、穗状、圆锥、束簇、头状或单花；有时茎生花。**花**两性或有时单性（部分含羞草亚科雌雄同株），辐射对称（羞草亚科），多少两侧对称（云实亚科）至完全两侧对称（蝶形花亚科）。**花萼**离生或合生成管状，稀退化。**花瓣**离生，覆瓦状或镊合状，稀退化；蝶形花冠中，上方 1 枚称为旗瓣，侧方 2 枚称为翼瓣，下方 2 枚常合生，称为龙骨瓣。**雄蕊** 10 枚或多数，花丝常合生成管状或 2 束（常为 9+1，偶为 5+5）；花药 2 室，基着或背着。**子房**上位；心皮常 1 枚或稀为少量离生；1 室；胚珠每室 1 至多枚；边缘胎座；花柱 1 个；柱头 1 个。**果实**常为荚果，干燥或肉质，开裂或不开裂，有时分节，成熟时断裂为仅具 1 枚种子的荚果段。**种子**不具或具瘠薄胚乳。

World ca.751/19, 500, worldwide.

China 167/1, 673.

This area 49/108 (22 tribes).

全世界共约 751 属 19 500 种，世界广布。

中国产 167 属 1 673 种。

本地区有 49 属 108 种（隶属 22 族）。

Canavalia rosea (Sw.) DC./ 海刀豆

1. Flowers actinomorphic, petals valvate in bud (subfam. Mimosoideae) / 花辐射对称，花瓣在芽中镊合状排列（含羞草亚科）
    2. Stamens 10 or fewer / 雄蕊 10 枚或更少 ....**I. Mimoseae/ 含羞草族**
    2. Stamens numerous, usually more than 10 / 雄蕊多数，通常多于 10 枚
        3. Filaments free or only connate at base / 花丝离生或仅在基部合生 ....**II. Acacieae/ 金合欢族**
        3. Filaments connate into a tube/ 花丝合生成管状 ....**III. Ingeae/ 印加树族**
1. Flowers ± zygomorphic, petals imbricate in bud / 花多少两侧对称，花瓣在芽中呈覆瓦状排列
    4. Flowers slightly zygomorphic; corolla not papilionaceous (subfam. Caesalpinioideae) / 花稍成两侧对称；花冠非蝶形（云实亚科）
        5. Leaves simple; palmately nerved / 单叶；掌状脉 ....**IV. Cercideae/ 紫荆族**
        5. Leaves once pinnate or bipinnate; pinnately nerved / 叶为一回或二回羽状复叶；羽状脉
            6. Leaves usually bipinnate/ 叶通常为二回羽状复叶 ....**V. Caesalpinieae/ 云实族**
            6. Leaves once pinnate; anthers basifixed/ 叶为一回羽状复叶；花药基着 ....**VI. Cassieae/ 决明族**
    4. Flowers very zygomorphic; corolla papilionaceous (subfam. Papilionoideae) / 花完全两侧对称；花冠蝶形（蝶形花亚科）
        7. Filaments all free or connate only at base / 花丝全部离生或仅基部合生
            8. Leaves odd-pinnate; calyxes usually subequally 5-dentate/ 叶为奇数羽状复叶；花萼常具不等大 5 锯齿 ....**VII. Sophoreae/ 槐族**
            8. Leaves palmately 3-foliolate; calyx usually deeply 5-lobed/ 叶为 3 枚小叶的掌状复叶；花萼常 5 深裂 ....**VIII. Thermopsideae/ 野决明族**
        7. Filaments partly or almost wholly united to one another / 花丝部分或几乎全部合生
            9. Anthers dimorphic, alternately dorsifixed and basifixed, either all equal or alternately longer and shorter/ 花药二型，背着与基着交互，等长或长短交互排列
                10. Legumes transversely septate and breaking up into 1-seeded joints/ 荚果横向断裂，每节具 1 粒种子 ....**IX. Aeschynomeneae/ 合萌族**
                10. Legumes not transversely septate / 荚果不横向断裂
                    11. Climbing plants; legumes usually thick, generally hairy, sometimes with stinging hairs/ 攀缘植物；荚果常厚实，通常被毛，有时被刺毛 ....**X. Phaseoleae/ 菜豆族**
                    11. Erect plants; legumes turgid /植株直立生长；荚果膨胀 ....**XI. Crotalarieae/ 猪屎豆族**
            9. Anthers uniform or nearly so, not alternately basifixed and dorsifixed, also not alternately longer and shorter/ 花药同型或近同型，不分成背着和基着，也不分成长短交互
                12. Free upper part of all or half of filaments dilated or expanded upward/ 上部离生花丝部分或全部肿大或向上扩展 ....**XII. Trifolieae/ 车轴草族**
                12. Free upper part of filaments ± filiform, not dilated upward / 上部离生花丝多少为线形，不肿大
                    13. Legumes breaking up into 1-seeded segments when maturity / 荚果成熟时开裂为仅含 1 粒种子的节荚
                        14. Leaves with stipels/ 叶片具小托叶 ....**XIII. Desmodieae/ 山蚂蝗族**
                        14. Leaves without stipels / 叶片无小托叶

15. Keel petals often obliquely truncate at apex; wings short or very small, rarely equaling keel petals / 龙骨瓣歪斜，先端截平，翼瓣短或极小，稀与龙骨瓣等长 .... **XIV. Hedysareae/ 岩黄芪族**

15. Keel petals obtuse or beaked, incurved; wings often transversely plicate/ 龙骨瓣钝头或喙状卷曲，翼瓣常具横皱褶纹 .... **IX. Aeschynomeneae/ 合萌族**

13. Legumes not breaking up into separate segments when maturity / 荚果成熟时不裂为节荚

16. Plants with indumentum composed mainly of T-shaped hairs / 植株主要被"丁"字毛

17. Legumes with septa between seeds, but not jointed; anthers tipped by a gland or apiculate/ 荚果种子间具隔，但不联合；花药先端具腺体或细尖 .... **XV. Indigofereae/ 木蓝族**

17. Legumes without septa between seeds; anthers not tipped by a gland or apiculate / 荚果种子间不具隔；花药先端无腺体也无细尖 .... **XVI. Galegeae/ 山羊豆族**

16. Plants glabrous or without indumentum composed of T-shaped hairs / 植株光滑或没有丁字毛

18. Stamens monadelphous and filaments partly connate a tube/ 雄蕊单体且花丝部分合生为管状 .... **XVII. Millettieae/ 崖豆藤族**

18. Stamens diadelphous, or if monadelphous then free at upper part or top / 雄蕊二体，若为单体则上部分离生

19. Leaves pinnately or digitately 3-foliolate / 叶为羽状 3 小叶或掌状 3 小叶

20. Leaves ± toothed/ 叶多少具锯齿 .... **XII. Trifolieae/ 车轴草族**

20. Leaves entire / 叶全缘

21. Legumes jointed/ 荚果分节 .... **XIII. Desmodieae/ 山蚂蝗族**

21. Legumes not jointed/ 荚果不分节 .... **X. Phaseoleae/ 菜豆族**

19. Leaves pinnate; leaflets generally numerous / 羽状复叶，小叶常多数

22. Leaves gland-dotted; corolla reduced to just standard/ 叶具腺点；花冠退化只剩旗瓣 .... **XVIII. Amorpheae/ 紫穗槐族**

22. Leaves not gland-dotted; corolla regular / 叶不具腺点；花冠正常

23. Legumes indehiscent/ 荚果不开裂 .... **XIX. Dalbergieae/ 黄檀族**

23. Legumes dehiscent, sometimes only so at apex / 荚果开裂，有时仅先端开裂

24. Rachis of leaves ending in a tendril or bristle/ 叶轴顶端具卷须或刚毛 .... **XX. Fabeae/ 野豌豆族**

24. Rachis of leaves not ending in a tendril or bristle / 叶轴顶端不具卷须或刚毛

25. Legumes inflated and bladder-like, sometimes dehiscent only at apex/ 荚果肿胀如膀胱状，有时先端开裂 .... **XVI. Galegeae/ 山羊豆族**

25. Legumes compressed / 荚果压扁

26. Inflorescence in terminal or leaf-opposed racemes, rarely axillary/ 花序顶生或与叶对生的总状花序，稀腋生 .... **XVII. Millettieae/ 崖豆藤族**

26. Inflorescence solitary flower, fasciculate, or in axillary racemes / 花序为单花簇生，或腋生总状花序

27. Leaves odd-pinnate; leaflets 2–12 pairs/ 奇数羽状复叶；小叶 2 ~ 12 对

....**XXI. Robinieae**/ 刺槐族

27. Leaves even-pinnate; leaflets 10 - 30 pairs/ 偶数羽状复叶；小叶 10 ~ 30 对
....**XXII. Sesbanieae**/ 田菁族

FABALES. Fabaceae. *Albizia julibrissin* Durazz./ 豆目 豆科 合欢

A. Flowering branch/ 花枝
B. Flowers in head/ 头状花序
C. Bipinnate/ 二回羽状复叶
D. Seeds/ 种子
E. Fruit (legume)/ 果实 (荚果)
F. Flower/ 花
G. Flowers in head, bud stage/ 头状花序，蕾期
(Scale/ 标尺： A–B 1 cm; C 1 mm; D–G 1 cm)

FABALES. Fabaceae. *Sesbania cannabina* (Retz.) Poir./ 豆目 豆科 田菁

A. Ventral view of flower/ 花腹面观

B. Calyx segregated/ 花萼离析

C. Racemes/ 总状花序

D. Wings/ 翼瓣

E. Pistil/ 雌蕊

F. Nine stamens fused/9 枚雄蕊合生

G. One stamen only basally fused/1 枚雄蕊仅基部联合

H. Vertical view of flower/ 花纵切面

I. Ventral view of standard/ 旗瓣腹面观

J. Vertical profile of flower/ 花纵剖面，除花瓣

K. Keels/ 龙骨瓣

(Scale/ 标尺： A–B 1 mm; C 1 cm; D–K 1 mm)

I. Mimoseae/ 含羞草族

1. Armed plant; legume dehiscent into several segments/ 植株具刺，荚果裂为数节 ....**1. *Mimosa*/ 含羞草属**
1. Unarmed plant; legume dehiscing longitudinal/ 植株不具刺；荚果纵裂 ....**2. *Leucaena*/ 银合欢属**

1. *Mimosa pudica* L./ 含羞草
（photo by WeiLiang MA/ 马炜梁拍摄）;

2. *Leucaena leucocephala* (Lam.) de Wit/ 银合欢
（Invasived/ 入侵）

II. Acacieae/ 金合欢族

Only *Acacia* Mill.

仅金合欢属 1 属

Acacia farnesiana (L.) Willd/ 金合欢（Widely cultivated in southern of this area / 本区域南部常见栽培）

III. Ingeae/ 印加树族

1. Legume dehiscent, curved/ 荚果开裂, 扭转 ....**1. *Archidendron*/ 猴耳环属**
1. Legume indehiscent, straight/ 荚果不裂, 劲直 ....**2. *Albizia*/ 合欢属**

1. *Archidendron lucidum* (Benth.) I.C. Nielsen/ 亮叶猴耳环;　　2. *Albizia julibrissin* Durazz./ 合欢

IV. Cercideae/ 紫荆族

1. Legume narrowly winged; leaves simple, entire/ 荚果具狭翅; 单叶, 全缘 ....**1. *Cercis*/ 紫荆属**
1. Legume without wings; leaves simple, usually bilobed/ 荚果不具翅; 单叶, 常2裂 ....**2. *Bauhinia*/ 羊蹄甲属**

1. *Cercis glabra* Pamp./ 湖北紫荆;　　2. *Bauhinia championii* (Benth.) Benth./ 龙须藤
（2a. Branch/ 枝条; 2b. Legume/ 荚果）

V. Caesalpinieae/ 云实族

Only *Caesalpinia* L.

仅云实属1属

Caesalpinia decapetala (Roth) Alston/ 云实
（a. Inflorescence/ 花序; b. Leaves bipinnate/ 二回羽状复叶）

VI. Cassieae/ 决明族

1. Bracteoles absent; petals subequal / 不具小苞片；花瓣近等大 .... **1. *Senna*/ 决明属**
1. Bracteoles present; petals unequal / 具小苞片；花瓣不等大 .... **2. *Chamaecrista*/ 山扁豆属**

1. *Senna occidentalis* (L.) Link/ 望江南；

2. *Chamaecrista mimosoides* (L.) Greene/ 山扁豆

VII. Sophoreae/ 槐族

Only *Styphnolobium* Schott

仅槐属 1 属

VIII. Thermopsideae/ 野决明族

Only *Thermopsis* R. Br.

仅野决明属 1 属

Thermopsis chinensis Benth. ex S. Moore/ 霍州油菜（小叶野决明）

Styphnolobium japonicum (L.) Schott/ 槐

IX. Aeschynomeneae/ 合萌族

1. Leaflets 8 to many per leaf; bracts small, shorter than flowers; bracteoles present / 小叶多于8枚；苞片小，短于花；具小苞片 ....**1. *Aeschynomene*/ 合萌属**
1. Leaflets usually 2 or 4 per leaf; bracts enlarged, usually enclosing flowers; bracteoles absent / 小叶常2或4枚；苞片增大，花常包藏于内；不具小苞片 ....**2. *Zornia*/ 丁癸草属**

1. *Aeschynomene indica* L./ 合萌（1a. Plants/ 植株；1b. Inflorescences/ 花序）；
2. *Zornia gibbosa* Span./ 丁癸草

X. Phaseoleae/ 菜豆族

1. Leaflets abaxially and calyxes generally with colored glands; bracteoles absent / 小叶叶背及萼片常被有色腺点；小苞片缺
   2. Legumes 3 or more seeded / 荚果具3粒以上种子 ....**1. *Dunbaria*/ 野扁豆属**
   2. Legumes 1 or 2 seeded / 荚果具1或2粒种子
      3. Twining herbs or shrublets; legumes compressed/ 缠绕草本或小灌木；荚果压扁 ....**2. *Rhynchosia*/ 鹿藿属**
      3. Erect shrubs or prostrate herbs; legumes turgid/ 直立灌木或匍匐草本；荚果肿胀 ....**3. *Flemingia*/ 千斤拔属**
1. Leaflets and calyxes without glands; bracteoles often present / 小叶及萼片无腺点；具小苞片
   4. Style flattened, thickened or twisted, mostly bearded / 花柱扁平，增厚或扭转，多数有须
      5. Style flattened laterally/ 花柱侧面扁平 ....**4. *Lablab*/ 扁豆属**
      5. Style terete or flattened dorsiventrally/ 花柱圆柱形或背腹压扁 ....**5. *Vigna*/ 豇豆属**
   4. Style generally terete and unbearded / 花柱常圆柱形，无须
      6. Petals generally unequal in length / 花瓣常不等大

7. Trees or shrubs; keels much shorter than standard/ 乔木或灌木；龙骨瓣远小于旗瓣 ....**6. *Erythrina*/ 刺桐属**

7. Twiners or climbers; keels usually longest petals/ 攀缘藤本；龙骨瓣为最大 ....**7. *Mucuna*/ 油麻藤属**

6. Petals subequal in length / 花瓣近等大

8. Inflorescence generally with nodes, swollen/ 花序轴具节，膨大 ....**8. *Canavalia*/ 刀豆属**

8. Inflorescence with nodes, not or only slightly swollen / 花序轴具节，不膨大或稍膨大

9. Flowers 2 or more per node of inflorescence/ 花序轴上每节具2至多朵花 ....**9. *Pueraria*/ 葛属**

9. Flower 1 per node of inflorescence/ 花序轴上每节仅具1朵花 ....**10. *Glycine*/ 大豆属**

1. *Dunbaria villosa* (Thunb.) Makino/ 野扁豆（1a. Legume/ 豆荚；1b. Flower/ 花）；

2. *Rhynchosia volubilis* Lour./ 鹿藿；

3. *Flemingia prostrata* Roxb./ 千斤拔（3a. Habitat/ 生境；3b. Infructescence/ 果序）；

4. *Lablab purpureus* (L.) Sweet/ 扁豆；

5. *Vigna unguiculata* (L.) Walp./ 豇豆（photo by WeiLiang MA/ 马炜梁拍摄）；

6. *Erythrina crista-galli* L./ 鸡冠刺桐（Naturalized/ 自然归化）；

7. *Mucuna sempervirens* Hemsl./ 常春油麻藤；

8. *Canavalia rosea* (Sw.) DC./ 海刀豆

9. *Pueraria montana* (Lour.) Merr. var. *lobata* (Willd.) Maesen & S.M. Almeida ex Sanjappa & Predeep/ 葛麻姆

10. *Glycine tomentella* Hayata/ 短绒野大豆

XI. Crotalarieae/ 猪屎豆族

Only *Crotalaria* L.

仅猪屎豆属 1 属

Crotalaria sessiliflora L./ 农吉利 (野百合)(a. Flowering plant/ 花期植株；b. Fruiting plant/ 果期植株)

XII. Trifolieae/ 车轴草族

1. Leaves palmately 3-foliolate; inflorescence generally an umbel or dense racemes, sometimes head-like/ 叶为 3 小叶掌状复叶；花序一般为伞形或密集总状，有时为头状 .... **1.** *Trifolium*/ 车轴草属
1. Leaves pinnately 3-foliolate / 叶为羽状 3 小叶
   2. Legumes straight/ 荚果劲直 .... **2.** *Melilotus*/ 草木犀属
   2. Legumes mostly spirally coiled, or curved/ 荚果多螺旋状卷曲，或弯曲 .... **3.** *Medicago*/ 苜蓿属

1. *Trifolium repens* L./ 白车轴草；
2. *Melilotus officinalis* (L.) Lam./ 草木犀；
3. *Medicago polymorpha* L./ 南苜蓿

XIII. Desmodieae/ 山蚂蝗族

1. Stipels absent, rarely present; legumes 1-jointed, 1-seeded, not glochidiate / 小托叶不存在，稀存在；荚果仅 1 节，每节 1 粒种子，不具倒刺毛
    2. Lateral veins of leaflets strict, extending to margin; stipules ovate/ 侧脉笔直，伸至叶缘；托叶卵圆形 ....**1. *Kummerowia*/ 鸡眼草属**
    2. Lateral veins of leaflets arcuate, not reaching to margin; stipules subulate / 侧脉弧曲，不达叶缘；托叶钻形
        3. Bracts 1-flowered, usually caducous/ 苞片腋内 1 朵花，早落 ....**2. *Campylotropis*/ 笐子梢属**
        3. Bracts 2-flowered, persistent/ 苞片腋内 2 朵花，宿存 ....**3. *Lespedeza*/ 胡枝子属**
1. Stipels present; legumes usually glochidiate, 2 to several jointed, rarely 1-jointed, 1-seeded / 小托叶存在；荚果常具倒刺毛，2 至数节，稀 1 节，每节 1 粒种子
    4. Calyxes glume-like, lobes striate/ 花萼颖片状，裂片具条纹 ....**4. *Alysicarpus*/ 链荚豆属**
    4. Calyxes not glume-like, lobes not striate / 花萼非颖片状，裂片不具条纹
        5. Legumes distinctly stipitate; stamens monadelphous / 荚果明显具柄；单体雄蕊 ....**5. *Hylodesmum*/ 长柄山蚂蝗属**
        5. Legumes not stipitate, or rarely shortly stipitate; stamens diadelphous, rarely monadelphous / 荚果不具柄，或稀具短柄；二体雄蕊，稀单体
            6. Joints of legume plicate-retrofracted/ 荚节反复折叠 ....**6. *Christia*/ 蝙蝠草属**
            6. Joints of legume not plicate-retrofracted / 荚节不反复折叠
                7. Petioles winged/ 叶柄具翅 ....**7. *Ohwia*/ 小槐花属**
                7. Petioles not winged/ 叶柄不具翅 ....**8. *Desmodium*/ 山蚂蝗属**

1. *Kummerowia striata* (Thunb.) Schindl. / 鸡眼草（1a.Flowering branch/ 花枝；1b. Legumes and seeds/ 荚果与种子）；
2. *Campylotropis macrocarpa* (Bunge) Rehder/ 笐子梢

3. *Lespedeza virgata* (Thunb.) DC./ 细梗胡枝子；
4. *Alysicarpus vaginalis* (L.) DC./ 链荚豆；
5. *Hylodesmum podocarpum* (DC.) H. Ohashi & R.R. Mill/ 长柄山蚂蝗；
6. *Christia obcordata* (Poir.) Bakh. f. ex Meeuwen/ 铺地蝙蝠草；
7. *Ohwia caudata* (Thunb.) H. Ohashi/ 小槐花；
8. *Desmodium heterophyllum* (Willd.) DC./ 异叶山蚂蝗

XIV. Hedysareae/ 岩黄芪族

Only *Caragana* Fabr.

仅锦鸡儿属 1 属

Caragana sinica (Buc'hoz) Rehder/ 锦鸡儿

XV. Indigofereae/ 木蓝族

Only *Indigofera* L.

仅木蓝属 1 属

Indigofera amblyantha Craib/ 多花木蓝

（a. Inflorescence, photo by XinXin ZHU/ 花序，朱鑫鑫拍摄；b. Legumes and seeds/ 荚果与种子）

XVI. Galegeae/ 山羊豆族

1. Keel petals half as long as wings; style shorter than or as long as ovary/ 龙骨瓣长为翼瓣的一半；花柱短于或与子房等长 ....**1. *Gueldenstaedtia*/ 米口袋属**
1. Keel petals subequal to or slightly shorter than wings; style longer than ovary/ 龙骨瓣与翼瓣近等长或稍短之；花柱长于子房 ....**2. *Astragalus*/ 黄芪属**

1. *Gueldenstaedtia verna* (Georgi) Boriss./ 少花米口袋； 2. *Astragalus sinicus* L./ 紫云英

XVII. Millettieae/ 崖豆藤族

1. Inflorescence racemes or panicles; flowers solitary in axil of a bract / 花序总状或圆锥状；花单生于苞腋
    2. Inflorescence racemes, pendent/ 总状花序，下垂 ....**1. *Wisteria*/ 紫藤属**
    2. Inflorescence panicles, not pendent/ 圆锥花序，不下垂 ....**2. *Callerya*/ 鸡血藤属**
1. Inflorescence pseudoracemes or pseudopanicles; flowers inserted on brachyblasts of floral axis/ 假总状花序或假圆锥花序；花着生于花序轴短枝上 ....**3. *Millettia*/ 崖豆藤属**

1. *Wisteria sinensis* (Sims) DC./ 紫藤；
2. *Callerya reticulata* (Benth.) Schot/ 网络鸡血藤；
3. *Millettia pachycarpa* Benth./ 厚果崖豆藤（photo by XinXin ZHU 朱鑫鑫拍摄）

XVIII. Amorpheae/ 紫穗槐族

Only *Amorpha* L.

仅紫穗槐属 1 属

Amorpha fruticosa L./ 紫穗槐（Naturalized/ 归化）

（a. Inflorescence/ 花序 photo by XinXin ZHU 朱鑫鑫拍摄；b. Infructescence/ 果序）

XIX. Dalbergieae/ 黄檀族

Only *Dalbergia* L.f.

仅黄檀属 1 属

Dalbergia assamica Benth./ 秧青（a. Flowering plant 花期植株；b. Flowers closeup 花部特写）

XX. Fabeae/ 野豌豆族

1. Staminal tube oblique at apex/ 雄蕊柱顶端倾斜 .... **1. *Vicia*/ 野豌豆属**
1. Staminal tube not oblique at apex/ 雄蕊柱顶端不倾斜 .... **2. *Lathyrus*/ 山黧豆属**

1. *Vicia sativa* L./ 救荒野豌豆；　　　　　　2. *Lathyrus japonicus* Willd./ 海滨山黧豆

XXI. Robinieae/ 刺槐族

Only *Robinia* L.

仅刺槐属 1 属

Robinia pseudoacacia L./ 刺槐

XXII. Sesbanieae/ 田菁族

Only *Sesbania* Scop.

仅田菁属 1 属

Sesbania cannabina (Retz.) Poir./ 田菁

142 Polygalaceae 远志科

Leaves simple, small nectar glands at petioles base / 单叶，叶柄基部具小腺体
Sepal unequal, inner 2 petaloid / 花萼不等大，内侧 2 枚花瓣状
Lower petal often boat-shaped, apically fimbricate / 中间花瓣为舟状，先端具流苏
Filaments fused, forming a tube / 花丝合生，呈管状

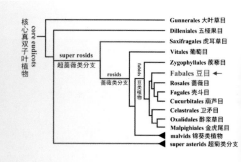

Perennial to annual **herbs**, **shrubs**, **trees** or **woody climbers**, rarely **saprophytic**. **Leaves** simple, alternate (usually spiral), opposite or whorled; margin entire; leaves rarely reduced and scale-like; sometimes small nectar glands at petioles base; stipules absent. **Inflorescence** panicles, racemes, spikes or rarely solitary flower; bract 1. **Flowers** bisexual or functionally unisexual, ±zygomorphic to less often actinomorphic; bracteoles 2. **Sepals** usually free or basally fused, deciduous, unequal, outer 3 green, often inner 2 petaloid. **Petals** basally fused or free, lower petal often boat-shaped, sometimes apically fimbricate, bright. **Stamen** filaments free, or fused and forming a sheath open on upper side and trough-like, attached to the petals; anthers basifixed; usually dehiscing by a single apical pore; rarely staminodes. **Ovary** superior; carpels fused; locules 2–8 or 1; ovule 1 or 4 to many per locule; placentation axile or rarely parietal. **Fruit** a loculicidal capsule (often winged), drupe, berry, nut or samara. **Seeds** sometimes hairy, often arillate.

多年生至一年生**草本**、**灌木**、**乔木**或**木质藤本**，稀为**腐生植物**。**叶**为单叶，互生（常螺旋状排列）、对生或轮生；全缘；叶稀退化为鳞片状；有时在叶柄基部有小腺体；托叶缺。**花序**为圆锥状、总状、穗状或稀单花；苞片 1 枚。**花**两性或功能上单性，多少为两侧对称，偶辐射对称；小苞片 2 枚。**萼片**常离生或基部合生，早落，不等大，外侧 3 枚绿色，内侧 2 枚常花瓣状。**花瓣**基部合生或离生，中间 1 枚常为舟状，有时先端具流苏，鲜艳。**雄蕊**花丝分离，或合生成鞘，上方开放成槽状，贴生于花瓣；花药基着；常顶部单孔开裂；稀具退化雄蕊。**子房**上位；心皮合生；2~8 或 1 室，胚珠每室 1 或 4 至多枚；中轴胎座或稀为侧膜胎座。**果实**为室背开裂蒴果（常具翅）、核果、浆果、坚果或翅果。**种子**有时被毛，常具假种皮。

World 26/965, worldwide, especially in tropical and subtropical regions.
China 6/53.
This area 1/2.

全世界共 26 属 965 种，世界广布，主产热带和亚热带地区。
中国产 6 属 53 种。
本地区有 1 属 2 种。

Polygala L. 远志属

FABALES. Polygalaceae. *Polygala japonica* Houtt./ 豆目 远志科 瓜子金

A. Whole plant/ 植株
B. Vertical profile of corolla/ 花冠纵剖面
C. Pistil/ 雌蕊
D. Androecium/ 雄蕊群
E. Vertical section of staminal sheath/ 雄蕊鞘纵切面
F. Dorsal view of leaf/ 叶背面观
G. Racemes/ 总状花序
H. Ventral view of leaf/ 叶正面观
I. Ventral view of flower/ 花顶面观 (腹面观)
J. Flowering branch/ 花枝
K. Vertical profile of flower/ 花纵剖面
L. Flower segregated/ 花离析

(Scale/ 标尺 : A 1 cm; B–C 1 mm; D–E 200 μm; F–L 1mm)

1-1. *Ploygala hongkongensis* Hemsl. var. *stenophylla* Migo/ 狭叶香港远志（1-1a. Flowering branch/ 花枝；1-1b. Fruit/ 果实）；

1-2. *P. japonica* Houtt./ 瓜子金（1-2a. Flowering branch/ 花枝；1-2b. Fruit/ 果实）

143 Rosaceae 蔷薇科

Leaves alternate; stipules paired / 叶互生；托叶成对
Flowers actinomorphic, bisexual, hypanthium / 花辐射对称，两性，具被丝托
Perianth 5-merous / 花 5 基数
Stamens numerous; filaments free / 雄蕊多数；花丝常分离

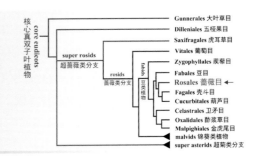

Trees, **shrubs**, or **herbs**, deciduous or evergreen. **Leaves** alternate, rarely opposite, simple or compound; stipules paired, free or adnate to petiole, rarely absent, persistent or deciduous; petiole usually with 2 glands at apical; often serrate, rarely entire. **Inflorescence** various, from single flower to umbellate, corymbs, racemes, panicles or fascicles. **Flowers** usually actinomorphic, bisexual, rarely unisexual and then plants dioecious; often with hypanthium (formed from basal parts of sepals, petals, and stamens), free from or adnate to ovary. **Sepals** usually 5, imbricate; epicalyx sometimes present. **Petals** as many as sepals, free, imbricate, sometimes absent. **Stamens** usually numerous, rarely few; filaments usually free, very rarely connate; anthers dorsifixed, didymous, 2-loculed. **Ovary** superior or part-inferior to inferior; carpels free to fused, 1 to many; locules 1-many; ovules 1–2(to many) per locule; placentation usually basal, less often axile or marginal, rarely apical. **Fruit** a follicle, drupe, achene, berry or pome, sometimes aggregated in compound fruits with fleshy receptacles.

乔木、灌木或草本，落叶或常绿。**叶**互生，稀对生，单叶或复叶；托叶成对，离生或与叶柄合生，稀缺，宿存或早落；叶柄先端常具 2 腺体；常有锯齿，稀全缘。**花序**多样，从单花至伞形、伞房、总状、圆锥状或簇生。**花**常辐射对称，两性，稀单性雌雄异株；常具被丝托（由萼片、花瓣和雄蕊的基部合生而成），与子房分离或合生。**花萼**常 5 枚，覆瓦状；有时具副萼。**花瓣**数同萼片，离生，覆瓦状，有时缺。**雄蕊**常多数，稀少量；花丝常分离，极稀合生；花药背着，双生，2 室。**子房**上位或半下位至下位，心皮离生或合生，1 至多数；1 至多室；每室胚珠 1~2 枚（至多枚）；常为基底胎座，偶为中轴或边缘胎座，稀为顶生胎座。**果实**为蓇葖果、核果、瘦果、浆果或梨果，有时在肉质花托上形成聚合果。

World 54–90/ca. 2, 950, worldwide, especially in northern hemisphere.
China 46/ca. 942.
This area 21/61.

全世界共 54~90 属约 2 950 种，世界广布，尤其是北半球。
中国产 46 属约 942 种。
本地区有 21 属 61 种。

1. Fruit dehiscent; stipules present or absent/ 果实开裂；托叶有或无
　　2. Flowers more than 2 cm in diam.; fruit a capsule/ 花较大，直径 2 cm 以上；果为蒴果 ....**1. *Exochorda*/ 白鹃梅属**
　　2. Flowers less than 2 cm in diam.; fruit a follicle/ 花较小，直径不超过 2 cm；果为蓇葖果
　　　　3. Carpels 1 or 2; stipules present/ 心皮 1 或 2 枚；具托叶 ....**2. *Stephanandra*/ 野珠兰属**
　　　　3. Carpels 5; stipules absent/ 心皮 5 枚；无托叶 ....**3. *Spiraea*/ 绣线菊属**
1. Fruit indehiscent; stipules present/ 果实不开裂；有托叶
　　4. Ovary inferior or semi-inferior/ 子房下位或半下位
　　　　5. Fruit with 1–5 pyrenes; carpels bony when mature/ 果实具 1~5 分核；心皮在成熟时变为坚硬骨质
　　　　　　6. Leaves pinnate/ 叶为羽状复叶 ....**4. *Osteomeles*/ 小石积属**
　　　　　　6. Leaves simple/ 叶为单叶
　　　　　　　　7. Plant evergreen; with 2 fertile ovules per carpel/ 叶常绿；每心皮具 2 枚可育胚珠 ....**5. *Pyracantha*/ 火棘属**
　　　　　　　　7. Plant deciduous; with 1 fertile ovule per carpel/ 叶凋落；每心皮具 1 枚可育胚珠 ....**6. *Crataegus*/ 山楂属**
　　　　5. Fruit a 1–5(–7)-loculed pome, each locule with 1 to many seeds; carpels leathery or papery when mature/ 果实为 1~5(~7) 室梨果，每室具 1 至多粒种子；心皮在成熟时变为革质或纸质
　　　　　　8. Inflorescence compound-corymbs or panicles, many flowered/ 花序复伞房或圆锥花序，多花
　　　　　　　　9. Ovary semi-inferior/ 子房半下位 ....**7. *Photinia*/ 石楠属**
　　　　　　　　9. Ovary inferior/ 子房下位 ....**8. *Rhaphiolepis*/ 石斑木属**
　　　　　　8. Inflorescence umbellate or racemes, sometimes flowers fascicled or solitary/ 花序伞形或总状，有时花簇生或单生
　　　　　　　　10. Styles free/ 花柱离生 ....**9. *Pyrus*/ 梨属**
　　　　　　　　10. Styles connate basally/ 花柱基部合生 ....**10. *Malus*/ 苹果属**
　　4. Ovary superior, rarely inferior/ 子房上位，稀下位
　　　　11. Carpels many; fruit an achene; leaves compound, very rarely simple/ 心皮多数；瘦果；复叶，极少单叶
　　　　　　12. Achenes enclosed in cupulate or urn-shaped torus/ 瘦果生于杯状或坛状花托内
　　　　　　　　13. Shrubs; stems usually prickly/ 灌木；茎常具刺 ....**11. *Rosa*/ 蔷薇属**
　　　　　　　　13. Perennial herbs or shrublets; stems not prickly/ 多年生草本或小灌木；茎不具刺
　　　　　　　　　　14. Petals present/ 有花瓣 ....**12. *Agrimonia*/ 龙芽草属**
　　　　　　　　　　14. Petals absent/ 无花瓣 ....**13. *Sanguisorba*/ 地榆属**
　　　　　　12. Achenes or drupelets borne on flat or convex torus/ 瘦果或小核果生于扁平或隆起花托上
　　　　　　　　15. Carpels borne at base of flat or slightly concave torus; stipules not adnate to petiole/ 心皮生于扁平或微凹的花托基部；托叶与叶柄分离 ....**14. *Kerria*/ 棣棠属**
　　　　　　　　15. Carpels borne on globose or conic torus; stipules adnate to petiole or not/ 心皮生于球形或圆锥形隆起的花托上；托叶与叶柄连合或分离
　　　　　　　　　　16. Fruit drupelets or drupaceous achenes; stems prickly, rarely unarmed/ 小核果或核果状瘦果；茎具刺，稀无刺 ....**15. *Rubus*/ 悬钩子属**

16. Fruit achenes; stems unarmed/ 瘦果；茎无刺
    17. Torus inflated and fleshy when maturity/ 花托成熟时膨大且肉质 .... **16. *Duchesnea*/ 蛇莓属**
    17. Torus dry when maturity/ 花托成熟时干燥 .... **17. *Potentilla*/ 委陵菜属**
11. Carpel 1(–5); fruit a drupe; leaves simple/ 心皮 1（~5）枚；核果；单叶
    18. Drupe grooved, hairy or glabrous with glaucous; endocarp conspicuously compressed/ 核果具沟，被毛或光滑而被蜡粉；内果皮明显压扁
        19. Axillary winter buds 3 with 2 lateral flower buds and 1 central leaf bud; endocarp often pitted, rarely smooth/ 冬芽侧芽 3 个，两侧花芽，中间叶芽；内果皮常有孔穴，稀光滑 .... **18. *Amygdalus*/ 桃属**
        19. Axillary winter buds single; endocarp usually smooth or inconspicuously pitted/ 冬芽侧芽 1 个；内果皮常光滑或具不明显孔穴 .... **19. *Armeniaca*/ 杏属**
    18. Drupe not grooved, glabrous but not glaucous; endocarp not or inconspicuously compressed/ 核果无沟，光滑不被蜡粉；内果皮不压扁或不明显
        20. Inflorescence usually with conspicuous bracts; flowers solitary to several in short racemes or corymbs/ 花序常具明显的苞片；花单生或数朵组成短总状或伞房花序 .... **20. *Cerasus*/ 樱属**
        20. Inflorescence with small bracts; flowers 10 or more in racemes/ 花序具小型苞片；总状花序花 10 朵以上 .... **21. *Padus*/ 稠李属**

ROSALES. Rosaceae. *Cerasus yedoensis* (Matsum.) A.N. Vassiljeva/ 蔷薇目 蔷薇科 东京樱花

A₁. Inflorescence/ 花序
A₂. Umbellate-racemose/ 花序伞形总状
B. Cross section of ovary/ 子房横切面
C. Vertical section of partly corolla, amplified/ 部分花冠纵切面，放大
D. Petals/ 花瓣
E. Vertical section of partly corolla/ 部分花冠纵切面
F. Dorsal view of calyx lobes/ 花萼裂片背面观
G. Calyx tube/ 花萼筒
H. Stamen/ 雄蕊
I. Pistil/ 雌蕊
J. Vertical profile of flower/ 花纵剖面
(Scale/ 标尺：A₁–A₂ 1 cm; B–C 1 mm; D–G 1 cm; H–J 1 mm)

1. *Exochorda racemosa* (Lindl.) Rehder/ 白鹃梅（1a. Infructescence/ 果序；1b. Flowering branch/ 花枝）；
2. *Stephanandra chinensis* Hance/ 华空木；
3. *Spiraea chinensis* Maxim./ 中华绣线菊；
4. *Osteomeles subrotunda* K. Koch/ 圆叶小石积；
5. *Pyracantha fortuneana* (Maxim.) H.L. Li/ 火棘（5a. Infructescence/ 果序；5b. Cross section of ovary, vertical section of flower, photo by WeiLiang MA/ 子房横切面，花纵切面，马炜梁拍摄）；
6. *Crataegus pinnatifida* Bunge/ 山楂（6a. Fruiting branch/ 果枝；6b. Cross section of ovary, vertical section of flower, photo by WeiLiang MA/ 子房横切面，花纵切面，马炜梁拍摄）；
7. *Photinia parvifolia* (E. Pritz.) C.K. Schneid./ 小叶石楠

8. *Rhaphiolepis indica* (L.) Lindl. ex Ker Gawl./ 石斑木;
9. *Pyrus calleryana* Decne./ 豆梨 (photo by WeiLiang MA/ 马炜梁拍摄) / 豆梨;
10. *Malus* sp./ 苹果属（not distributed in this area, photo by XinXin ZHU/ 非本区域分布，朱鑫鑫拍摄）;
11. *Rosa henryi* Boulenger/ 软条七蔷薇;
12. *Agrimonia pilosa* Ledeb./ 龙芽草;
13. *Sanguisorba officinalis* L./ 地榆;
14. *Kerria japonica* (L.) DC./ 棣棠花 (14a. Flowers/ 花; 14b. Vertical section of flower/ 花纵切面（photo by WeiLiang MA/ 马炜梁拍摄）;
15-1. *Rubus buergeri* Miq./ 寒莓

15-2. *R. hirsutus* Thunb./ 蓬虆；

15-3. *R. adenophorus* Rolfe/ 腺毛莓；

16. *Duchesnea indica* (Andrews) Teschem./ 蛇莓；

17. *Potentilla* sp./ 委陵菜属；

18. *Amygdalus persica* L./ 桃；

19. *Armeniaca mume* Siebold/ 梅；

20. *Cerasus discoidea* T.T. Yu & C.L. Li/ 迎春樱桃；

21. *Padus obtusata* (Koehne) T.T. Yu & T.C. Ku/ 细齿稠李

146 Elaeagnaceae 胡颓子科

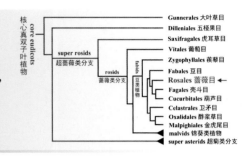

Usually xerophytes or halophytes / 常为旱生或盐生植物
Branches spine-tipped / 小枝先端常呈刺状
Flowers actinomorphic, often fleshy hypanthium / 花辐射对称，常具肉质托杯
Fruit enclosed in calyx tube / 果实包裹于萼筒内

Shrubs, small **trees** or rarely **woody climbers**, usually xerophytes or halophytes. **Leaves** often deciduous, simple, alternate (usually spiral) or opposite; margin entire; petioles present, sometimes branches spine-tipped; stipules absent; usually with conspicuous silver or golden stellate hairs throughout. **Inflorescence** axillary solitary flower, racemes, spikes or fascicles on short shoots. **Flowers** bisexual or rarely polygamous, actinomorphic, often fleshy hypanthium. **Perianth** 1-whorled, fused (at least basally), usually valvate. **Stamen** filaments attached to hypanthium. **Ovary** superior, but tightly enclosed in basal part of calyx and apparently inferior; carpel 1; 1-loculed ; ovule 1; placentation basal. **Fruit** pseudo-drupe, indehiscent. **Seed** 1.

灌木、小**乔木**或稀为**木质藤本**，常为旱生或盐生植物。**叶**多落叶性，单叶，互生（常螺旋状排列）或对生；全缘；具叶柄，有时小枝先端呈刺状；托叶缺；全株被明显的银色或金色星状毛。**花序**为腋生单花、总状、穗状或簇生于短枝。**花**两性或稀杂性，辐射对称，常有肉质托杯。**花被片** 1 轮，合生（至少基部如此），常镊合状。**雄蕊**花丝贴生于托杯。**子房**上位，但由于紧紧包裹于花萼筒基部而呈下位状；心皮 1 枚；1 室；胚珠 1 枚；基底胎座。**果实**为假核果，不裂。**种子** 1 粒。

World 3/90, Northern temperate and tropical regions.
China 2/74.
This area 1/7.

全世界共 3 属 90 种，分布于北温带和热带地区。
中国产 2 属 74 种。
本地区有 1 属 7 种。

Elaeagnus L. 胡颓子属

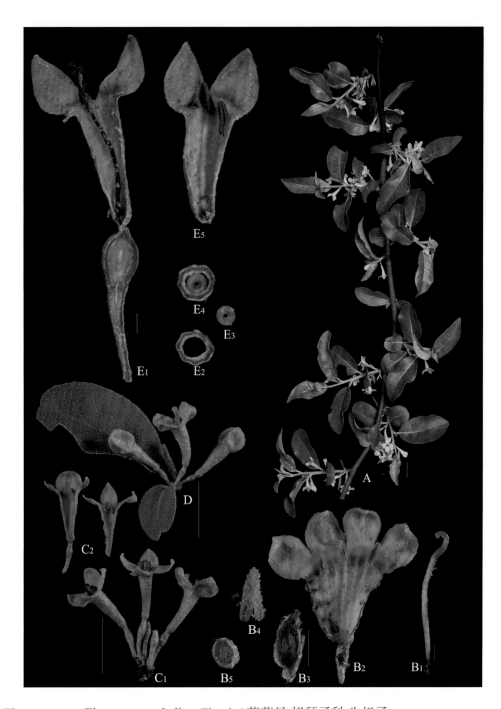

ROSALES. Elaeagnaceae. *Elaeagnus umbellata* Thunb./ 蔷薇目 胡颓子科 牛奶子

A. Flowering branch/ 花枝

B_1. Stigma and style/ 柱头与花柱

B_2. Inside of perianth tube/ 花被筒内侧

B_3. Vertical section of ovary/ 子房纵切面

B_4. Anther/ 花药

B_5. Cross section of ovary/ 子房横切面

C_1. Flowers clustered/ 花簇生

C_2. Vertical profile of corolla/ 花冠纵剖面

D. Inflorescence clustered in axillary/ 花序簇生叶腋

E_1. Vertical section of flower/ 花纵切面

E_2. Cross section of perianth tube, ovary removed/ 花被筒横切面，除子房

E_3. Cross section of ovary/ 子房横切面

E_4. Cross section of perianth tube/ 花被筒横切面

E_5. Vertical profile of corolla, amplified/ 花冠纵剖，放大

(Scale/ 标尺： A 1 cm; B_1–B_5 1 mm; C_1–D 1 cm; E_1–E_5 1 mm)

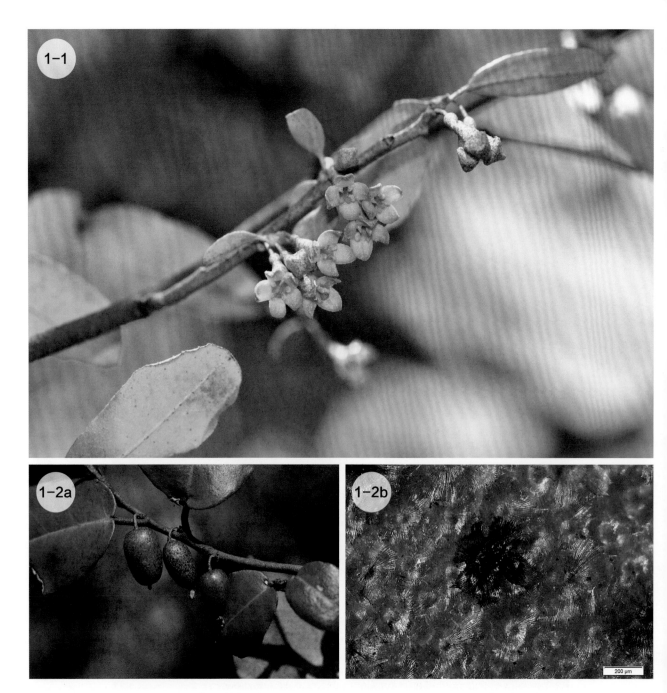

1-1. *Elaeagnus argyi* H. Lév./ 佘山羊奶子；

1-2. *E. pungens* Thunb./ 胡颓子（1-2a. Fruits/ 果实；1-2b. Scales/ 鳞片）（Scale/ 标尺：200μm）

147 Rhamnaceae 鼠李科

Plant often thorny / 植株常具刺
Flowers not bright, with nectar disk / 花色不显，具蜜腺盘
Sepals valvate, inside fleshy layer forming a keel / 花萼镊合状，肉质内层成龙骨状
Stamens 1-whorled, opposite the petals / 雄蕊 1 轮，与花瓣对生

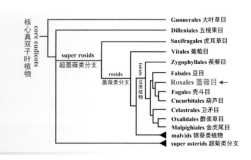

Trees, **shrubs**, **woody climbers** or **herbs**, often thorny, rarely succulent. **Leaves** sometimes deciduous, simple, alternate or less often opposite; margin toothed or entire; venation pinnate or distinctly 3-nerved; petioles present to absent; stipules small, often turned to spines or thorns. **Inflorescence** cymes or fascicles or solitary flower. **Flowers** yellowish to greenish, rarely bright; bisexual to unisexual (plants monoecious, dioecious or androdioecious), actinomorphic, nectar disk present or absent. **Sepals** valvate, fused or free, often with a fleshy layer inside usually forming a keel, alternate with petals or petaloid when petals are absent. **Petals** free or rarely absent, clawed or sessile, hooded or often keeled, often smaller than sepals. **Stamens** 1-whorled, opposite the petals, filaments attached to the base of the petals; anthers dorsifixed, introrse, dehiscing by longitudinal slits. **Ovary** superior, part-inferior or inferior; carpels fused; locules (1–)2–3(–5); ovules usually 1 or 2 per locule; placentation basal or axile. **Fruit** a capsule, drupe, samara or nut.

乔木、灌木、木质藤本或**草本**，常多刺，稀肉质。**叶**多落叶性，单叶，互生或稀对生；叶缘具齿或全缘；羽状脉或 3 出脉；叶柄有或无；托叶小，或有时变为刺或角状。**花序**聚伞状或簇生或单花。**花**淡黄色至淡绿色，稀鲜艳；两性至单性（雌雄同株、雌雄异株或雄全异株），辐射对称，蜜腺盘有或无。**萼片**镊合状，合生或离生，内侧常肉质成龙骨状，与花瓣互生或花瓣缺失时呈花瓣状。**花瓣**离生或稀缺失，具爪或无柄，兜帽状或具龙骨突，多小于花萼。**雄蕊** 1 轮，与花瓣对生，花丝贴生于花瓣基部；花药背着，内向，纵裂。**子房**上位、半下位或下位；心皮合生;(1~)2~3(~5)室；胚珠每室 1 或 2 枚；基底胎座或中轴胎座。**果实**为蒴果、核果、翅果或坚果。

World 58/1, 000, worldwide, especially in the tropics and warm temperate regions.
China 13/137.
This area 5/11.

全世界共 58 属 1 000 种，世界广布，特别是热带和暖温带地区。
中国产 13 属 137 种。
本地区有 5 属 11 种。

1. Leaves pinnately veined; fruits not winged/ 叶为羽状脉；果实不具翅
   2. Fruit 1-stoned drupe/ 核果具 1 核
      3. Leaves margin entire; inflorescences terminal or both axillary / 叶全缘；

Sageretia thea (Osbeck) M.C. Johnst./ 雀梅藤

花序顶生或兼具腋生 .... **1. *Berchemia*/** 勾儿茶属
   3. Leaves margin denticulate; inflorescence axillary/ 叶具细牙齿；花序腋生 .... **2. *Rhamnella*/** 猫乳属
 2. Fruit with 2-4 pyrenes/ 果具 2~4 分核
   4. Flowers nearly sessile/ 花近无梗 .... **3. *Sageretia*/** 雀梅藤属
   4. Flowers distinctly pedicellate/ 花明显具梗 .... **4. *Rhamnus*/** 鼠李属
1. Leaves triplinerved; fruits winged/ 叶为基出 3 脉；果实具翅 .... **5. *Paliurus*/** 马甲子属

ROSALES. Rhamnaceae. *Sageretia rugosa* Hance/ 蔷薇目 鼠李科 皱叶雀梅藤 (not distributed here/ 非本区域分布)

A. Flowering branch/ 花枝
B. Inflorescence/ 花序
C. Frontal view of corolla, amplified/ 花冠正面观，放大
D$_1$. Flower segregated/ 花离析
D$_2$. Pistil/ 雌蕊
D$_3$. Sepal/ 萼片
D$_4$. Stamens opposite with petals/ 雄蕊与花瓣对生
D$_5$. Vertical section of corolla/ 花冠纵切面
D$_6$. Frontal view of corolla/ 花冠正面观
D$_7$. Sepal adaxially keeled/ 萼片腹面具中肋
(Scale/ 标尺： A 1 cm； B-D$_7$ 1 mm)

1. *Berchemia floribunda* (Wall.) Brongn./ 多花勾儿茶；
2. *Rhamnella franguloides* (Maxim.) Weberb./ 猫乳；
3. *Sageretia thea* (Osbeck) M.C. Johnst./ 雀梅藤
   （3a. Infructescence/ 果序；3b. Inflorescence/ 花序）；
4. *Rhamnus crenata* Siebold & Zucc./ 长叶冻绿；
5. *Paliurus ramosissimus* (Lour.) Poir./ 马甲子

148 Ulmaceae 榆科

Bark notably ridged or spiny / 树皮明显具脊或刺
Leaves simple, alternate, usually toothed / 单叶互生，常具锯齿
Perianth 1-whorled, persistent / 花被片 1 轮，宿存
Fruit usually a samara / 果实常为翅果

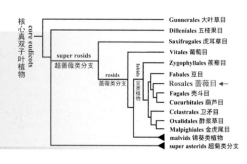

Trees or **shrubs**; bark notably ridged or spiny. **Leaves** deciduous or evergreen, simple, usually alternate; margin entire or toothed (singly or doubly); petioles present; stipules present; cystoliths rare. **Inflorescence** solitary flower, cymes, fascicles or panicles. **Flowers** bisexual or unisexual (monoecious, andromonoecious or polygamomonoecious), actinomorphic. **Perianth** 1-whorled, fused or free, persistent. **Stamens** opposite the perianth; anthers dorsifixed, extrorse; staminodes present. **Ovary** superior; carpels fused; 1(–2)-loculed; ovule 1; placentation apical; stigmas often 2-lobed. **Fruit** usually a dry samara, unwinged in *Zelkova*.

乔木或**灌木**；树皮明显具脊或刺。**叶**为落叶或常绿，单叶，常互生；全缘或有锯齿（单锯齿或重锯齿）；具叶柄；托叶存在；少有钟乳体。**花序**为单花、聚伞状、簇生或圆锥状。**花**两性或单性（雌雄同株、雄全同株或杂性同株），辐射对称。**花被片** 1 轮，合生或离生，宿存。**雄蕊**与花被片对生；花药背着，外向；退化雄蕊存在。**子房**上位；心皮合生；1(~2)室；胚珠 1 枚；顶生胎座；柱头常 2 裂。**果实**常为干燥翅果，榉属无翅。

World 8/35, mostly in northern temperate regions, especially in Asia.
China 3/25.
This area 2/3.

全世界共 8 属 35 种，主要分布于北温带，特别是亚洲。
中国产 3 属 25 种。
本地区有 2 属 3 种。

1. Fruit dry, broadly winged/ 果实干燥，具宽翅 .... **1. *Ulmus*/ 榆属**
1. Fruit a drupe, not winged/ 核果，无翅 .... **2. *Zelkova*/ 榉属**

1. *Ulmus parvifolia* Jacq./ 榔榆；
2. *Zelkova serrata* (Thunb.) Makino/ 榉树

149 Cannabaceae 大麻科

Leaves often aromatic / 叶常具芳香味
Perianth 1-whorled, sometimes persistent / 花被片 1 轮，有时宿存
Carpels fused; 1-loculed; ovule 1 / 心皮合生；1 室 1 胚珠
Fruit usually a drupe or achene / 果实常为核果或瘦果

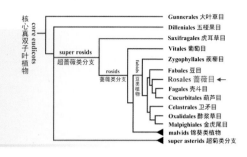

Trees, **shrubs**, **woody climbers** or **herbs**. **Leaves** often aromatic, deciduous, simple or compound, usually alternate or opposite; margin entire or toothed; petioles present; stipules present, often membranous and deciduous; cystoliths present or absent. **Inflorescence** racemes, cymes or panicles; bracteate, sometimes leafy, petaloid. **Flowers** small, unisexual (plants dioecious, monoecious or polygamomonoecious) or less often bisexual (then often functionally unisexual), actinomorphic. **Perianth** 1-whorled, 4 or 5; free or often partly fused, sometimes persistent. **Stamens** opposite tepals; 4 or 5; anthers dorsifixed; sometimes staminodes in female. **Ovary** superior; carpels fused; 1-loculed; ovule 1; placentation apical; styles 2. **Fruit** a drupe, samara or achene with persistent perianth.

乔木、灌木、木质藤本或草本。**叶**常具芳香味，落叶性，单叶或复叶，常互生或对生；全缘或有锯齿；具叶柄；托叶存在，常膜质或早落；有钟乳体或缺。**花序**总状、聚伞状或圆锥状；有苞片，有时为叶状或花瓣状。**花**小，单性（雌雄异株、雌雄同株或杂性同株）或稀为两性（则为功能性单性花），辐射对称。**花被片** 1 轮，4 或 5 枚；离生或常部分合生，有时宿存。**雄蕊**与花被片对生；4 或 5 枚；花药背着；有时雌花中具退化雄蕊。**子房**上位；心皮合生；1 室；胚珠 1 枚；顶生胎座；花柱 2 个。**果实**为核果、翅果或具宿存花被片的瘦果。

World ca. 10/180, disjunct between New World and Old World, from tropics to temperate regions.
China 7/25.
This area 4/8.

全世界共约 10 属 180 种，新世界与旧世界间断分布，热带至温带地区均产。
中国产 7 属 25 种。
本地区有 4 属 8 种。

1. Twining herbs; leaves opposite/ 缠绕草本，叶对生 .... **1.** *Humulus*/ 葎草属
1. Trees, or shrubs/ 乔木或灌木
  2. Lateral veins extending to margin, each ending in a tooth/ 侧脉直行，直达齿端 .... **2.** *Aphananthe*/ 糙叶树属
  2. Lateral veins anastomosing before reaching margin/ 侧脉未达边缘就弯曲闭合
    3. Flowers unisexual; fruit 1.5–4 mm in diam./ 花单性；果径 1.5~4 mm .... **3.** *Trema*/ 山黄麻属
    3. Flowers polygamous; fruit 5–15 mm in diam./ 花杂性；果径 5~15 mm .... **4.** *Celtis*/ 朴属

1. *Humulus scandens* (Lour.) Merr./ 葎草；
2. *Aphananthe aspera* (Thunb.) Planch./ 糙叶树（2a. Male flowers/ 雄花；2b. Leaves/ 叶）；
3. *Trema cannabina* Lour. var. *dielsiana* (Hand.-Mazz.) C.J. Chen/ 山油麻（3a. Male flowers/ 雄花；3b. Female flowers/ 雌花）；
4-1. *Celtis sinensis* Pers./ 朴树（Drupe/ 核果）；
4-2. *C. chekiangensis* W.C. Cheng/ 天目朴树（Female flowers/ 雌花）

150 Moraceae 桑科

Conspicuous milky exudate / 具明显乳汁
Stipules caducous, leaves circular scars / 托叶早落，具托叶环痕
Inflorescence axillary, syncarp / 花序腋生，聚花果
Flowers minute, perianth 1-whorled / 花微小，单轮花被

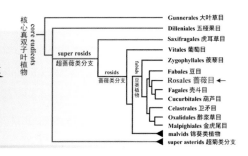

Trees, **shrubs**, **woody climbers**, **hemi-epiphytes** or rarely **herbs**, conspicuous milky exudate (except *Fatoua*), sometimes spiny. **Leaves** simple or pinnately compound, alternate or opposite; margin entire, rarely toothed or palmately lobed; sometimes with cystoliths; stipules interpetiolar or intrapetiolar, conspicuous terminal stipule that leaves circular scars. **Inflorescence** axillary, individual flowers usually congested to capitulum, spikes, panicles or hypanthodium. **Flowers** minute, unisexual (plant monoecious or dioecious) or bisexual, actinomorphic. **Perianth** 1-whorled, sepaloid, reduced or absent. **Male flowers**: stamens opposite the perianth; anthers 1 or 2-loculed, crescent-shaped to top-shaped. **Female flowers**: ovary superior to inferior; carpels free or fused; 1(–2)-loculed; ovule 1 per locule; placentation apical or basal; stigmas 1–2. **Fruit** usually a drupe or achene, often joined into a syncarp, associated with fleshy perianth, bracts or receptacle.

乔木、灌木、木质藤本、半附生或稀为**草本**，具明显乳汁（除水蛇麻属），有时多刺。**叶**为单叶或羽状复叶，互生或对生；全缘，稀有锯齿或掌状分裂；有时具钟乳体；具叶柄间或叶柄内托叶，具明显托叶环痕。**花序**腋生，常由多朵花聚合成头状、穗状、圆锥状或隐头花序。**花**微小，单性（雌雄同株或雌雄异株）或两性，辐射对称。**花被片**1轮，萼片状，退化或缺。**雄花**：雄蕊与花被片对生；花药1或2室，新月形至陀螺形。**雌花**：子房上位至下位；心皮离生或合生；1(~2)室；胚珠每室1枚；顶生或基底胎座；柱头1~2个。**果实**常为核果或瘦果，常与肉质花被、苞片或花托合生成聚花果。

World 39/1,125, tropical to warm temperate regions.
China 9/144.
This area 5/22.

全世界共39属1 125种，分布于热带至暖温带地区。
中国产9属144种。
本地区有5属22种。

1. Herbs, without exudate/ 草本，无乳汁 .... **1. *Fatoua*/ 水蛇麻属**
1. Trees, shrubs, or climbers, with exudate/ 乔木、灌木或藤本，有乳汁
  2. Inflorescence a hypanthodium/ 隐头花序 .... **2. *Ficus*/ 榕属**
  2. Inflorescence a capitulum, spikes, or racemes/ 头状、穗状或总状花序
    3. The key reference material in fruit or with female flowers/ 按照果实或雌花材料检索
      4. Leaf margin clearly toothed/ 叶缘明显具锯齿

5. Calyx lobes imbricate/ 萼裂片覆瓦状排列 ....**4. *Morus*/ 桑属**
5. Calyx lobes valvate/ 萼裂片镊合状排列 ....**5. *Broussonetia*/ 构属**
4. Leaf margin entire or shallowly crenate/ 叶缘全缘或具浅圆齿 ....**3. *Maclura*/ 橙桑属**
3. The key reference material with male flowers/ 按照雄花材料检索
6. Stamens straight in flower buds/ 雄蕊在花芽中直立 ....**3. *Maclura*/ 橙桑属**
6. Stamens inflexed in flower buds/ 雄蕊在芽中内折
7. Male flower calyx lobes imbricate/ 雄花萼裂片覆瓦状排列 ....**4. *Morus*/ 桑属**
7. Male flower calyx lobes valvate/ 雄花萼裂片镊合状排列 ....**5. *Broussonetia*/ 构属**

1. *Fatoua villosa* (Thunb.) Nakai/ 水蛇麻；
2. *Ficus pumila* L. var. *awkeotsang* (Makino) Corner/ 爱玉子（2a. Flowering branch/ 花枝；2b. Gall flowers and male flowers/ 瘿花与雄花）；
3. *Maclura tricuspidata* Carrière/ 柘（3a. Male plants/ 雄株；3b. Syncarps/ 聚花果）；
4. *Morus australis* Poir./ 鸡桑；
5. *Broussonetia kazinoki* Siebold & Zucc./ 楮

151　Urticaceae 荨麻科

Bark with clear or white exudate / 植株具清澈或白色乳汁
Leaves sometimes with stinging hairs, stipules present / 叶有时具螯毛，具托叶
Perianth 1-whorled, sepaloid, often persistent / 花被片 1 轮，萼片状，常宿存
Stigma 1, diverse / 柱头 1 个，形状多样

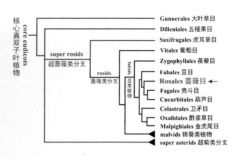

Annual or perennial **herbs**, **trees**, **shrubs** or **woody climbers**; bark with clear or white exudate; stems often fibrous, sometimes succulent. **Leaves** simple or compound, alternate or opposite; margin usually entire or pinnatisect to lobed; stipules interpetiolar or intrapetiolar, or absent; sometimes hairs stinging; usually cystoliths present (except Cecropieae). **Inflorescence** spikes, cymes, racemes, panicles or heads, sometimes crowded on common enlarged cup-like or discoid receptacle, or rarely solitary flower; often bracteate. **Flowers** unisexual (plants dioecious or monoecious) or rarely bisexual, actinomorphic or zygomorphic. **Perianth** 1-whorled, sepaloid, often persistent, sometimes reduced to absent in female flowers. **Stamens** equal in number to tepals, opposite the tepals; anthers basifixed, sometimes explosively dehiscent; staminodes reduced or scale-like if present. **Ovary** usually superior or part-inferior, reduced in male flowers; carpel 1 or carpels fused (2 carpels but one really reduced); 1-loculed; ovule 1; placentation basal; style 1; stigma 1, diverse. **Fruit** an achene, nut or rarely a drupe.

一年生或多年生**草本**、**乔木**、**灌木或木质藤本**；树皮具清澈或白色乳汁；茎纤维发达，有时肉质。**叶**为单叶或复叶，互生或对生；常全缘或羽状全裂至浅裂；具叶柄间或叶柄内托叶，或缺；有时被螯毛；常具钟乳体（除号角树族）。**花序**穗状、聚伞状、总状、圆锥状或头状，有时簇生于扩大的杯状或盘状花托上，稀为单花；常具苞片。花单性（雌雄异株或雌雄同株）或稀为两性，辐射对称或两侧对称。**花被片** 1 轮，萼片状，常宿存，雌花中有时退化至无。**雄蕊**与花被同数，与花被片对生；花药基着，有时炸裂；退化雄蕊缩减，如存在呈鳞片状。**子房**常上位或半下位，在雄花中退化；心皮 1 枚或合生（心皮 2 枚，其中 1 枚退化）；1 室；胚珠 1 枚；基底胎座；花柱 1 个；柱头 1 个，形状多样。**果实**为瘦果、坚果或稀为核果。

World 55/2,626, worldwide, except Antarctica, mainly in the tropical and subtropical humid habitats, extending into temperate regions.
China 26/ca. 430.
This area 6/12.

全世界共 55 属 2 626 种，除南极洲外，世界广布，但主要分布于热带与亚热带潮湿地区，延伸至温带地区。中国产 26 属约 430 种。本地区有 6 属 12 种。

Oreocnide frutescens (Thunb.) Miq./ 紫麻

1. Plants armed with stinging hairs/ 植株具螫毛 ··· **1. *Nanocnide*/ 花点草属**
1. Plants without stinging hairs/ 植株无螫毛
    2. Style absent, stigma penicillate/ 花柱缺，柱头画笔头状 ································· **2. *Pilea*/ 冷水花属**
    2. Style present, stigma various type/ 具花柱，柱头多样
        3. Leaves opposite/ 叶对生
            4. Leaf blade margin entire/ 叶全缘
                5. Male perianth with annular crown formed from transverse crests of lobes/ 雄花花被顶端横折，形成环绕花被的冠状物 ·· **3. *Gonostegia*/ 糯米团属**
                5. Male flowers without this characters/ 雄花无此特征 ················ **4. *Pouzolzia*/ 雾水葛属**
            4. Leaf blade margin serrate/ 叶具锯齿 ·· **5. *Boehmeria*/ 苎麻属**
        3. Leaves alternate/ 叶互生
            6. Stigma filiform/ 柱头丝状
                7. Achene not lustrous; stigma persistent/ 瘦果无光泽；柱头宿存 ············ **5. *Boehmeria*/ 苎麻属**
                7. Achene lustrous; stigma usually caducous/ 瘦果有光泽；柱头常早落 ········· **4. *Pouzolzia*/ 雾水葛属**
            6. Stigma peltate/ 柱头盾状 ··· **6. *Oreocnide*/ 紫麻属**

1. *Nanocnide japonica* Blume/ 花点草；
2. *Pilea angulata* (Blume) Blume/ 圆瓣冷水花（showing stigma penicillate, not distributed here, photo by WeiLiang MA/ 仅展示画笔头状柱头，非本区域分布，马炜梁拍摄）；
3. *Gonostegia hirta* (Blume) Miq./ 糯米团；
4. *Pouzolzia zeylanica* (L.) Benn. & R. Br. / 雾水葛；
5. *Boehmeria macrophylla* Hornem./ 水苎麻；
6. *Oreocnide frutescens* (Thunb.) Miq./ 紫麻

153 Fagaceae 壳斗科

Woody plants, monoecious / 木本植物，雌雄同株
Stipules often triangular / 托叶常三棱形
Flowers unisexual, perianth bract-like / 花单性，花被片苞片状
Fruit with subtending cupule / 果包于壳斗中

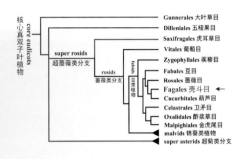

Trees or **shrubs**; monoecious, evergreen or deciduous. **Leaves** ±gland-doted, simple, alternate or whorled; margin entire, toothed or deeply lobed; petioles present, swollen at base; stipules often triangular, caducous; sometimes resins. **Flowers** unisexual (plants monoecious, rarely dioecious), actinomorphic. **Perianth** fused or free, bract-like. **Male flowers** in heads or catkins; stamen filaments free, rarely basally fused; sometimes pistillodes. **Female flowers** in spikes or solitary, each flower with bracts (= cupule); staminodes absent or 6–12; ovary inferior; carpels fused; locules 2–6(–9); ovules 2 per locule; placentation axile to apical; styles 3–6. **Fruit** a nut embedding in cupule.

乔木或**灌木**；雌雄同株，常绿或落叶。**叶**多少有腺点，单叶，互生或轮生；全缘、有锯齿或深裂；具叶柄，基部膨大；托叶常三棱形，早落；有时具树脂。**花**单性（雌雄同株，稀雌雄异株），辐射对称。**花被片**合生或离生，苞片状。**雄花**组成头状或柔荑花序；花丝分离，稀基部合生；有时具退化雌蕊。**雌花**成穗状花序或单生，每朵花具总苞(壳斗)；退化雄蕊缺或6~12枚；子房下位；心皮合生；2~6(~9)室；胚珠每室2枚；中轴胎座至顶生胎座；花柱3~6个。**果实**为壳斗包被的坚果。

World 7/900+, widely in Northern tropics, subtropics and temperate regions, especially in Asia.
China 6/ca. 295.
This area 4/16.

Quercus phillyreoides A. Gray/ 乌冈栎

全世界共7属900余种，广布于北半球热带、亚热带和温带地区，尤其是亚洲。
中国产6属约295种。
本地区有4属16种。

1. Male inflorescence erect/ 雄花序直立
   2. Leaves deciduous/ 叶凋落 ....**1.** *Castanea*/ 栗属
   2. Leaves evergreen/ 叶常绿
      3. Cupules solitary on rachis/ 壳斗单生于花序轴 ....**2.** *Castanopsis*/ 锥属
      3. Cupules in cymes on rachis/ 壳斗在花序轴上呈聚伞状 ....**3.** *Lithocarpus*/ 柯属

1. Male inflorescence pendulous/ 雄花序下垂 ....**4. *Quercus*/ 栎属**

FAGALES. Fagaceae. *Quercus variabilis* Blume/ 壳斗目 壳斗科 栓皮栎
A. Male flowering branch/ 雄花枝
B. Male catkin/ 雄柔荑花序
C. Leaves/ 叶
D$_1$. Male flower, anther not open/ 雄花，花药未开裂
D$_2$. Male flower, anther opened/ 雄花，花药开裂
D$_3$. Stamen/ 雄蕊
D$_4$. Perianth in male flower/ 雄花花被片
D$_5$. Dorsal view in male flowers/ 雄花背面观
(Scale/ 标尺：A–C 1 cm; D$_1$–D$_5$ 1 mm)

1. *Castanea seguinii* Dode/ 茅栗；
2. *Castanopsis carlesii* (Hemsl.) Hayata/ 米槠；
3. *Lithocarpus glaber* (Thunb.) Nakai/ 柯；
4. *Quercus acutissima* Carruth./ 麻栎；
5. *Quercus ciliaris* C.C. Huang & Y.T. Chang/ 细叶青冈

154　Myricaceae 杨梅科

Leaves often aromatic, with peltate glands / 叶常芳香，具盾状腺体
Inflorescence spikes or catkin-like / 花序穗状或柔荑花序
Flowers unisexual, perianth absent / 花单性，花被缺
Drupe often waxy and warty / 核果常具蜡质和疣突

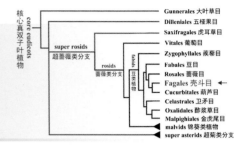

Trees or **shrubs**. **Leaves** often aromatic (sweet odour), often with peltate glands; simple, alternate (spiral); margin entire to toothed or rarely pinnatifid; petioles present; stipules absent or present. **Inflorescence** spikes or catkin-like; bracts 1(–3) per flower. **Flowers** unisexual (plants dioecious or monoecious) or rarely bisexual, actinomorphic. **Perianth** absent or with 6 tiny lobes. **Male flowers** with usually 2 bracteoles; stamen filaments free or fused; anthers dorsifixed, extrorse, dehiscing longitudinally. **Female flowers** with bracteoles 2–4; ovary superior to inferior; carpels fused; 1-loculed; ovule 1; placentation basal; styles 2. **Fruit** a drupe (often waxy and warty) or nut-like.

乔木或**灌木**。**叶**常具芳香味（甜味），多具盾状腺体；单叶，互生（螺旋状排列）；全缘至有锯齿或稀羽裂；有叶柄；托叶无或有。**花序**穗状或柔荑花序；每朵花具 1(~3) 枚苞片。**花**单性（雌雄异株或雌雄同株）或稀为两性，辐射对称。**花被片**缺或具 6 枚细小裂片。**雄花**常具 2 枚小苞片；花丝分离或合生；花药背着，外向，纵裂。**雌花**具 2~4 枚小苞片；子房上位至下位；心皮合生；1 室；胚珠 1 枚；基底胎座；花柱 2 个。**果实**为核果（常具蜡质和疣突）或坚果状。

World 4/50, temperate and subtropical regions.
China 1/4.
This area 1/1.

全世界共 4 属 50 种，分布于温带和亚热带。
中国产 1 属 4 种。
本地区有 1 属 1 种。

Morella Lour. 杨梅属

Morella rubra Lour. / 杨梅（a. Fruits/ 果实；b. Male inflorescence / 雄花序）

155 Juglandaceae 胡桃科

Plant woody, terminal buds often naked / 木本植物，顶芽常为裸芽
Leaves aromatic, often deciduous, compound / 叶具芳香味，多落叶性，复叶
Flowers unisexual, bracts often lobed or wing-like / 花单性，苞片常分裂或呈翅状
Cotyledons 4-lobed, much contorted / 子叶 4 裂，极度扭曲

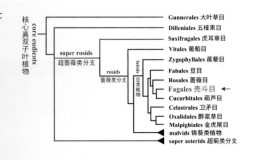

Trees or rarely **shrubs**; terminal buds naked or with scales. **Leaves** aromatic, often deciduous, compound (even or odd-pinnate to trifoliate), alternate (usually spiral) or opposite; leaflet margin toothed or entire; stipules absent or present; peltate scales, often with aromatic glands. **Inflorescence** pendulous or sometimes erect, lateral or terminal on reduced shoots. **Flowers** unisexual (plants monoecious or rarely dioecious), actinomorphic; bracts often lobed or wing-like (conspicuous in fruit). **Perianth** reduced to small lobes or absent and ± fused to bracts. **Male flowers** in catkins or racemes; stamen filaments free, short; anthers basifixed, glabrous or pubescent, 2- loculed, dehiscing longitudinally. **Female flowers** solitary or spikes; ovary inferior or superior; carpels fused; 1(–2)-loculed ; ovule 1; placentation basal or axile; styles 2; stigmas 2, feathery. **Fruit** a drupe-like nut or winged samara. **Seed** solitary, without endosperm; cotyledons 4-lobed, much contorted.

乔木或稀为**灌木**；顶芽为裸芽或被鳞片。**叶**具芳香味，多落叶性，复叶（偶数或奇数羽状复叶至 3 出复叶），互生（常螺旋状排列）或对生；小叶有锯齿或全缘；托叶无或有；被盾状鳞片，常具芳香腺体。**花序**下垂或有时直立，侧生或顶生于短枝上。**花**单性（雌雄同株或稀雌雄异株），辐射对称；苞片常分裂或为翅状（果期明显）。**花被片**退化至小裂片或缺，多少与苞片合生。**雄花**为柔荑或总状花序；花丝分离，短；花药基着，光滑或被毛，2 室，纵裂。**雌花**单生或穗状；子房下位或上位；心皮合生；1(~2) 室；胚珠 1 枚；基底胎座或中轴胎座；花柱 2 个；柱头 2 个，羽毛状。**果实**为核果状坚果或翅果。**种子**单生，无胚乳；子叶 4 裂，极度扭曲。

World 9/ca. 71, Northern tropics to temperate regions.
China 8/27.
This area 2/2.

全世界共 9 属约 71 种，分布于北半球热带至温带地区。
中国产 8 属 27 种。
本地区有 2 属 2 种。

Platycarya strobilacea Siebold & Zucc./ 化香树

Pterocarya stenoptera C. DC./ 枫杨

1. Fruiting spike cone-like; branchlets with solid pith/ 果序球果状；小枝髓实心 .... **1. *Platycarya*/ 化香树属**
1. Fruiting spike in catkins; branchlets with chambered pith/ 果序柔荑状；小枝髓片状分隔 .... **2. *Pterocarya*/ 枫杨属**

1. *Platycarya strobilacea* Siebold & Zucc./ 化香树
 (1a. Infructescence/ 果序；1b. Inflorescence/ 花序)；
2. *Pterocarya stenoptera* C. DC./ 枫杨
 (2a. Infructescence/ 果序；2b. Male inflorescence/ 雄花序)

156 Casuarinaceae 木麻黄科

Trees or shrubs evergreen, equisetoid / 常绿乔木或灌木，似木贼
Branches with longitudinal furrows / 枝条具纵沟槽
Inflorescence spike-like or head-like, flowers unisexual / 花序穗状或头状，花单性
Fruits multiple cone-like structures / 果实聚合成球果状

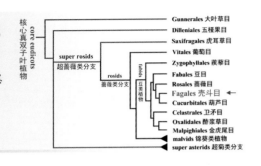

Trees or shrubs evergreen, equisetoid or conifer-like; branches with longitudinal furrows, deep and closed or shallow and open. Leaves tooth-like, minutely whorled; petioles absent; stipules absent. Inflorescence spike-like or head-like. Flowers unisexual (plants dioecious or monoecious), actinomorphic. Male flowers in terminal spikes; bracteoles 2; perianth 1–2-lobed; stamen 1; anthers basifixed, 2-loculed, dehiscing longitudinally. Female flowers in ±heads; bract 1, leaf-like; perianth absent; ovary superior; carpels 2, fused; 2-loculed ; ovules absent in posterior locule and 2 in the anterior one; placentation axile to rarely basal; stigmas 2, reddish, linear. Fruits multiple cone-like structures with samaras to winged nuts. Seed 1; embryo straight, often more than 1.

常绿**乔木**或**灌木**，似木贼或针叶树；小枝具纵沟槽，深且闭合或浅而开放。**叶**成鞘齿，细小轮生；叶柄缺；托叶缺。**花序**穗状或头状。**花**单性（雌雄异株或雌雄同株），辐射对称。**雄花**成顶生穗状花序；小苞片2枚；花被1~2裂；雄蕊1枚；花药基着，2室，纵裂。**雌花**近头状；苞片1枚，叶状；花被缺；子房上位；心皮2枚，合生；2室；后方1室败育无胚珠，前方1室具2枚胚珠；中轴胎座，稀基底胎座；柱头2个，红色，线形。**果实**为翅果或具翅坚果，聚合为球果状。**种子**1粒；胚伸直，常多于1粒。

World 4/96, Australia, Southeast Asia and Pacific islands.
China 3/20+ (all introduced).
This area 3/11 (personal communication).

全世界共4属96种，分布于澳大利亚、东南亚和太平洋群岛。
中国引种3属20余种（全部为引入）。
本地区有3属11种（个人交流）。

1. Furrows of branchlets shallow and open, exposing the stomates; cone with a broad bract beneath each pair of bracteoles/ 小枝的沟槽浅而开放，气孔裸露；球果的每对小苞片下具1枚宽阔的苞片 .... **1. *Gymnostoma*/ 方木麻黄属**
1. Furrows of branchlets deep and narrow, the stomates concealed; cone with more or less thin and inconspicuous bracts/ 小枝的沟槽深而狭，气孔隐藏；球果的苞片多少瘦薄而不显
   2. Mature samaras grey or yellow-brown, dull; cone bracteoles thin and without any dorsal protuberances; teeth 5 to many/ 成熟翅果灰白或黄褐色，无光泽；球果小苞片瘦薄，背部无凸起；鞘齿5至多枚 ....

....**2. *Casuarina*/ 木麻黄属**

2. Mature samaras reddish brown to black, shining; cone bracteoles thickly woody and convex, mostly with an angular, divided or spiny dorsal protuberances; teeth 4 to many/ 成熟翅果红褐色至黑色，有光泽；球果小苞片厚木质，背部具角状，分裂或刺状凸起物；鞘齿 4 至多枚 ....**3. *Allocasuarina*/ 异木麻黄属**

2. *Casuarina glauca* Sieber ex Spreng. / 粗枝木麻黄
(2a. Plant/ 植株；2b. Male inflorescence/ 雄花序；
2c. Infructescence/ 果序；2d. Infructescence dehiscent/ 果序开裂)

158　Betulaceae 桦木科

Trees deciduous, bark smooth, brightly coloured / 落叶乔木，树皮光滑，颜色鲜艳
Leaves often doubly toothed; stipules intrapetiolar / 叶常具重锯齿；叶柄内托叶
Inflorescence catkins, perianth absent / 柔荑花序，花被缺
Infructescence woody and strobilus-like / 果序木质孢子叶球状

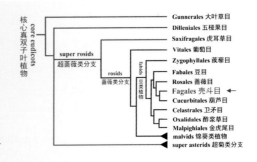

Trees or **shrubs**; bark often smooth, brightly coloured and peeling. **Leaves** often deciduous, simple, alternate; margin toothed, often doubly toothed; petioles present; stipules intrapetiolar, often deciduous. **Flowers** unisexual (plants usually monoecious), actinomorphic. **Male flowers** in catkins with many overlapping bracts; each bract usually subtending a small dichasium with 1–3 male flowers; perianth absent or minute; stamen filaments short; anthers dorsifixed, 2-loculed, extrorse, dehiscing longitudinally. **Female flowers** in spikes or catkins with many overlapping bracts; each bract subtending a small dichasium with 2 or 3 female flowers; perianth absent or scale-like lobes; ovary superior to inferior; carpels fused; 2(–3)-loculed ; ovules 2 or 1 by abortion, per locule; placentation axile to apical. **Fruit** a nut (sometimes winged) in woody and strobilus-like infructescence.

乔木或**灌木**；树皮常光滑，鲜艳，剥落。**叶**多落叶性，单叶，互生；有锯齿，常重锯齿；具叶柄；叶柄内托叶，常早落。**花**单性（常雌雄同株），辐射对称。**雄花**柔荑花序，具多数重叠苞片；每个苞片内常包含一个具1~3朵雄花的二歧聚伞花序；花被缺或细小；花丝短；花药背着，2室，外向，纵裂。**雌花**穗状或柔荑花序，具多数重叠苞片；每个苞片内包含一个具2或3朵雌花的二歧聚伞花序；花被缺或鳞片状；子房上位至下位；心皮合生；2(~3)室；每室胚珠2或1枚败育；中轴胎座至顶生胎座。**果实**为坚果（有时具翅），组成木质孢子叶球状果序。

World 6/150–200, Asia, Europe and North America temperate regions, few to South America.
China 6/89.
This area 1/1.

全世界共6属150~200种，分布于亚洲、欧洲和北美洲，南美洲少量分布。
中国产6属89种。
本地区有1属1种。

Carpinus L. 鹅耳枥属

Carpinus putoensis W.C. Cheng/ 普陀鹅耳枥

163 Cucurbitaceae 葫芦科

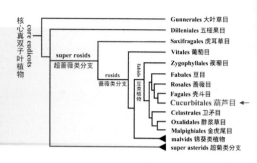

Creeping, herbaceous climbers / 匍匐草质藤本
Axillary tendrils usually present / 常具腋生卷须
Ovary inferior, placentation parietal / 子房下位，侧膜胎座
Usually a pepo / 常为瓠果

Perennial or annual, creeping, herbaceous **climbers**, woody climbers, rarely **shrubs** or **trees**. **Leaves** simple, compound (palmate, 3-foliate), alternate (spiral); margin entire or toothed; venation usually palmate; petioles usually present, axillary tendrils present or absent; stipules absent; often coarse hairs. **Inflorescence** solitary flower, cymes, racemes, panicles, spikes or umbels. **Flowers** unisexual (plants dioecious or monoecious) or very rarely bisexual, actinomorphic or rarely zygomorphic; hypanthium, often basally necteriferous. **Sepals** fused, imbricate. **Petals** basally fused or free, sometimes with coarse hairs; often yellow or red. **Stamens** alternating with petals, usually 5 or 3, of which one often 1-loculed and others 2-loculed; filaments separate or variously united into a column; anthers separate or coherent into a head; anther cells straight to conduplicate, extrorse; sometimes staminodes. **Ovary** inferior to part-inferior; carpels fused or rarely 1; locules 1–5; ovules 1 to many per locule; placentation usually parietal; style free or fused; stigma enlarged or 2-lobed. **Fruit** usually a hard-shelled berry (= pepo), capsule, samara or rarely an achene. **Seeds** often compressed; embryo with leaf-like cotyledons.

多年生或一年生匍匐草质**藤本**，木质藤本或稀为**灌木**或**乔木**。**叶**为单叶、复叶（掌状或 3 枚小叶），互生（螺旋状排列）；全缘或有锯齿；常为掌状脉；常具叶柄，卷须有或无；托叶缺；常被糙毛。**花序**为单花、聚伞状、总状、圆锥状、穗状或伞形。**花**单性（雌雄异株或雌雄同株）或极稀为两性，辐射对称或稀两侧对称；具托杯，基部常具花蜜。**花萼**合生，覆瓦状。**花瓣**基部合生或离生，有时被糙毛；常黄色或红色。**雄蕊**与花瓣互生，常 5 枚或 3 枚，其中 1 枚花药 1 室，其他 2 枚 2 室；花丝分离或联合成各式合蕊柱；花药分离或连着成头状；药室通直至对折，外向；有时具退化雄蕊。**子房**下位至半下位；心皮合生或稀为 1 枚；1~5 室；每室胚珠 1 至多枚；常侧膜胎座；花柱分离或合生；柱头增大或 2 裂。**果实**常为硬壳浆果（瓠果）、蒴果、翅果或稀为瘦果。**种子**常压扁；胚具叶状子叶。

World 95/960, tropics to subtropics, few to temperate regions.

China 30/147.

This area 5/5.

全世界共 95 属 960 种，分布于热带和亚热带，少数至温带。

中国产 30 属 147 种。

本地区有 5 属 5 种。

1. Corolla segments fimbriate/ 花冠裂片流苏状 ....**1. *Trichosanthes*/** 栝楼属

1. Corolla segments not fimbriate/ 花冠裂片非流苏状
    2. Stamens 5/ 雄蕊 5 枚
        3. Leaves palmately compound/ 叶为掌状复叶 .... **2. *Gynostemma*/ 绞股蓝属**
        3. Leaves simple/ 叶为单叶 .... **3. *Actinostemma*/ 盒子草属**
    2. Stamens 3 or 1/ 雄蕊 3 或 1 枚
        4. Anther cells incurved or reflexed/ 药室弧曲或之字形折曲 .... **4. *Solena*/ 茅瓜属**
        4. Anther cells straight/ 药室通直 .... **5. *Zehneria*/ 马㼎儿属**

CUCURBITALES. Cucurbitaceae. *Trichosanthes cucumerina* L./ 葫芦目 葫芦科 瓜叶栝楼 (not distributed here/ 非本区域分布)

A. Ventral view of male flower/ 雄花花冠腹面观
B. Dorsal view of male flower/ 雄花花冠背面观
C. Corolla segregated/ 花冠离析
D. Stamens/ 雄蕊
E₁. Vertical section of corolla (inside)/ 花冠纵切面(内侧)
E₂. Vertical section of calyx tube (partly)/ 花萼筒纵切面(局部)
E₃. Vertical section of corolla (outside)/ 花冠纵切面(外侧)

(Scale/ 标尺: A–C 1 cm; D 1 mm; E₁–E₃ 1 cm)

1. *Trichosanthes kirilowii* Maxim./ 栝楼；
2. *Gynostemma pentaphyllum* (Thunb.) Makino/ 绞股蓝；
3. *Actinostemma tenerum* Griff./ 盒子草；
4. *Solena heterophylla* Lour./ 茅瓜；
5. *Zehneria japonica* (Thunb.) H.Y. Liu/ 马㼎儿

166 Begoniaceae 秋海棠科

Semi-succulent herbs / 近肉质草本
Blade often oblique / 叶偏斜
Flowers unisexual, anthers and stigma twisted / 花单性，花药与花柱扭曲
Capsule winged; seeds numerous, testa reticulate / 具翅蒴果；种子多数，种皮具网纹

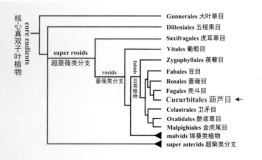

Perennial or rarely annual **herbs**, usually semi-succulent, shrubs or rarely woody **climbers**; usually rhizomatous, or tuberous. **Leaves** simple, rarely compound (palmate to odd-pinnate), alternate or all basal, or rarely opposite; petioles present to absent; stipules intrapetiolar, conspicuous, usually deciduous; blade often oblique; margin irregularly serrate and divided, occasionally entire, venation usually palmate. **Inflorescence** axillary, many to paired cymes or solitary flower. **Flowers** unisexual (plants monoecious or rarely dioecious), actinomorphic or zygomorphic; bracteoles deciduous, leaving scar. **Male flowers**: tepals (2–) 4(–5), unequal; stamen filaments free or rarely fused; anthers basifixed, twisted and yellow, 2-loculed, apical or lateral. **Female flowers**: tepals 4–5(–9) or 10, free or rarely fused; ovary inferior or part-inferior, pendulous or ascending; carpels fused; locules (1–)3(–8); ovules many; placentation axile or parietal; styles 2–6, forked once or more; stigma often twisted or U-shaped, papillose. **Fruit** usually a capsule, often winged, or less often a berry. **Seeds** numerous, testa reticulate.

多年生或稀一年生**草本**，常近肉质，灌木或稀木质**藤本**；多具根状茎或球茎。**叶**为单叶，稀为复叶（掌状至奇数羽状），互生或全部基生，或稀为对生；叶柄有或无；叶柄内托叶，显著，常早落；叶片常偏斜；叶缘具不规则锯齿或分裂，偶全缘，多掌脉。**花序**腋生，多个或成对聚伞花序或单花。**花**单性（雌雄同株或稀雌雄异株），辐射对称或两侧对称；苞片早落，留痕。**雄花**：花被片(2~) 4 (~5)枚，不等大；花丝分离或稀合生；花药基着，扭曲，黄色，2室，顶裂或侧裂。**雌花**：花被片4~5 (~9)或10枚，分离或稀合生；子房下位或半下位，下垂或上升；心皮合生；(1~) 3 (~8)室；胚珠多数；中轴胎座或侧膜胎座；花柱2~6个，一次或多次分叉；柱头常扭曲或呈U形，具乳头状毛。**果实**常为蒴果，多具翅，或偶为浆果。**种子**极多，种皮具网纹。

World 2(–3)/ca. 1, 400, widely in tropics and subtropics.
China 1/180+.
This area 1/1.

全世界共 2（~3）属约 1 400 种，分布于热带和亚热带。
中国产 1 属 180 余种。
本地区有 1 属 1 种。

Begonia L. 秋海棠属

Begonia grandis Dryand./ 秋海棠

CUCURBITALES. Begoniaceae. *Begonia reniformis* Dryand./ 葫芦目 秋海棠科 肾叶秋海棠（not distributed here/ 非本区域分布）

A. Leaves/ 叶
B. Dichotomous cymes/ 二岐聚伞花序
C_1. Cross section of ovary, amplified/ 子房横切面，放大
C_2. Cross section of ovary/ 子房横切面
D. Stigma forked once/ 柱头分叉
E. Perianth lobes/ 花被裂片
F. Styles 3/ 花柱 3 枚
G. Lateral view of pistillate flower, perianth lobes removed/ 雌花侧面观，除花被裂片
(Scale/ 标尺： A–B 1 cm； C_1–G 1 mm)

168 Celastraceae 卫矛科

Leaves simple, often opposite / 单叶，常对生
Flowers small, conspicuous nectar disk / 花小，具明显腺盘
Stamens 1-whorled, alternate with petals / 雄蕊 1 轮，与花瓣互生
Seeds often arillate / 种子常具假种皮

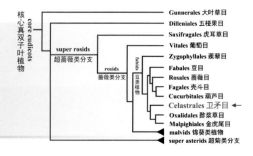

Trees, **shrubs**, woody **climbers**, **subshrubs** or annual to perennial **herbs**; sometimes thorny. **Leaves** simple, alternate or opposite, rarely whorled; margin entire or toothed; petioles usually present or absent; often stipules, small, deciduous. **Inflorescence** axillary or terminal, few to many flowered, cymes, thyrsoid, racemoses, fasciculate, or solitary flower, rarely panicles. **Flowers** small (usually < 1 cm in diam.), bisexual or unisexual (plants often dioecious, also monoecious, andromonoecious, gynodioecious or polyagmous), actinomorphic or rarely zygomorphic, conspicuous nectar disk or rarely absent. **Perianth** (3 or) 4- or 5-merous. **Sepals** fused or free, usually imbricate. **Petals** free or rarely absent. **Stamens** 1-whorled or many, alternate with petals; anther (1 or) 2-loculed, basifixed to dorsifixed, dehiscing longitudinally or obliquely, introrse, extrorse, or latrorse; sometimes staminodes (2–) 3–5. **Ovary** usually superior or rarely part-inferior; carpels fused; locules (1–) 2–5 (–10); ovules 1 to many per locule; placentation axile, rarely parietal or basal. **Fruit** a capsule, berry, drupe, samara or nut. **Seeds** often arillate.

乔木、**灌木**、**木质藤本**、**亚灌木**或一年生至多年生**草本**；有时多刺。**叶**为单叶，互生或对生，稀轮生；全缘或有锯齿；叶柄常存在或缺；常具托叶，小，脱落。**花序**腋生或顶生，稀少至多花，聚伞状、聚伞圆锥状、总状、簇生或为单花，稀为圆锥状。**花**小（直径常小于 1 cm），两性或单性（常雌雄异株或雌雄同株、雄全同株、雌全异株或杂性），辐射对称或稀两侧对称，具明显腺盘，稀缺。**花被片** (3) 4 或 5 基数。**花萼**合生或离生，常覆瓦状。**花瓣**分离或稀缺。**雄蕊** 1 轮或多轮，与花瓣互生；花药 (1) 或 2 室，基着至背着，纵裂或斜向开裂，内向、外向或侧向；有时具 (2~) 3~5 枚退化雄蕊。**子房**常上位或稀半下位；心皮合生；(1~) 2~5 (~10) 室；胚珠每室 1 至多枚；中轴胎座，稀侧膜胎座或基底胎座。**果实**为蒴果、浆果、核果、翅果或坚果。**种子**常具假种皮。

World 94/1, 400, tropics and subtropics, few to temperate regions.
China 15/257.
This area 3/16.

全世界共 94 属 1 400 种，分布于热带和亚热带地区，少数至温带地区。
中国产 15 属 257 种。
本地区有 3 属 16 种。

Euonymus alatus (Thunb.) Siebold/ 卫矛

1. Leaves opposite/ 叶对生 .... **1.** ***Euonymus*/** 卫矛属

1. Leaves alternate/ 叶互生
    2. Seeds completely covered by arils / 种子全为假种皮包被 ....**2. *Celastrus*/ 南蛇藤属**
    2. Seeds only basally or lower half covered by arils / 假种皮仅基部包被或半包种子
    ....**3. *Gymnosporia*/ 美登木属**

CELASTRALES. Celastraceae. *Euonymus alatus* (Thunb.) Siebold/ 卫矛目 卫矛科 卫矛

A. Flowering branch/ 花枝
B. Branch with corky wings/ 木栓翅
C. Cross section of young branch, with corky wing/ 小枝横切面，具木栓翅
D. Dichotomous cymes/ 二歧聚伞花序
E_1. Anther obliquely dehiscent/ 花药斜向开裂
E_2. Stamen/ 雄蕊
F_1. Vertical section of flower/ 花纵切面
F_2. Cross section of ovary/ 子房横切面
G_1. Dorsal view of corolla/ 花冠背面观
G_2. Ventral view of corolla/ 花冠腹面观
G_3. Flower segregated/ 花离析
H. Fruiting branch/ 果枝
(Scale/ 标尺： A–B 1 cm； C 1 mm； D 1 cm； E_1–F_2 1 mm； G_1–G_3 1 cm； H 1 mm)

1. *Euonymus centidens* H. Lév./ 百齿卫矛（Flowering branch/ 花枝）；
2. *Celastrus gemmatus* Loes./ 大芽南蛇藤 (2a. Inflorescence/ 花序；2b. Infructescence/ 果序)；
3. *Gymnosporia diversifolia* Maxim./ 变叶裸实（3a. Flower/ 花；3b. Arils/ 假种皮）

171 Oxalidaceae 酢浆草科

Leaves compound, alternate / 复叶，互生
Flower heteromorphic, heterostylous / 花异形，花柱异长
Flowers 5-merous, tepals free / 花 5 基数，花被离生
Fruit a loculicidal capsule / 蒴果室背开裂

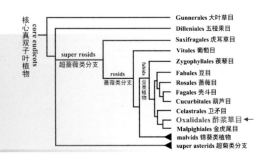

Perennial or annual **herbs**, sometimes **shrubs** or **trees**, or woody **climbers**; bulbous. **Leaves** alternate, pinnate or palmate; leaflets often swollen at base, margin entire, stipules absent or small. **Inflorescence** umbellate, cymes, or racemoses, or solitary flower. **Flowers** bisexual or rarely unisexual, actinomorphic, 5-merous, usually heteromorphic, heterostylous. **Sepals** 5, free, imbricate, persistent. **Petals** 5, free or rarely fused or absent, imbricate, often clawed. **Stamens** 10, in 2-whorled; outer whorl usually with shorter filaments, opposite petals; filaments basally fused; anthers dorsifixed, introrse, 2-loculed, with longitudinal slits. **Ovary** superior; carpels 5 and fused, or free; locules 5; ovules (1 or) 2 to several per locule; placentation axile or rarely parietal; styles (3–) 5, distinct; stigmas capitate or shortly 2-lobed. **Fruit** a loculicidal capsule or a berry. **Seeds** often with basal aril involved in explosive ejection of seed from capsule; endosperm fleshy.

多年生或一年生**草本**，有时为**灌木**或**乔木**，或木质**藤本**；常有球茎。**叶**互生，羽状或掌状；小叶基部常肿大，全缘，托叶缺或小。花序伞形、聚伞、总状或为单花。**花**两性或稀为单性，辐射对称，5 基数，常异形，花柱异长。**花萼** 5 枚，分离，覆瓦状，宿存。**花瓣** 5 枚，分离或稀合生，或缺，覆瓦状，常具爪。**雄蕊** 10 枚，2 轮；外轮花丝较短，与花瓣对生；花丝基部合生；花药背着，内向，2 室，纵裂。**子房**上位；心皮 5 枚，合生，或分离；5 室；胚珠每室（1 或）2 至数枚；中轴胎座或稀侧膜胎座；花柱 (3~) 5 个，分离；柱头头状或浅 2 裂。**果实**为室背开裂蒴果，或浆果。**种子**基部常具假种皮，可辅助将种子从蒴果中弹射出；胚乳肉质。

World 5/780, tropics and subtropics, also in temperate regions, mainly in South America.
China 3/ca.13.
This area 1/3.

全世界共 5 属 780 种，分布于热带和亚热带地区，温带亦产，主要分布于南美洲。
中国产 3 属约 13 种。
本地区有 1 属 3 种。

Oxalis L. 酢浆草属

Oxalis corniculata L./ 酢浆草

OXALIDALES. Oxalidaceae. *Oxalis corniculata* L./ 酢浆草目 酢浆草科 酢浆草

A. Whole plant/ 植株
B₁. Inflorescence, bud stage/ 花序，蕾期
B₂. Young fruit/ 幼果
B₃. Inflorescence umbellate/ 伞形花序
B₄. Vertical section of flower/ 花纵切面
C₁. Pistil/ 雌蕊
C₂. Androecium/ 雄蕊群
C₃. Cross section of ovary/ 子房横切面
D. Capsule/ 蒴果
E. Seeds (Seeds red–brown, aril yellowish)/ 种子（种子红褐色，假种皮淡黄色）
F. Embryo/ 胚
G. Flower segregated/ 花离析
H. Leaves/ 叶
I. Stamen/ 雄蕊
(Scale/ 标尺： A–B₄ 1 cm; C₁–F 1 mm; G–H 1 cm; I 1 mm)

173 Elaeocarpaceae 杜英科

Leaves simple, petioles swollen at ends / 单叶，叶柄两端膨大
Flower actinomorphic, often pendulous / 花辐射对称，常下垂
Tepals valvate, petals margin laciniate / 花被镊合状，花瓣边缘流苏状
Stamens connective with awn or tip at apices / 药隔先端具芒或突尖

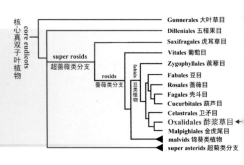

Trees, shrubs or **subshrubs**, sometimes **ericoidous**. **Leaves** simple or rarely compound, often alternate or rarely opposite or whorled, margin entire or usually toothed; petioles swollen at base or both ends; stipules present or absent. **Inflorescence** axillary or terminal, racemes, cymes, panicles or sometimes fascicled or solitary flower. **Flowers** bisexual or unisexual, or polygamous, 4–5-merous, actinomorphic, often pendulous, usually nectar disk. **Sepals** usually free or basally fused, valvate. **Petals** free or sometimes absent, valvate or imbricate, margin laciniate or rarely entire. **Stamens** 8 to numerous; filaments free and borne on disks; anthers basifixed, 2-loculed, dehiscing by apical or longitudinal slits, awned or tipped with hairs at apex. **Ovary** superior; carpels fused; 2 to several locules; placentation various; ovules 1 to several per locule; styles connate or free. **Fruit** a capsule, drupe or berry. **Seeds** with copious endosperm; sometimes with aril.

乔木、**灌木**或**亚灌木**，有时**石南状**。**叶**为单叶或稀复叶，常互生或稀对生或轮生，全缘或常有锯齿；叶柄基部或两端均膨大；托叶有或无。**花序**腋生或顶生，总状、聚伞状、圆锥状或有时簇生或为单花。**花**两性或单性，或杂性，4~5基数，辐射对称，常下垂，多具腺盘。**花萼**常分离或基部合生，镊合状。**花瓣**分离或有时缺，镊合状或覆瓦状，边缘具流苏或稀全缘。**雄蕊**8枚至多数；花丝分离，生于花盘上；花药基着，2室，顶孔开裂或纵裂，先端药隔伸出为芒状或突尖，被毛。**子房**上位；心皮合生；2至多室；胎座类型多样；胚珠每室1至数枚；花柱合生或分离。**果实**为蒴果、核果或浆果。**种子**胚乳丰富；有时具假种皮。

World 12/605, tropics and subtropics, absent from Africa.
China 2/52.
This area 1/3.

全世界共12属605种，分布于热带和亚热带地区，非洲除外。
中国产2属52种。
本地区有1属3种。

Elaeocarpus L. 杜英属

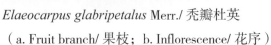

Elaeocarpus glabripetalus Merr./ 秃瓣杜英
（a. Fruit branch/ 果枝；b. Inflorescence/ 花序）

179 Rhizophoraceae 红树科

Woody plants with aerial roots / 木本植物，具气生根
Leaves simple, stipules interpetiolar / 单叶，具叶柄间托叶
Petals often lobed, lobes laciniate / 花瓣分裂，具流苏
Viviparous seedling / 种子胎生

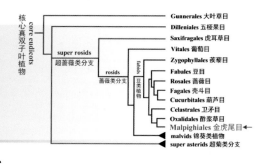

Small trees or **shrubs**, with aerial roots. **Leaves** simple, opposite or whorled; margin entire to toothed; petioles present; stipules interpetiolar, deciduous, always with colleters. **Inflorescence** axillary, cymes, or rarely solitary flowers. **Flowers** bisexual or rarely unisexual (plants dioecious), actinomorphic, sometimes a nectary disk. **Sepals** free or deeply 5-lobed, valvate, persistent. **Petals** often lobed, lobes laciniate; with long setae in sinus between petals. **Stamens** numerous, on a disk; anthers 4-loculed, dehiscing longitudinally. **Ovary** superior to inferior, 2 to many locules; ovules 2 to many per locule; placentation axile to apical; style 1; stigma entire, capitate, or lobed. **Fruit** a berry, drupe or capsule, sometimes attached to plant until falling with viviparous seedling.

小乔木或**灌木**，具气生根。**叶**为单叶，对生或轮生；全缘至有锯齿；具叶柄；有叶柄间托叶，脱落，常具黏液毛。**花序**腋生，聚伞状或稀为单花。**花**两性或稀单性（雌雄异株），辐射对称，有时具蜜腺盘。**花萼**分离或5深裂，镊合状，宿存。**花瓣**常分裂，裂片有流苏；花瓣间缺刻处常具长刚毛。**雄蕊**多数，着生于花盘上；花药4室，纵裂。**子房**上位至下位，2至多室；胚珠每室2至多枚；中轴胎座至顶生胎座；花柱1个；柱头不裂，头状或分裂。**果实**为浆果、核果或蒴果，有时果实随胎生种苗一起离开母株。

World 16/149, mainly in tropical seacoast and inland.
China 6/14.
This area 1/1 (introduced).

全世界共16属149种，主要分布于热带海岸和内陆。
中国产6属14种。
本地区有1属1种（引入）。

Kandelia (DC.) Wight & Arm.
秋茄树属

Kandelia obovata Sheue, H.Y. Liu & J. Yong/ 秋茄树（a. Inflorescence, in bud stage/ 花序，蕾期；b. Young hypocoty/ 幼嫩的下胚轴）

186 Hypericaceae 金丝桃科

Leaves simple, opposite, with gland-doted / 单叶对生，具腺点
Stamens many, sometimes in bundles / 雄蕊多数，有时成束
Flowers bisexual, ovary superior / 花两性，子房上位

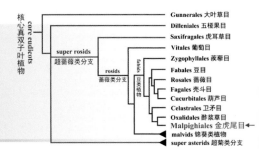

Trees, **shrubs** or annual **herbs**. **Leaves** sometimes black or red gland-doted, simple, opposite, rarely whorled or alternate; margin entire, rarely toothed; petioles present or absent; stipules absent. **Inflorescence** usually cymes, thyrsus or solitary flower. **Flowers** bisexual, actinomorphic. **Sepals** free, imbricate. **Petals** free, imbricate, often contorted. **Stamens** many, filaments free or fused, sometimes in 4 or 5 bundles; anthers dehiscing longitudinally, extrorse; staminodes alternating with petals or absent. **Ovary** superior; carpels fused; locules 3–5; ovules 1 to many per locule; placentation parietal or axile; styles 1–5, free or united; stigmas free and branched. **Fruit** a capsule, berry or rarely a drupe. **Seeds** 1 to many, without or almost without endosperm.

乔木、**灌木**或一年生**草本**。**叶**有时具黑色或红色腺点，单叶，对生，稀轮生或互生；全缘，稀有锯齿；叶柄有或无；托叶缺。**花序**常聚伞状，聚伞圆锥状或单花。**花**两性，辐射对称。**花萼**分离，覆瓦状。**花瓣**分离，覆瓦状，常扭曲。**雄蕊**多数，花丝分离或合生，有时成4或5束；花药纵裂，外向；退化雄蕊与花瓣互生或缺。**子房**上位；心皮合生；3~5室；胚珠每室1至多枚；侧膜胎座或中轴胎座；花柱1~5个，分离或合生；柱头分离并分枝。**果实**为蒴果、浆果或稀为核果。**种子**1粒至多数，无或几无胚乳。

World 9/540, worldwide.
China 4/69.
This area 1/6.

全世界共9属540种，世界广布。
中国产4属69种。
本地区有1属6种。

Hypericum L. 金丝桃属

1-1. *Hypericum ascyron* L./ 黄海棠；
1-2. *H. japonicum* Thunb./ 地耳草；
1-3. *H. sampsonii* Hance/ 元宝草

MALPIGHIALES. Hypericaceae. *Hypericum patulum* Thunb./ 金虎尾目 金丝桃科 金丝梅

A. Flowering branch/ 花枝
B. Flower segregated/ 花离析
C. Cross section of ovary/ 子房横切面
D. Pistil and stamens/ 雌蕊与雄蕊
E. Lateral view of flower, petals removed/ 花侧面观，除花瓣
F. Stamens in 5 bundles/ 雄蕊 5 束
G. Pistil, amplified/ 雌蕊，放大
H. Cross section of ovary, amplified/ 子房横切面，放大

(Scale/ 标尺： A 1 dm; B 1 cm; C 1 mm; D-G 1 cm; H 1 mm)

200 Violaceae 堇菜科

Leaves simple, stipules small or leaf-like / 单叶，托叶小或叶状
Tepals imbricate, usually unequal / 花被覆瓦状，常不等大
Sepals persistent; anterior petal spurred at base / 花萼宿存；前侧花瓣基部有距
Stamens connectives with membranous appendages / 药隔常具膜质附属物

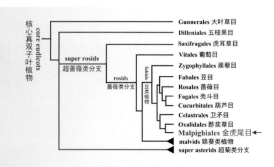

Herbs annual or perennial, **shrubs**, or **subshrubs**, rarely **trees**. **Leaves** simple, alternate or opposite; margin toothed, entire or pinnatisect; petioles present to almost absent; stipules small or leaf-like, sometimes deciduous. **Inflorescence** solitary flower, cymes, heads, racemes, panicles or fascicles; bracts persistent. **Flowers** bisexual or rarely unisexual (plants dioecious), rarely polygamous, actinomorphic or zygomorphic; bracteoles 2. **Sepals** 5, free or basally fused, imbricate, sometimes unequal, persistent. **Petals** 5, free, imbricate, unequal, anterior one saccate, gibbous or spurred at base. **Stamens** 5, connectives often dilated into membranous appendages, anterior 2 stamens with spur-like nectary at base. **Ovary** superior; 3–5 carpels fused; 1-loculed; ovules 1 to many; placentation parietal; style 1; stigmas variously shaped. **Fruit** a 3-valved capsule, berry, nut or follicle-like. **Seeds** often carunculate.

一年生或多年生**草本**，**灌木**或**亚灌木**，稀**乔木**。**叶**为单叶，互生或对生；有锯齿、全缘或羽状全裂；具叶柄或几无；托叶小或叶状，有时脱落。**花序**为单花、聚伞状、头状、总状、圆锥状或簇生；苞片宿存。**花**两性或稀单性（雌雄异株），稀杂性，辐射对称或两侧对称；小苞片 2 枚。**花萼** 5 枚，离生或基部合生，覆瓦状，有时不等大，宿存。**花瓣** 5 枚，离生，覆瓦状，不等大，囊状，基部凸起或为距。**雄蕊** 5 枚，药隔常增大形成膜质附属物，前侧 2 枚基部有距状蜜腺。**子房**上位；3~5 枚心皮合生；1 室；胚珠 1 至多枚；侧膜胎座；花柱 1 个；柱头形态多样。**果实**为 3 瓣裂蒴果、浆果、坚果或蓇葖果状。**种子**常有种阜。

World 22/ca.1, 100, worldwide.

China 3/101.

This area 1/12.

全世界共 22 属约 1 100 种，全球广布。

中国产 3 属 101 种。

本地区有 1 属 12 种。

1-1. *Viola betonicifolia* Sm./ 戟叶堇菜；
1-2. *V. diffusa* Ging./ 七星莲

***Viola* L. 堇菜属**

MALPIGHIALES. Violaceae. *Viola cornuta* Cultivar/ 金虎尾目 堇菜科 三色堇品种

A. Whole plant/ 植株

B₁. Vertical section of ovary, amplified/ 子房纵切面，放大

B₂. Vertical section of ovary/ 子房纵切面

B₃. Calyx segregated/ 花萼离析

B₄. Petals segregated/ 花瓣离析

B₅. Pistil, amplified/ 雌蕊，放大

C₁. Lateral view of flower, petals removed/ 花侧面观，除花瓣

C₂. Dorsal view of flower/ 花背面观

C₃. Frontal view of flower/ 花正面观

C₄. Vertical section of flower/ 花纵切面

C₅. Vertical section of flower, with pedicel/ 花纵切面，带花梗

C₆. Pistil/ 雌蕊

(Scale/ 标尺：A 1 cm; B₁ 1 mm; B₂–B₄ 1 cm; B₅ 1 mm; C₁–C₆ 1 cm)

202 Passifloraceae 西番莲科

Leaves simple, usually with glands on petiole / 单叶，叶柄常有腺体
Filamentous corona, conspicuous / 有花丝状副花冠，显著
Androgynophore present / 具雌雄蕊柄
1-loculed, berry or capsule / 子房1室，浆果或蒴果

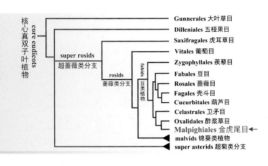

Climbers, woody to herbaceous, rarely **shrubs** or **trees**, often with tendrils. **Leaves** simple or rarely palmately compound, alternate; margin entire or dissected, usually with glands on petiole and/or blade; stipules linear to leaf-like, often glandular. **Inflorescences** axillary, or rarely terminal; cymes, racemes, panicles, fascicles or rarely solitary flower; bracts usually 3. **Flowers** bisexual or rarely unisexual, actinomorphic. **Sepals** fused or free, imbricate, persistent. **Petals** usually free, basally fused or rarely absent. **Stamens** attached to hypanthium or androgynophore or attached to perianth tube; anthers introrse; staminodes many, forming a filamentous or scale-like conspicuous corona. **Ovary** superior to part-inferior; carpels fused into gynophore; 1-loculed; ovules 3 to many; placentation parietal; styles 3(–5), free; stigmas capitate. **Fruit** a berry.

藤本，木质至草质，稀**灌木**或**乔木**，常具卷须。**叶**为单叶或稀掌状复叶，互生；全缘或分裂，叶柄和/或叶片常具腺体；托叶线形至叶状，常具腺体。**花序**腋生，或稀顶生；聚伞状、总状、圆锥状、簇生或稀为单花；苞片常3枚。**花**两性或稀单性，辐射对称。**花萼**合生或离生，覆瓦状，宿存。**花瓣**常离生，基部合生或稀缺失。**雄蕊**着生于托杯或雌雄蕊柄上，或着生于花被筒内；花药内向；退化雄蕊多，组成显著的副花冠，呈花丝状或鳞片状。**子房**上位至半下位；心皮合生形成雌蕊柄；1室；胚珠3至多枚；侧膜胎座；花柱3 (~5)个，离生；柱头头状。**果实**为浆果。

World 27/975, mainly in tropics, especially in America and Africa, also in warm temperate regions.
China 2/23.
This area 1/1.

全世界共27属975种，主要分布于热带，特别是美洲和非洲，暖温带也产。
中国产2属23种。
本地区有1属1种。

***Passiflora* L. 西番莲属**

Passiflora edulis Sims/ 鸡蛋果（a. Fruits/ 果实；b. Flower/ 花）

204   Salicaceae 杨柳科

Leaves simple, often with salicoid teeth / 单叶，常有柳型锯齿
Flowers actinomorphic; pedicel often articulated / 花辐射对称；花梗常具关节
Tepals free, petals similar to sepals / 花被片分离，花瓣与花萼近似
Seeds often arillate or hairs / 种子常具假种皮或被毛

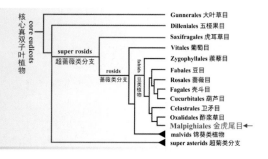

Trees or **shrubs**; pseudoterminal buds in *Salix*. **Leaves** usually deciduous, gland-doted, simple, alternate or rarely opposite; margin entire or toothed, often with salicoid teeth; petioles present; stipules small and intrapetiolar or absent. **Inflorescence** racemes, spikes, catkins, panicles, corymbs or solitary flower; often bracteate. **Flowers** bisexual or sometimes unisexual (dioecious), actinomorphic; pedicel often articulated. **Perianth** present or absent, often an annular or glandular disk present. **Sepals** free or basally fused, valvate or imbricate. **Petals** absent or free, similar to sepals. **Stamens** attached singly or in bundles, filaments free or basally fused. **Ovary** superior or part-inferior; carpels fused; 1-loculed ; ovules 2 to many; placentation usually parietal, rarely basal or axile; styles 1–8. **Fruit** a capsule, berry or rarely a drupe or samara. **Seeds** often arillate or hairs.

乔木或**灌木**；柳属具假顶芽。**叶**常为落叶性，具腺点，单叶，互生或稀对生；全缘或有锯齿，常有柳型锯齿；具叶柄；托叶小，为叶柄内托叶或缺失。**花序**总状、穗状、柔荑花序、圆锥状、伞房状或为单花；常具苞片。**花**两性或有时单性（雌雄异株），辐射对称；花梗常具关节。**花被片**存在或缺，花盘常为环形或腺状。**花萼**离生或基部合生，镊合状或覆瓦状。**花瓣**缺或离生，与花萼近似。**雄蕊**单生或成束，花丝分离或基部合生。**子房**上位或半下位；心皮合生；1室；胚珠2至多数；常侧膜胎座，稀基底胎座或中轴胎座；花柱1~8个。**果实**为蒴果、浆果或稀为核果或翅果。**种子**常具假种皮或被毛。

World 50+/1, 800+, worldwide, especially in Northern temperate regions.
China 13/ca. 385.
This area 4/5.

全世界共 50 余属 1 800 余种，世界广布，尤其是北温带。
中国产 13 属约 385 种。
本地区有 4 属 5 种。

1. Inflorescence a catkin/ 柔荑花序 .... **1.** *Salix*/ 柳属
1. Inflorescence not a catkin/ 非柔荑花序
  2. Petals present/ 有花瓣 .... **2.** *Scolopia*/ 箣柊属
  2. Petals absent/ 无花瓣
    3. Flowers bisexual/ 花两性 .... **3.** *Casearia*/ 脚骨脆属
    3. Flowers unisexual/ 花单性 .... **4.** *Xylosma*/ 柞木属

1. *Salix babylonica* L./ 垂柳；
2. *Scolopia chinensis* (Lour.) Clos/ 箣柊；
3. *Casearia glomerata* Roxb./ 球花脚骨脆；
4. *Xylosma congesta* (Lour.) Merr./ 柞木（photo by WeiLiang MA/ 马炜梁拍摄）

207 Euphorbiaceae 大戟科

Plants often with exudates / 植株常具乳汁
Leaves usually simple, sometimes glands present on leaf base /
多为单叶，叶基常具腺体
Flowers unisexual, actinomorphic / 花单性，辐射对称
Ovary superior, 3 carpels / 上位子房，3 枚心皮

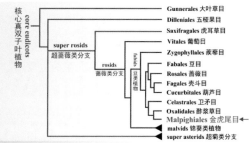

Trees, **shrubs**, **herbs** or rarely woody or herbaceous **lianas**, sometimes succulent and/or spiny. **Leaves** usually simple or compound (often palmate), alternate or opposite, rarely whorled; margin entire to toothed; sometimes glands present on leaf base and/or petiole apex; petioles present or absent; usually stipules present, deciduous, sometimes reduced to spines or glands; often exudates (clear to various color); hairs, if present, simple, stellate, lepidote or T-shaped. **Inflorescence** cymes, racemes, panicles to thyrses, fascicle, spikes or solitary flower; sometimes reduced in bisexual cyathium; floral bracts often with nectary glands. **Flowers** unisexual (plants monoecious or dioecious), actinomorphic or rarely zygomorphic, disk present or absent; sometimes bracteolate. **Sepals** fused into a calyx tube, free or very rarely absent. **Petals** free to reduced or absent. **Stamen** filaments free or fused; sometimes staminodes; anthers 2 (–4)-loculed, mostly dehiscing longitudinally. **Ovary** superior; carpels fused (often carpels 3); locules (1–) 2–5 (or many); ovule 1 per locule; placentation axile or apical; stigmas usually 3. **Fruit** usually an explosive schizocarp or capsule, rarely a drupe or berry.

乔木、**灌木**、**草本**或稀为木质或草质**藤本**，有时肉质，或多刺。**叶**常为单叶或复叶（常掌状），互生或对生，稀轮生；全缘至有锯齿；有时叶基和/或叶柄先端具腺体；叶柄有或无；托叶常存在，脱落，有时退化成刺或腺体；常有乳汁（清澈至各色）；毛被如存在，则为单毛、星状毛、鳞片状或丁字毛。**花序**为聚伞状、总状、圆锥状至聚伞圆锥状、簇生、穗状或为单花；有时简化为两性的杯状聚伞花序；苞片常具蜜腺。**花**单性（雌雄同株或雌雄异株），辐射对称或稀两侧对称，花盘存在或缺失；有时具小苞片。**花萼**合生成萼筒，或离生，极稀缺失。**花瓣**离生至退化或缺。**雄蕊**花丝分离或合生；有时具退化雄蕊；花药 2 (~4) 室，多为纵裂。**子房**上位；心皮合生（多3枚心皮）；3室，稀2或4室或更多或更少；胚珠每室1枚；中轴胎座或顶生胎座；柱头常为3个。**果实**常为炸裂式分果或蒴果，稀为核果或浆果。

World 217/6, 745, Tropical to temperate regions.
China 56/253.
This area 6/19.

全世界共 217 属 6 745 种，热带至温带分布。
中国产 56 属 253 种。
本地区有 6 属 19 种。

Euphorbia maculata L./ 斑地锦

1. Inflorescence not a cyathium/ 花序不呈杯状聚伞花序
    2. Filaments incurved in bud; pistillode absent/ 花丝在芽内弯曲；无退化雌蕊 ....**1. *Croton*/ 巴豆属**
    2. Filaments erect in bud; pistillode present or not/ 花丝在芽内直立；退化雌蕊存在或无
        3. Male sepals valvate/ 雄花萼片镊合状
            4. Male petals present/ 雄花有花瓣 ....**2. *Vernicia*/ 油桐属**
            4. Male petals absent/ 雄花无花瓣
                5. Filaments connate into several bundles/ 花丝合生成数枚雄蕊束 ....**3. *Ricinus*/ 蓖麻属**
                5. Filaments free or connate only at base/ 花丝分离或仅基部合生
                    6. Anthers globose or oval/ 花药球形或椭圆形 ....**2. *Vernicia*/ 油桐属**
                    6. Anthers vermiform/ 花药蠕虫形 ....**4. *Acalypha*/ 铁苋菜属**
        3. Male sepals not valvate/ 雄花萼片非镊合状 ....**5. *Triadica*/ 乌桕属**
1. Inflorescence a cyathium/ 花序为杯状聚伞花序 ....**6. *Euphorbia*/ 大戟属**

1. *Croton cascarilloides* Raeusch./ 银叶巴豆；
2. *Vernicia fordii* (Hemsl.) Airy Shaw/ 油桐；
3. *Ricinus communis* L./ 蓖麻；
4. *Acalypha australis* L./ 铁苋菜；
5. *Triadica sebifera* (L.) Small/ 乌桕；
6. *Euphorbia sieboldiana* C. Morren & Decne./ 钩腺大戟

211 Phyllanthaceae 叶下珠科

Sap absent / 无乳汁
Inflorescence axillary / 花序腋生
Flowers unisexual, actinomorphic / 花单性，辐射对称
Ovules 2 per locule / 胚珠每室 2 枚

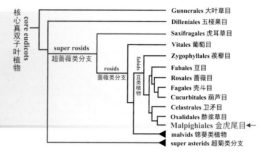

Trees, **shrubs** or **herbs**, rarely climbers. **Leaves** simple or rarely compound, alternate or rarely opposite; margin entire or rarely toothed; petioles present; stipules usually present; sap absent. **Inflorescence** axillary or rarely terminal, cymes, racemes, panicles, fascicles or solitary flower; rarely cyathia. **Flowers** unisexual (plants monoecious or dioecious) or rarely bisexual, actinomorphic; disk present or absent; usually yellow to green or pink to maroon; bracteate. **Sepals** usually free or fused, usually imbricate or valvate. **Petals** free or absent. **Stamen** filaments free or fused; anthers usually extrorse, dehiscing longitudinally or horizontally; staminodes rarely present. **Ovary** superior; carpels fused, sometimes pistillodes in male flowers; locules (1–) 2–5 (–20); ovules 2 per locule; placentation apical to axile; styles usually 2–3, apex 2-lobed, rarely entire. **Fruit** usually a capsule to schizocarp, less often a drupe or berry; columella persistent.

乔木、灌木或草本，稀藤本。**叶**为单叶或稀复叶，互生或稀对生；全缘或稀有锯齿；具叶柄；常有托叶；无乳汁。**花序**腋生或稀顶生，聚伞状、总状、圆锥状、簇生或为单花；稀为杯状聚伞花序。**花**单性（雌雄同株或雌雄异株）或稀两性，辐射对称；花盘有或无；常为黄色至绿色，或粉色至红褐色；具苞片。**花萼**常分离或合生，常覆瓦状或镊合状。**花瓣**离生或缺。**雄蕊**花丝分离或合生；花药常外向，纵裂或横裂；退化雄蕊稀存在。**子房**上位；心皮合生，有时雄花中具退化雌蕊；(1~) 2~5 (~20) 室；胚珠每室 2 枚；顶生胎座至中轴胎座；花柱常 2~3 个，先端 2 裂，稀全缘。**果实**常为蒴果或分果，少为核果或浆果；中柱宿存。

World 59/1, 700, widely distributed in tropical regions, especially in tropical Asia.
China 15/128.
This area 3/10.

全世界共 59 属 1 700 种，广布于热带，主要分布于热带亚洲，少数可至温带。
中国产 15 属 128 种。
本地区有 3 属 10 种。

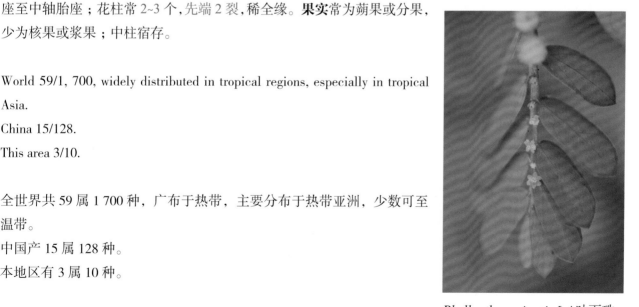

Phyllanthus urinaria L./ 叶下珠

1. Flowers without a disk/ 无花盘
　　2. Styles connate/ 花柱合生 ....**1. *Glochidion*/ 算盘子属**

2. Styles free or connate at base only/ 花柱离生或仅基部合生 .... **2. *Breynia*/ 黑面神属**
1. Flowers with a fleshy disk/ 有肉质花盘 .... **3. *Phyllanthus*/ 叶下珠属**

1. *Glochidion puber* (L.) Hutch./ 算盘子；
2. *Breynia rostrata* Merr./ 喙果黑面神；
3. *Phyllanthus leptoclados* Benth./ 细枝叶下珠

212 Geraniaceae 牻牛儿苗科

Leaves gland-doted, sometimes aromatic / 叶片具腺点，常具芳香味
Flowers bisexual, 5 nectary glands / 两性花，具5枚蜜腺
Ovary superior, style with 5 branches / 上位子房，花柱分枝5个
Schizocarp with awned mericarps / 分果，果爿具芒

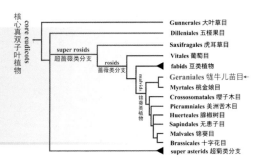

Perennial or annual **herbs**, rarely **shrublets** or **shrubs**; stems sometimes succulent. **Leaves** gland-doted, sometimes aromatic, simple or compound, alternate or opposite; margin dissected or lobed; petioles present; stipules present or absent. **Inflorescence** cymes, pseudoumbels, or rarely solitary flower. **Flowers** bisexual, actinomorphic or zygomorphic, 5 nectary glands alternating with outer stamens; bracteolate. **Sepals** free or basally fused, ±imbricate, persistent. **Petals** free, clawed, imbricate. **Stamens** 5 or 10, usually in 2 whorls, opposite the petals, sometimes a whorl reduced to staminodes; filaments free or ±fused basally; anthers 2-loculed, dorsifixed, introrse; dehiscing longitudinally. **Ovary** superior; carpels 5, fused; 5-loculed; ovules 1–2 (to many) per locule; placentation axile; styles with 5 stigmatic branches or style 1 with capitate stigma. **Fruit** a schizocarp with 5 one-seeded awned mericarps which separate elastically from a central beak, or a loculicidal capsule. **Seeds** usually with little or no endosperm.

多年生或一年生**草本**，稀**小灌木**或**灌木**；茎有时肉质。**叶**具腺点，有时芳香，单叶或复叶，互生或对生；边缘有锯齿或分裂；具叶柄；托叶有或无。**花序**聚伞状、假伞形或稀为单花。**花**两性，辐射对称或两侧对称，具5枚蜜腺，与外轮雄蕊互生；具小苞片。**花萼**离生或基部合生，多少覆瓦状，宿存。**花瓣**离生，具爪，覆瓦状。**雄蕊**5或10枚，常2轮，与花瓣对生，有时1轮退化为退化雄蕊；花丝分离或基部多少合生；花药2室，背着，内向；纵裂。**子房**上位；心皮5枚，合生；5室；胚珠每室1~2（至多）枚；中轴胎座；花柱具5个柱头状分枝或为单个头状柱头。**果实**为分果，分果爿5个，每个分果爿具1粒种子，具芒，由喙状中轴弹射，或为室背开裂蒴果。**种子**胚乳常稀薄或无。

World 6/780, widely distributed in temperate regions and tropical montane habitats.
China 2/59.
This area 1/1.

全世界共6属780种，广布于温带、亚热带和热带山地。
中国产2属59种。
本地区有1属1种。

Geranium L. 老鹳草属

Geranium carolinianum L./ 野老鹳草
（a. Flower/ 花；b. Fruit branch/ 果枝）

GERANIALES. Geraniaceae. *Geranium carolinianum* L./ 牻牛儿苗目 牻牛儿苗科 野老鹳草

A. Whole plant/ 植株
B. Androecuim segregated/ 雄蕊群离析
C. Perianth segregated/ 花被片离析
D₁. Young fruit/ 幼果
D₂. Pistil/ 雌蕊
D₃. Cross section of ovary/ 子房横切面
D₄. Cross section of young fruit, ovules removed/ 幼果横切面，除胚珠
E. Flowering branch/ 花枝
F. Cross section of ovary, amplified/ 子房横切面，放大
G. Infructescence/ 果序

(Scale/ 标尺： A 5 cm; B–D₄ 1 mm; E 1 cm; F 1 mm; G 1cm)

215 Lythraceae 千屈菜科

Tetragonal branch with edges / 方茎具棱
Leaves simple, opposite / 单叶，对生
Hypanthium ribbed or spurred / 被丝托具棱或距
Petals often clawed, wrinkled at maturity / 花瓣常具爪，成熟时皱褶

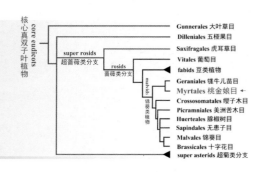

Trees, **shrubs**, often with flaky bark, or **herbs**; sometimes aquatic or in marshes, branch sometimes tetragonal. **Leaves** simple, often opposite, margin entire or toothed; petioles present or absent; stipule present or absent. **Inflorescence** solitary flower, cymes, panicles, racemes or spikes. **Flowers** bisexual or rarely unsexual (plants dioecious or monoecious), actinomorphic or zygomorphic; hypanthium associated with an epicalyx, ribbed or spurred, tubular or campanulate, persistent. **Sepals** usually partly fused or free, valvate. **Petals** free or absent, often clawed, crumpled in bud and wrinkled at maturity. **Stamens** usually 2-whorled, attached to hypanthium; anthers dorsifixed or rarely basifixed, introrse; filaments unequal in length. **Ovary** often superior or inferior to part-inferior; carpels fused; locules 1-several; ovules many per locule; placentation usually axile or parietal. **Fruit** a leathery or membranous capsule, or berry.

乔木、**灌木**，常具薄片状易脱落的树皮，或为**草本**；有时水生或沼生，枝条有时四棱形。**叶**为单叶，常对生，全缘或有锯齿；叶柄有或无；托叶有或无。**花序**为单花、聚伞状、圆锥状、总状或穗状。**花**两性或稀单性（雌雄异株或雌雄同株），辐射对称或两侧对称；被丝托常与副萼合生，具棱或距，管状或钟状，宿存。**花萼**常部分合生或离生，镊合状。**花瓣**分离或缺，常具爪，在花芽中皱褶，成熟时具皱纹。**雄蕊**常2轮，着生于被丝托；花药背着或稀基着，内向；花丝不等长。**子房**常上位或下位至半下位；心皮合生；1至数室；胚珠每室多枚；中轴胎座或侧膜胎座。**果实**为革质或膜质蒴果，或浆果。

World 32/ca. 602, widely distributed in tropics and subtropics.

China 11/45.

This area 4/6.

全世界共32属约602种，广布于热带与亚热带地区。

中国产11属45种。

本地区有4属6种。

1. Ovary superior/ 子房上位
    2. Herbs/ 草本
        3. Flowers usually 3 or more per axil/ 花每腋常3朵以上 .... **1. *Ammannia*/ 水苋菜属**
        3. Flower usually solitary, axillary/ 花常单生于腋间 .... **2. *Rotala*/ 节节菜属**
    2. Trees or shrubs/ 乔木或灌木 .... **3. *Lagerstroemia*/ 紫薇属**
1. Ovary inferior or part-inferior/ 子房下位或半下位 .... **4. *Trapa*/ 菱属**

1. *Ammannia multiflora* Roxb./ 多花水苋菜;
2. *Rotala rotundifolia* (Ruch.-Ham. ex Roxb.) Koehne/ 圆叶节节菜;
3. *Lagerstroemia indica* L./ 紫薇;
4. *Trapa natans* L./ 欧菱

216 Onagraceae 柳叶菜科

Leaves simple, intrapetiolar / 单叶，具叶柄内托叶
Well-developed hypanthium / 具发达被丝托
Sepals fused, valvate; petals free, often clawed / 花萼合生，镊合状；花瓣分离，常具爪
Loculicidal capsule / 室背开裂的蒴果

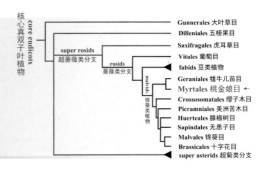

Perennial to annual **herbs** or **shrubs**, rarely **trees**, sometimes epiphytic. **Leaves** simple, opposite, alternate or whorled; margin entire to often toothed to pinnatifid; petioles present to absent; sometimes small stipules, intrapetiolar, or absent. **Inflorescence** solitary flower, panicles, racemes or spikes. **Flowers** bisexual or sometimes unisexual (gynodioecious or dioecious), actinomorphic or zygomorphic; often well-developed hypanthium, nectariferous within or absent. **Sepals** fused to rarely free, valvate; green to petaloid. **Petals** free or rarely absent, often clawed. **Stamens** usually 2-whorled (4+4 or 6+6) or reduced to 1-2 within or inserted on the hypanthium; anthers versatile or basifixed; 2-loculed; dehiscing by longitudinal slits; introrse; sometimes staminodes 1 or 2-4. **Ovary** inferior or part-inferior; carpels fused; locules (2-) 4 (-7); ovules 1 to many per locule; placentation axile or parietal; stigma globose, peltate or lobed. **Fruit** a loculicidal capsule, or indehiscent berry or nut. **Seeds** small; sometimes with a coma; endosperm absent.

多年生至一年生**草本**或**灌木**，稀**乔木**，有时附生。**叶**为单叶，对生、互生或轮生；全缘、有锯齿或羽裂；叶柄有或无；有时托叶小，叶柄内托叶，或缺。**花序**为单花、圆锥状、总状或穗状。**花**为两性或有时单性（雌全异株或雌雄异株），辐射对称或两侧对称；常具发达被丝托，内有花蜜或无。**萼片**合生，稀离生；镊合状；绿色至花瓣状。**花瓣**离生至稀缺失，常具爪。**雄蕊**常 2 轮（4+4 或 6+6），或退化成 1~2 枚着生于被丝托内；花药丁字状着生或基部着生；2 室；纵裂；内向；有时具 1 或 2~4 枚退化雄蕊。**子房**下位或半下位；心皮合生；(2~) 4 (~7) 室；胚珠每室 1 至多枚；中轴胎座或侧膜胎座；柱头球形、盾状或分裂。**果实**为室背开裂的蒴果、不裂的浆果或坚果。**种子**小，有时具种缨；胚乳缺。

World 15/ca. 650, widely in temperate regions and tropics, especially in Western North America.
China 6/64 (3 genera introduced and naturalized, 1 genus cultivated).
This area 3/8.

全世界共 15 属 650 种，广布于温带与热带地区，特别是北美西部。
中国产 6 属 64 种（其中 3 属引种并逸生，1 属引种栽培）。
本地区有 3 属 8 种。

1. Seeds with coma/ 种子具种缨 .... **1. *Epilobium*/ 柳叶菜属**
1. Seeds without coma/ 种子无种缨
　　2. Sepals persistent after anthesis/ 萼片花后宿存 .... **2. *Ludwigia*/ 丁香蓼属**
　　2. Sepals deciduous after anthesis/ 萼片花后脱落 .... **3. *Oenothera*/ 月见草属**

MYRTALES. Onagraceae. *Oenothera speciosa* Nutt./ 桃金娘目 柳叶菜科 美丽月见草

A. Flowering branch/ 花枝

B. Ventral and dorsal view of leaves/ 叶腹面观与背面观

C_1. Vertical section of ovary/ 子房纵切面

C_2. Cross section of ovary, most of ovules removed/ 子房横切面，大多胚珠移除

C_3. Cross section of ovary/ 子房横切面

D. Lateral view of flower, petals removed/ 花侧面观，除花瓣

E. Flower segregated/ 花离析

F_1. Stamen/ 雄蕊

F_2. Anther, pollen possess viscin thread/ 花药，花粉具黏丝

(Scale/ 标尺： A 5 cm; B 1 cm; C_1–C_3 1 mm; D–E 1 cm; F_1–F_2 1 mm)

1. *Epilobium pyrricholophum* Franch. & Sav./ 长籽柳叶菜;
2. *Ludwigia prostrata* Roxb./ 丁香蓼;
3. *Oenothera laciniata* Hill/ 裂叶月见草

218 Myrtaceae 桃金娘科

Woody plants, usually with essential oils / 木本，常具芳香精油
Leaves usually pellucid gland-dots, simple, opposite / 叶常具透明腺点，单叶，对生
Sepals and petals fused into calyptra independently / 花萼与花瓣分别合生为帽状体
Stamens showy, often many / 雄蕊极显著，常多数

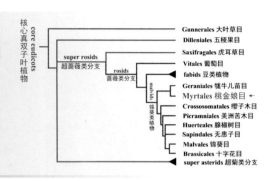

Trees, **shrubs** or rarely **subshrubs**; usually with essential oils; bark sometimes flaking. **Leaves** usually pellucid gland-dotted, often aromatic, rarely deciduous, simple, opposite, sometimes alternate or rarely whorled; margin entire; petioles present or absent; stipules absent or rarely colleters. **Inflorescence** solitary flower, cymes, spikes, racemes, corymbs, panicles, thyrses or heads; sometimes cauliflorous and/or ramiflorous; sometimes bracteate. **Flowers** bisexual, sometimes polygamous or rarely unisexual; conspicuous hypanthium; sometimes bracteolate. **Sepals** fused into a calyptra or free, often imbricate. **Petals** fused, reduced into calyptra or free, imbricate, usually 4–5. **Stamens** often showy, few to usually many, free, fused or in bundles; anthers 2-loculed, usually dorsifixed, dehiscing longitudinally or rarely terminally, introrse; connectives usually terminating in 1 or more apical glands; sometimes staminodes. **Ovary** inferior, part-inferior or rarely superior; carpels fused; locules (1–) 2–5 (–12); ovules 1 to many per locule; placentation axile, less often apical, basal or parietal; style 1; stigma 1. **Fruit** a fleshy berry, capsule or rarely a nut. **Seeds** without endosperm or endosperm sparse and thin.

乔木、**灌木**或稀为**亚灌木**；常具芳香油；树皮常剥落状。**叶**常具透明腺点，多具芳香味，稀脱落，单叶，对生，有时互生或稀轮生；全缘；叶柄有或无；托叶缺或稀具黏液。**花序**为单花、聚伞状、穗状、总状、伞房状、圆锥状、聚伞圆锥状或头状；有时具老茎生花和/或枝花现象；有时具苞片。**花**两性，有时杂性或稀为单性；被丝托显著；有时具小苞片。**花萼**合生成帽状体或离生，常覆瓦状。**花瓣**合生，退化为帽状体或离生，覆瓦状，常4~5枚。**雄蕊**常显著，少数至多数，分离、合生或成束；花药2室，常背着，纵裂或稀顶裂，内向；药隔顶端常延伸成1或多个腺体；有时具退化雄蕊。**子房**下位、半下位或稀上位；心皮合生；(1~) 2~5 (~12) 室；胚珠每室1至多枚；中轴胎座，稀为顶生胎座、基底胎座或侧膜胎座；花柱1个；柱头1个。**果实**为肉质浆果、蒴果，或稀为坚果。**种子**无胚乳或胚乳稀少而薄。

World 131/4, 620, worldwide, mainly in warm tropics.
China 10/121.
This area 2/3.

全世界共131属4 620种，世界广布，主要分布于热带温暖地区。
中国产10属121种。
本地区有2属3种。

1. Leaves pinnately veined; seed usually 1/ 叶为羽状脉；种子常1粒 ....**1. *Syzygium*/ 蒲桃属**
1. Leaves tripliveined; seeds usually numerous/ 叶为离基3出脉；种子常多数 ....**2. *Rhodomyrtus*/ 桃金娘属**

1. *Syzygium buxifolium* Hook. & Arn./ 赤楠；

2. *Rhodomyrtus tomentosa* (Aiton) Hassk./ 桃金娘（2a. Branch/ 枝条；2b. Flowering branch/ 花枝）

219 Melastomataceae 野牡丹科

Leaves simple, opposite and decussate / 单叶，交互对生
Bracts conspicuous and persistent / 苞片常明显且宿存
Flowers bisexual, hypanthium well-developed / 花两性，被丝托发达
Stamens twice as petals, connective appendaged / 雄蕊数 2 倍于花瓣数，药隔具附属物

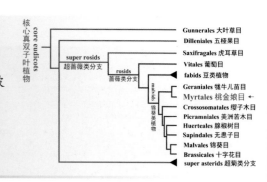

Trees, **shrubs**, **woody climbers** or **herbs**, rarely epiphytic. **Leaves** simple, usually opposite and decussate, rarely whorled or alternate; margin entire or toothed; petioles present; stipules absent or rarely present. **Inflorescence** cymes, umbel, racemes, panicles, or a scorpioid cyme, rarely solitary flower; bracts sometimes conspicuous and persistent. **Flowers** bisexual, actinomorphic but androdioecious often slightly zygomorphic, usually 4 or 5-merous; hypanthium well-developed; often bracteolate. **Sepals** fused, inserted on the rim of hypanthium, somewhat lobbed, valvate. **Petals** equal to sepals, imbricate. **Stamens** 2 × as many as petals, 2-whorled; isomorphic or dimorphic; connective elongated or variously appendaged; anthers basifixed, introrse, dehiscing by 1 or 2 apical pores or by short longitudinal slits. **Ovary** inferior or part-inferior, carpels fused, locule 1–5; ovules 2 to many per locule; placentation axile, free central, basal or parietal. **Fruit** a dry or fleshy capsule or a berry, dehiscing loculicidally or indehiscent. **Seeds** usually small, curved through half a circle or wedge-shaped.

乔木、**灌木**、**木质藤本**或**草本**，稀为附生植物。**叶**为单叶，常交互对生，稀轮生或互生；全缘或具齿；具叶柄；托叶缺或稀存在。**花序**聚伞状、伞形、总状、圆锥状或蝎尾状，稀单花；苞片有时明显且宿存。**花**常**两性**，辐射对称但在雄全异株类群中常稍呈两侧对称，多 4 或 5 基数；被丝托发达；常具小苞片。**花萼**合生，嵌生于托杯边缘，多少分裂，镊合状。**花瓣**与花萼同数，覆瓦状。**雄蕊**数 2 倍于花瓣，2 轮；同型或异型；药隔延伸或具各式附属物；花药基着，内向，1 或 2 孔裂或短纵裂。**子房**下位或半下位，心皮合生，1~5 室；胚珠每室 2 至多数；中轴胎座、特立中央胎座、基底胎座或侧膜胎座。**果实**为干燥或肉质蒴果或浆果，室背开裂或不开裂。**种子**小，弯曲成半圆或呈楔形。

World 188/5, 055, worldwide, mainly in the tropics and subtropics.
China 21/114.
This area 2/3.

全世界共 188 属 5 055 种，世界广布，主要分布于热带和亚热带地区。
中国产 21 属 114 种。
本地区有 2 属 3 种。

Melastoma dodecandrum Lour. / 地菍

1. Stamen whorls equal or subequal in length and shape; fruit a dry capsule/ 两轮雄蕊同形，近等长；果实为干燥蒴果 ....**1. *Osbeckia*/ 金锦香属**
1. Stamen whorls unequal in length and shape; fruit a fleshy capsule/ 两轮雄蕊异形，不等长；果实为肉质蒴果 ....**2. *Melastoma*/ 野牡丹属**

1. *Osbeckia stellata* Buch.-Ham. ex Ker Gawl./ 星毛金锦香；　　2. *Melastoma dodecandrum* Lour./ 地菍

226 Staphyleaceae 省沽油科

Leaves opposite odd-pinnately compound / 奇数羽状复叶，对生
Flowers bisexual, sepals often petaloid, persistent / 两性花，花萼花瓣状，宿存
Ovary superior, follicles or capsules/ 子房上位，蓇葖果或蒴果

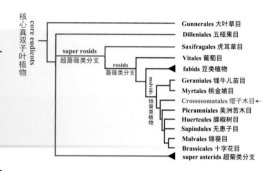

Trees or **shrubs**. **Leaves** often opposite odd-pinnately compound, or trifoliate, rarely simple; leaflets margin slightly incised, leaflets often with petiolules; stipules and stipels, sometimes reduced glands or absent. **Inflorescence** panicles or racemes. **Flowers** bisexual or rarely unisexual (plants monoecious, polygamomonoecious or dioecious), actinomorphic, nectar disk present. **Sepals** fused or free, imbricate, unequal, often petaloid, persistent. **Petals** free or fused (becoming cup-like), imbricate. **Stamens** alternating with corolla lobes; anthers dorsifixed, dehiscing by longitudinal slits; introrse. **Ovary** superior to rarely part-inferior; carpels 3(–2), slightly united at the base; locules 2–3 (–4); ovules 2 to many per locule; placentation axile. **Fruits** inflated capsules or 1–3 follicle.

乔木或**灌木**。**叶**常为对生的奇数羽状复叶，或三出羽状复叶，稀单叶；小叶稍有缺刻，具小叶柄；具托叶和小托叶，有时退化为腺体或缺。**花序**圆锥状或总状。**花**两性或稀单性（雌雄同株、杂性同株或雌雄异株），辐射对称，有腺盘。**花萼**合生或离生，镊合状，不等大，常花瓣状，宿存。**花瓣**离生或合生（呈杯状），覆瓦状。**雄蕊**与花冠裂片互生；花药背着，纵裂，内向。**子房**上位至稀半下位；心皮 3(~2) 枚，仅基部稍联合；2–3 (~4) 室；胚珠每室 2 至多枚；中轴胎座。**果实**为膨大的蒴果或 1~3 枚蓇葖果。

World 3/40–60, tropical Asia, America and northern temperate regions.
China 3/20.
This area 1/1.

全世界共 3 属 40~60 种，分布于热带亚洲、美洲和北温带。
中国产 3 属 20 种。
本地区有 1 属 1 种。

Euscaphis Siebold & Zucc. 野鸦椿属

Euscaphis japonica (Thunb.) Kanitz/ 野鸦椿
（a. Follicle/ 蓇葖果；b. Inflorescence/ 花序）

239 Anacardiaceae 漆树科

Leaves often compound, aromatic / 常为复叶，具芳香味
Exudate turning black when exposed to air / 树脂暴露于空气后变黑
Flowers unisexual, ovary superior / 花单性，子房上位
Ovule 1 per locule / 胚珠每室 1 枚

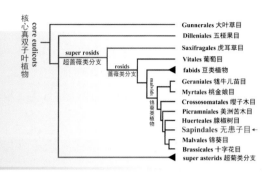

Trees, **shrubs**, woody **climbers** or rarely **subshrubs**. **Leaves** often aromatic, sometimes deciduous, compound or simple, alternate (spiral) or rarely opposite; margin entire to less often toothed; petioles often swollen at the base; stipules absent; often resinous exudate, white, clear or coloured exudate turning black when exposed to air. **Inflorescence** panicles, thyrses, spikes, racemes or rarely solitary flower; sometimes epiphyllous in female flowers or cauliflorous; bracteate. **Flowers** unisexual (monoecious, andromonoecious, polygamomonoecious, dioecious, or gynodioecious) or sometimes bisexual, small, usually actinomorphic, often nectar disk inside the stamens; bracteoles deciduous. **Sepals** usually fused basally, free or absent. **Petals** free or basally fused, imbricate, lobes longer than tube or absent. **Stamens** opposite the sepals, filaments usually free, filaments attached to base of nectar disk when present; anthers usually dorsifixed or basifixed; dehiscing longitudinally, 2-loculed; staminodes in female flowers. **Ovary** superior or rarely inferior; carpels fused or 1; locules 1–5; ovule 1 per locule; placentation basal or parietal. **Fruit** usually an indehiscent drupe, samara or rarely achene-like.

乔木、**灌木**、**木质藤本**或稀**亚灌木**。**叶**常具芳香味，有时早落，复叶或单叶，互生（螺旋状）或稀为对生；全缘，稀具锯齿；叶柄基部常膨大；托叶缺；常有树脂，白色、清澈或有色，暴露于空气后变黑。**花序**为圆锥状、聚伞圆锥状、穗状、总状或稀为单花；雌花有时叶面着生或茎生；具苞片。**花**单性（雌雄同株、雄全同株、杂性同株、雌雄异株或雌全同株）或有时两性，小，常辐射对称，雄蕊内侧常具腺盘；小苞片脱落。**花萼**常基部合生，离生或缺失。**花瓣**离生或基部合生，覆瓦状，裂片长于花冠筒或缺。**雄蕊**与花萼对生，花丝常分离，若具腺盘则着生于基部腺盘上；花药常背着或基着；纵裂，2 室；雌花中具退化雄蕊。**子房**上位或稀下位；心皮合生或为 1；1~5 室；胚珠每室 1 枚；基底胎座或侧膜胎座。**果实**常为不裂的核果、翅果或稀瘦果状。

World 81/800+, mainly in tropics and subtropics, few in temperate regions.
China 17/55.
This area 4/6.

全世界共 81 属 800 余种，主要分布于热带与亚热带地区，少数到温带地区。
中国产 17 属 55 种。
本地区有 4 属 6 种。

1. Perianth reduced to 1-whorled or absent/ 花为单被花或无被花 .... **1. *Pistacia*/ 黄连木属**

1. Perianth of 2 distinct whorls/ 花被 2 轮
   2. Ovary 1-loculed; stamens 5/ 子房 1 室；雄蕊 5 枚
      3. Inflorescence axillary/ 花序腋生 ....**2. *Toxicodendron*/ 漆树属**
      3. Inflorescence terminal/ 花序顶生 ....**3. *Rhus*/ 盐麸木属**
   2. Ovary 5-loculed; stamens 10/ 子房 5 室；雄蕊 10 枚 ....**4. *Choerospondias*/ 南酸枣属**

1. *Pistacia chinensis* Bunge/ 黄连木；
2. *Toxicodendron verniciflum* (Stokes) F.A. Barkley/ 漆树；
3. *Rhus chinensis* Mill./ 盐麸木；
4. *Choerospondias axillaris* (Roxb.) B.L. Burtt & A.W. Hill/ 南酸枣

240 Sapindaceae 无患子科

Leaves often gland-doted, compound / 复叶，多具腺点
Terminal leaflet often aborted and formed a tip / 顶小叶退化，残留突尖
Flowers functionally unisexual / 花为功能性单性
Fruit often 1-seeded and 1-loculed by abortion / 果实常退化仅剩 1 室 1 种子

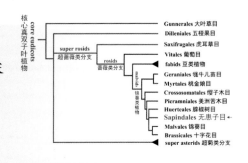

Trees, shrubs, woody climbers or herbs. Leaves often gland-doted, sometimes deciduous, compound (odd- or even-pinnate, palmate, 3-foliate or bi-pinnate) or rarely simple, alternate or opposite, sometimes an aborted terminal leaflet on rachis forming a tip; leaflet margin entire or toothed; petioles often swollen at base; stipules absent. Inflorescence cymes, panicles, thyrses or rarely solitary flower; sometimes cauliflorous to ramiflorous; usually bracteate. Flowers functionally unisexual or bisexual, actinomorphic or zygomorphic, nectar disk often outside the stamens, rarely absent. Sepals usually free or basally fused, imbricate or less often valvate. Petals free, usually clawed, often with scales or hair-tufted basal appendages. Stamen filaments usually attached to nectar disk, filaments often hairy; anthers dorsifixed; dehiscing longitudinally; introrse; staminodes in female flowers, but filaments shorter. Ovary superior, pistillodes in male flowers; carpels fused; locules 1–8; ovules 1-several per locule; placentation axile to basal, rarely parietal; stigma entire or 2 or 3(or 4)-lobed. Fruit a loculicidal capsule, berry, drupe, nut, samara or schizocarp, often 1-seeded and 1-loculed by abortion. Seeds often with a conspicuous fleshy aril or sarcotesta.

乔木、灌木、木质藤本或草本。叶多具腺点，有时早落，复叶（奇数或偶数羽状、掌状、3 枚小叶或二回羽状复叶）或稀为单叶，互生或对生，有时顶小叶退化，残留突尖；小叶全缘或具齿；叶柄基部常膨大；托叶缺。花序聚伞状、圆锥状、聚伞圆锥状或稀为单花；有时具老茎生花或枝花现象；常具苞片。花为功能性单性或两性，辐射对称或两侧对称，腺盘常位于雄蕊外侧，稀缺失。花萼常离生或基部合生，覆瓦状或稀为镊合状。花瓣离生，常具爪，基部附属物多具鳞片或毛簇。雄蕊花丝常贴生于腺盘，花丝常被毛；花药背着；纵裂；内向；雌花中有退化雄蕊，但花丝较短。子房上位，雄花中具退化雌蕊；心皮合生；1~8 室；胚珠每室 1 至数枚；中轴胎座至基底胎座，稀侧膜胎座；柱头光滑或 2 或 3（或 4）裂。果实为室背开裂蒴果、浆果、核果、坚果、翅果或分果，常退化为 1 室 1 粒种子。种子常有显著的肉质假种皮或肉质外种皮。

World ca. 141/1, 900, mainly in tropics and subtropics, few to temperate regions.
China 25/158.
This area 4/6.

全世界共约 141 属 1 900 种，主要分布于热带与亚热带，少数至温带。

Sapindus saponaria L./ 无患子

中国产 25 属 158 种。

本地区有 4 属 6 种。

1. Fruit indehiscent, baccarium or samara/ 果实不裂，分浆果或翅果
    2. Fruit a samara/ 果实为翅果 ....**1. *Acer*/ 槭属**
    2. Fruit a baccarium/ 果实为分浆果 ....**2. *Sapindus*/ 无患子属**
1. Fruit loculicidal capsule/ 蒴果，室背开裂
    3. Leaves odd-pinnate; capsule not winged/ 奇数羽状复叶；蒴果不具翅 ....**3. *Koelreuteria*/ 栾树属**
    3. Leaves simple; capsule winged/ 单叶；蒴果具翅 ....**4. *Dodonaea*/ 车桑子属**

1. *Acer tataricum* L. subsp. *ginnala* (Maxim.) Wesm./ 茶条枫；

2. *Sapindus saponaria* L./ 无患子；

3. *Koelreuteria bipinnata* Franch./ 复羽叶栾树；

4. *Dodonaea viscosa* Jacq./ 车桑子

241 Rutaceae 芸香科

Thorns always present / 多具刺
Often with pellucid gland-dots / 常有透明腺点
Leaves 1-foliolate or compound / 单身复叶或复叶
Hesperidium / 柑果

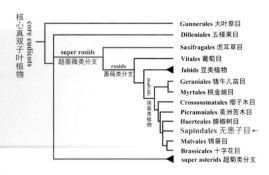

Trees or **shrubs**, rarely **herbs** or **woody climbers**, often with thorns, spines, or prickles; usually with pellucid gland-dots throughout the plant. **Leaves** aromatic, 1-foliolate or variously compound, mostly alternate; margin entire or toothed; petioles present; stipules absent. **Inflorescence** cymes, panicles, racemes, corymbs or solitary flower; rarely cauliflorous or epiphyllous; bracteate. **Flowers** bisexual or less often unisexual; usually 3–5-merous; actinomorphic or rarely zygomorphic, usually a prominent nectar disk inside the stamens or absent. **Sepals** imbricate or valvate, free or basally fused, sometimes inconspicuous. **Petals** free or fused. **Stamens** usually 8–10; filaments free or basally to completely fused. **Ovary** superior, part-inferior or inferior, carpels free, fused or 1; placentation usually axile or parietal. **Fruit** a hesperidium, follicles, drupe or berry.

乔木或**灌木**，稀**草本**或**木质藤本**，常具枝刺或皮刺；全株常具透明腺点。**叶**具芳香味，单身复叶或多种复叶类型，绝大部分互生；全缘或具齿；叶柄存在；托叶缺。**花序**为聚伞状、圆锥状、总状、伞房状或为单花；稀老茎生花或叶面着生；具苞片。**花**两性或稀单性；常3~5基数；辐射对称或稀两侧对称，常具显著花盘，位于雄蕊内侧，或缺。**花萼**覆瓦状或镊合状，离生或基部合生，有时不显著。**花瓣**离生或合生。**雄蕊**常8~10枚；花丝分离，或基部至全部合生。**子房**上位、半下位或下位，心皮离生、合生或仅为1枚；常为中轴胎座或侧膜胎座。**果实**为柑果、蓇葖果、核果或浆果。

World 155/1, 600, worldwide, mainly in the tropics and subtropics, few in temperate regions.
China 23/127.
This area 6/15.

全世界共155属1600种，世界广布，主要分布于热带和亚热带，少数至温带。
中国产23属127种。
本地区有6属15种。

Zanthoxylum armatum DC./ 竹叶花椒

1. Leaves opposite/ 叶对生
  2. Leaves mostly odd-pinnate/ 叶多为奇数羽状复叶 .... **1. *Tetradium*/** 吴茱萸属
  2. Leaves digitately 3-foliolate or 1-foliolate/ 叶为3出掌状复叶或单身复叶 .... **2. *Melicope*/** 蜜茱萸属
1. Leaves alternate/ 叶互生
  3. Fruit follicular or drupaceous/ 果实为蓇葖果或核果状

4. Fruit a follicles/ 果为蓇葖果 ....**3. *Zanthoxylum*/ 花椒属**
4. Fruit a berry-like drupe/ 果为浆果状的核果 ....**4. *Toddalia*/ 飞龙掌血属**
3. Fruit berry/ 果实为浆果
  5. Stems often with spines; leaves 1-foliolate, simple, or rarely digitately 3-foliolate/ 茎常具刺；叶为单身复叶、单叶, 稀为 3 出掌状复叶 ....**5. *Citrus*/ 柑橘属**
  5. Stems unarmed; leaves odd-pinnate/ 茎不具刺；叶为奇数羽状复叶 ....**6. *Murraya*/ 九里香属**

1. *Tetradium glabrifolium* (Champ. ex Benth.) T.G. Hartley/ 楝叶吴萸；
2. *Melicope pteleifolia* (Champ. ex Benth.) T.G. Hartley/ 三桠苦；
3. *Zanthoxylum schinifolium* Siebold & Zucc./ 青花椒；
4. *Toddalia asiatica* (L.) Lam./ 飞龙掌血；
5. *Citrus japonica* Thunb./ 金柑；
6. *Murraya exotica* L./ 九里香

242 Simaroubaceae 苦木科

Woody plants with bitter bark and twigs / 木本植物，树皮和枝条味苦
Inflorescence axillary; flowers small / 花序腋生；花小
Filaments often with hairy basal appendages / 花丝基部附属物被毛
Ovary superior, carpels usually free / 子房上位，心皮常离生

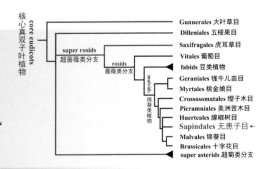

Trees or **shrubs** usually with bitter bark and twigs (poisonous). **Leaves** usually compound (even or odd-pinnate or 1-3-foliate) or rarely simple, alternate (spiral) or rarely opposite, leaflets opposite or alternate; rarely reduced to scales or absent; leaflets margin entire or toothed; petioles present; stipules rarely present. **Inflorescence** axillary; cymes, racemes, spikes, panicles or catkins. **Flowers** usually small; unisexual or bisexual, actinomorphic, nectar disk usually inside the stamens; bracteate. **Sepals** basally fused, free or absent, usually imbricate or valvate. **Petals** free or absent, usually imbricate or contorted to valvate, usually smaller than the sepals. **Stamens** usually 2-whorled, filaments free, often with hairy basal appendages; anthers usually dorsifixed, 2-loculed, with a longitudinal slit. **Ovary** superior; carpels weakly united at base, free, usually 2-5 or 1; locules (1-) 2-5; ovule 1 per locule; placentation axile; styles free or fused; stigmas 2-5. **Fruit** a druparium or samarium.

乔木或**灌木**，树皮与枝条味苦（有剧毒）。**叶**常为复叶（偶数或奇数羽状或1~3枚小叶）或稀为单叶，互生（螺旋状）或稀对生，小叶对生或互生；稀退化为鳞片状或缺；小叶全缘或具齿；叶柄存在；托叶稀存在。**花序腋生**；聚伞状、总状、穗状、圆锥状或柔荑花序。**花**常小；单性或两性，辐射对称，腺盘常位于雄蕊内侧；具苞片。**花萼**基部合生、离生或缺，常覆瓦状或镊合状。**花瓣**离生或缺，常覆瓦状或扭曲为镊合状，常小于花萼。**雄蕊**常2轮，花丝分离，基部附属物被毛；花药常背着，2室，纵裂。**子房**上位；心皮基部稍联合或离生，常2~5或1枚；(1~) 2~5室；胚珠每室1枚；中轴胎座；花柱分离或合生；柱头2~5个。**果实**为核果状或翅果状。

World 22/ca. 109, tropical and subtropical regions.
China 3/10.
This area 1/1.

全世界共22属约109种，分布于热带至亚热带地区。
中国产3属10种。
本地区有1属1种。

***Ailanthus* Desf. 臭椿属**

Ailanthus altissima (Mill.) Swingle/ 臭椿

SAPINDALES. Simaroubaceae. *Picrasma quassioides* (D. Don) Benn./ 无患子目 苦木科 苦树（not distributed here/ 非本区域分布）

A. Fruiting branch/ 果枝

B_1. Vertical section of flower/ 花纵切面

B_2. Vertical section of flower, with pedicel/ 花纵切面，具花梗

C. Vertical section of fruit/ 果实纵切面

D. Lateral view of fruit/ 果实侧面观

E. Flower segregated/ 花离析

F_{1-3}. Ventral view of flower, 3–5-merous/ 花腹面观，依次为 3~5 基数

F_{4-6}. Dorsal view of flower, 3–5-merous/ 花背面观，依次为 3~5 基数

(Scale/ 标尺：A 5 cm; B_1–C 1 mm; D 1 cm; E–F_6 1 mm)

243 Meliaceae 楝科

Leaves usually compound, alternate / 常为复叶，互生
Flowers usually functionally unisexual, actinomorphic / 花常功能性单性，辐射对称
Stamen filaments usually fused into a tube / 花丝常合生成管

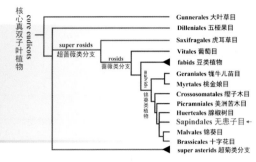

Trees, **shrubs** or rarely **herbs**, trees often buttressed; often bark bitter, sometimes aromatic or rarely with milky exudate. **Leaves** sometimes deciduous, usually compound or less often simple, alternate (spiral) or rarely opposite (decussate), leaflets often opposite, sometimes with black gland-dotted; leaflet blade with base somewhat oblique; margin usually entire, toothed or lobed; petioles present, often swollen, sometimes spiny; stipules absent; sometimes hairs including simple, T-shaped, stellate and peltate types. **Inflorescence** axillary; usually panicles to thyrses or cymes, racemes, spikes or solitary flower; sometimes cauliflorous to ramiflorous or epiphyllous. **Flowers** usually functionally unisexual (monoecious, polygamomonoecious or dioecious) or bisexual, actinomorphic, sometimes a nectar disk; bracteolate. **Sepals** usually fused or free, cup-shaped or tubular; imbricate or valvate. **Petals** free or fused; sometimes adnate to staminal tube. **Stamen** filaments usually fused into a tube, often androgynophore; anthers usually sessile on stamen tube; 2-loculed, dehiscing longitudinally; ±dorsifixed, introrse; sometimes staminodes 5-100. **Ovary** superior; carpels fused; locules (1-) 2-6(-20); ovules 1 to many per locule; placentation usually axile or parietal. **Fruit** a capsule, berry or drupe.

乔木、**灌木**或稀为**草本**，乔木常具板根；树皮味苦，有时具芳香味或稀具乳汁。**叶**有时脱落，常为复叶或稀单叶，互生（螺旋状）或稀对生（交互对生），小叶常对生，有时具黑色腺点；小叶片基部不对称；全缘、具齿或分裂；具叶柄，常膨大，有时多刺；托叶缺；有时被单毛、丁字状毛、星状毛或盾状毛。**花序**腋生；常圆锥状至聚伞圆锥状，或聚伞状、总状、穗状或为单花；有时具老茎生花至枝花现象，或叶面着生。花常功能性单性（雌雄同株、杂性同株或雌雄异株）或两性，辐射对称，有时具腺盘；具小苞片。**花萼**常合生或离生，萼筒为杯状或管状；萼裂片覆瓦状或镊合状。**花瓣**离生或合生；有时贴生于雄蕊管。**雄蕊**花丝常合生成管，常具雌雄蕊柄；花药常无柄，着生于雄蕊管；2室，纵裂；多少背着，内向；有时具5~100枚退化雄蕊。**子房**上位；心皮合生；(1~) 2-6(~20)室；胚珠每室1至多枚；常中轴胎座或侧膜胎座；**果实**为蒴果、浆果或核果。

World 50/ca. 575, tropical and subtropical regions, few in temperate regions.
China 17/40.
This area 3/3.

全世界共50属约575种，分布于热带和亚热带地区，少数至温带。

Melia azedarach L./ 楝

中国产 17 属 40 种。

本地区有 3 属 3 种。

1. Seeds winged/ 种子具翅
    2. Filaments distinct/ 花丝分离 .... **1. *Toona*/ 香椿属**
    2. Filaments connate/ 花丝合生 .... **2. *Chukrasia*/ 麻楝属**
1. Seeds not winged/ 种子不具翅 .... **3. *Melia*/ 楝属**

1. *Toona sinensis* (A. Juss.) M. Roem./ 香椿；
2. *Chukrasia tabularis* A. Juss./ 麻楝；
3. *Melia azedarach* L./ 楝

247 Malvaceae 锦葵科

Plants with peltate scales or stellate hairs / 植株被盾状鳞片或星状毛
Perianth 5-merous, epicalyx present / 花萼5枚，具副萼
Stamen filaments often connate into tube / 常为单体雄蕊
Pollen spiny / 花粉具刺

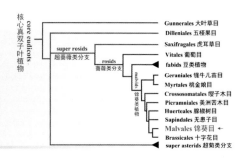

Herbs, **subshrubs**, **shrubs** or **trees**, rarely woody **climbers**. **Leaves** simple or compound, usually alternate; margin entire, or various lobed; petioles, often swollen at both ends; stipules usually present, usually well-developed, deciduous; often with peltate scales or stellate hairs, sometimes simple or glandular. **Inflorescence** solitary flower or cymes; sometimes cauliflorous to ramiflorous; epicalyx often present, forming an involucre, 3 to many lobed or absent. **Flowers** bisexual or rarely unisexual, usually conspicuous; actinomorphic or rarely zygomorphic, nectaries often present. **Sepals** 5, basally fused or free, usually valvate or sometimes imbricate, sometimes petaloid. **Petals** 5, free or absent, contorted or imbricate; basally adnate to base of filament tube. **Stamen** filaments often connate into tube, usually androgynophore; anthers 1-loculed; pollen spiny. **Ovary** superior, sometimes gynophore; carpels usually fused or free; locules many; ovules 1 to many per locule; placentation axile or rarely parietal or separating from one another and from axis; styles apex branched or capitate. **Fruit** a loculicidal capsule or schizocarp, rarely a berry or samara.

草本、**亚灌木**、**灌木**或**乔木**，稀木质藤本。**叶**为单叶或复叶，常互生；全缘或缺刻多样；具叶柄，两端常膨大；托叶存在，常发达，脱落；常被盾状鳞片或星状毛，有时为单毛或腺毛。**花序**为单花，或聚伞状；有时具老茎生花至枝花现象；副萼常存在，形成总苞状，3至多裂或缺失。**花**两性或稀单性，常显著；辐射对称或稀为两侧对称，常具蜜腺盘。**花萼**5枚，基部合生或离生，常镊合状或偶为覆瓦状，有时呈花瓣状。**花瓣**5枚，离生或缺，扭曲或覆瓦状；基部与花丝管底部联合。**雄蕊**常为单体雄蕊，常具雌雄蕊柄；花药1室；**花粉具刺**。**子房**上位，有时具雌蕊柄；心皮常合生或离生；多室；胚珠每室1至多枚；中轴胎座或稀侧膜胎座或自中轴处彼此分离；花柱顶端分枝或头状。**果实**为室背开裂的蒴果或分果，稀为浆果或翅果。

World ca.243/4, 300, tropics and temperate regions.
China 51/246.
This area 15/23.

全世界共约243属4 300种，主要分布在热带和温带。
中国产51属246种。
本地区有15属23种。

1. Stamens 1–4 times as many as tepals; anthers 2-loculed/ 雄蕊为花被片数目的1~4倍；花药2室
  2. Flowers without petals/ 花无花瓣 .... **1. *Firmiana*/ 梧桐属**

2. Flowers with petals/ 花有花瓣
　　3. Ovary with elongate androgynophore/ 子房生于伸长的雌雄蕊柄上 ....**2. *Helicteres*/** 山芝麻属
　　3. Ovary sessile/ 子房无柄
　　　　4. Flowers without staminodes/ 花无退化雄蕊
　　　　　　5. Styles 5/ 花柱 5 个 ....**3. *Melochia*/** 马松子属
　　　　　　5. Style 1/ 花柱 1 个 ....**4. *Waltheria*/** 蛇婆子属
　　　　4. Flowers with staminodes/ 花有退化雄蕊 ....**5. *Corchoropsis*/** 田麻属
1. Stamens numerous; anthers 1-loculed/ 雄蕊多数；花药 1 室
　　6. Flowers solitary/ 花单生
　　　　7. Fruit a capsule/ 果为蒴果
　　　　　　8. Seeds glabrous and smooth/ 种子光滑 ....**6. *Abelmoschus*/** 秋葵属
　　　　　　8. Seeds hairy/ 种子被毛 ....**7. *Hibiscus*/** 木槿属
　　　　7. Fruit a schizocarp/ 果为分果
　　　　　　9. Filament tube with anthers inserted along sides/ 雄蕊柱上只外部生药 ....**8. *Urena*/** 梵天花属
　　　　　　9. Filament tube with anthers inserted to apex/ 雄蕊柱上花药着生至顶部
　　　　　　　　10. Epicalyx absent/ 不具副萼
　　　　　　　　　　11. Ovule 1 per locule; mericarps often indehiscent/ 每室具 1 枚胚珠；分果爿不开裂 ....**9. *Sida*/** 黄花棯属
　　　　　　　　　　11. Ovules 2 or more per locule; mericarps eventually dehiscent/ 每室具 2 至多枚胚珠；分果爿开裂 ....**10. *Abutilon*/** 苘麻属
　　　　　　　　10. Epicalyx present/ 具副萼
　　　　　　　　　　12. Epicalyx lobes 6–9/ 副萼 6~9 枚 ....**11. *Althaea*/** 药葵属
　　　　　　　　　　12. Epicalyx lobes 3/ 副萼 3 枚
　　　　　　　　　　　　13. Stigmas capitate/ 柱头头状 ....**12. *Malvastrum*/** 赛葵属
　　　　　　　　　　　　13. Stigmas filiform/ 柱头丝状 ....**13. *Malva*/** 锦葵属
　　6. Inflorescence cymes or thyrses/ 花排成聚伞花序或聚伞圆锥花序
　　　　14. Each inflorescence with a bract partially connate with peduncle/ 花序梗一部分贴生于苞片上 ....**14. *Tilia*/** 椴属
　　　　14. Inflorescence without such a bract/ 花序梗上无贴生苞片 ....**15. *Grewia*/** 扁担杆属

1. *Firmiana simplex* (L.) W. Wight/ 梧桐；
2. *Helicteres angustifolia* L./ 山芝麻；
3. *Melochia corchorifolia* L./ 马松子

10. *Abutilon theophrasti* Medik./ 苘麻（10a. Plant/ 植株；10b. Cross section of ovary/ 子房横切面）；

11. *Althaea officinalis* L./ 药葵；

12. *Malvastrum coromandelianum* (L.) Garcke/ 赛葵；

4. *Waltheria indica* L./ 蛇婆子；

5. *Corchoropsis crenata* Siebold & Zucc./ 田麻；

6. *Abelmoschus esculentus* (L.) Moench/ 咖啡黄葵；

7. *Hibiscus mutabilis* L./ 木芙蓉；

8. *Urena lobata* L./ 地桃花；

9. *Sida rhombifolia* L./ 白背黄花稔；

13. *Malva pusilla* Sm./ 圆叶锦葵；

14. *Tilia henryana* Szyszył. var. *subglabra* V. Engl/ 糯米椴；

15. *Grewia biloba* G. Don var. *parviflora* (Bunge) Hand.–Mazz./ 小花扁担杆

249 Thymelaeaceae 瑞香科

Bark tough and fibrous / 树皮坚韧，纤维发达
Leaves simple, margin entire; stipules absent / 单叶全缘；无托叶
Sepals fused and tubular / 花萼合生，管状
Ovary superior, ovule 1 per locules / 子房上位，胚珠每室 1 枚

Trees, **shrubs**, woody **climbers** or **herbs**; bark tough and fibrous, twigs often with unpleasant odour. **Leaves** sometimes gland-dotted, sometimes deciduous, simple, alternate (often clustered at branch ends) or opposite, rarely some ternate; margin entire; petioles present or absent; stipules usually absent; often silky hairs. **Inflorescence** often terminal; solitary flower, racemes, heads, umbels or fascicles; sometimes cauliflorous; sometimes involucral bracts. **Flowers** usually bisexual, sometimes unisexual, usually actinomorphic, sometimes a nectar disk. **Sepals** fused and tubular or rarely free, imbricate, petaloid; mostly caducous. **Petals** free, reduced or absent, imbricate. **Stamens** opposite the sepals, 1–2-whorled, 1–2 or many, filaments free or attached to perianth; anthers basifixed or rarely dorsifixed; staminodes in female flowers. **Ovary** superior; sometimes shortly stalked; carpels fused; 1–12-loculed; ovule 1 per locule; styles filiform, caducous; placentation parietal, apical or axile. **Fruit** mostly indehiscent, dry or fleshy, sometimes a loculicidal capsule.

乔木、灌木、木质藤本或草本；树皮坚韧，纤维发达，枝条常具难闻气味。**叶**有时具腺点，有时脱落，单叶，互生（常簇生枝顶）或对生，稀 3 出；全缘；叶柄有或无；托叶常缺；常被绢毛。**花序**常顶生；单花、总状、头状、伞状或簇生；有时老茎生花；偶具总苞。**花**常两性，有时单性，常辐射对称，有时具腺盘。**花萼**合生，管状，或稀离生，覆瓦状，呈花瓣状；大多早落。**花瓣**离生，退化或缺，覆瓦状。**雄蕊**与花萼对生，1~2 轮，1~2 枚或多枚，花丝分离或贴生于花被片；花药基着或稀背着；雌花中具退化雄蕊。**子房**上位；有时具短柄；心皮合生；1~12 室；胚珠每室 1 枚；花柱丝状，早落；侧膜胎座、顶生胎座或中轴胎座。**果实**基本不裂，干燥或肉质，偶为室背开裂蒴果。

World 49/892, widely distributed in tropical and temperate regions, especially in Africa and Oceania.

China 9/115.

This area 3/6.

全世界共 49 属 892 种，广布于热带至温带，尤其是非洲和大洋洲。

中国产 9 属 115 种。

本地区有 3 属 6 种。

Wikstroemia indica (L.) C.A. Mey./ 了哥王

1. Leaves alternate/ 叶互生

   2. Styles very short or obscure; stigma capitate/ 花柱极短或近无；柱头头状 ....**1. *Daphne*/ 瑞香属**

2. Styles long; stigmas terete/ 花柱长；柱头圆筒形 .... **2. *Edgeworthia*/ 结香属**
1. Leaves opposite/ 叶对生 .... **3. *Wikstroemia*/ 荛花属**

1. *Daphne kiusiana* Miq. var. *atrocaulis* (Rehder) F. Maek./ 毛瑞香；
2. *Edgeworthia chrysantha* Lindl./ 结香；
3. *Wikstroemia indica* (L.) C.A. Mey./ 了哥王

268 Capparaceae 山柑科

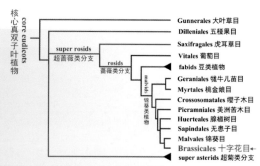

Stipules usually spiny / 托叶常刺状
Flower bisexual, 4-merous / 花两性，4 基数
Stamens opposite the sepals, usually exserted / 雄蕊与花萼对生，常伸出
Ovary superior, often stalked / 子房上位，多具柄

Shrubs, **trees**, sometimes woody **climbers** or very rarely **herbs**; usually evergreen. **Leaves** rarely absent, sometimes gland–dotted, simple or compound, alternate or rarely opposite; petioles present; stipules usually spiny or absent; hairs various or absent. **Inflorescence** axillary; solitary flower, racemes, panicles corymbs or umbels; rarely bracteate. **Flowers** bisexual or rarely unisexual, actinomorphic to zygomorphic, often floral nectar scales outside stamens. **Sepals** 4 (–8), free or fused, imbricate or valvate, 1 or 2-whorled, sometimes petaloid. **Petals** 4 (–8); free or absent, imbricate or rarely valvate, sometimes clawed. **Stamens** usually opposite the sepals, usually exserted, usually short to long androgynophore; anthers basifixed, 2-loculed, introrse, dehiscing longitudinally; staminodes 0–7. **Ovary** superior; carpels fused, often stalked, ± as long as stamens; locules 1–2 or 4–6; ovules few–many per locule; placentation parietal; styles short to absent. **Fruit** usually a fleshy berry or a capsule.

灌木、乔木，有时为木质藤本或极稀为草本；多常绿。叶稀缺失，有时具腺点，单叶或复叶，互生或稀对生；具叶柄；托叶常刺状或缺；毛被多样或缺。花序腋生；单花、总状、伞房花序圆锥状或伞形；稀具苞片。花两性或稀单性，辐射对称至两侧对称，雄蕊外侧常具腺鳞。花萼 4 (~8) 枚，离生或合生，覆瓦状或镊合状，1 或 2 轮，有时呈花瓣状。花瓣 4 (~8) 枚；离生或缺，覆瓦状或稀镊合状，有时具爪。雄蕊常与花萼对生，多伸出，常具短至长的雌雄蕊柄；花药基着，2 室，内向，纵裂；退化雄蕊 0~7 枚。子房上位；心皮合生，常具柄，多少与雄蕊等长；1~2 室或 4~6 室；胚珠每室数枚至多枚；侧膜胎座；花柱短至缺失。果实常为肉质浆果或蒴果。

World ca. 28/650, widely in tropics and subtropics, few to temperate regions.
China 2/42.
This area 1/1.

全世界共约 28 属 650 种，分布于热带和亚热带，少量分布至温带地区。
中国产 2 属 42 种。
本地区有 1 属 1 种。

Capparis L. 山柑属

Capparis cantoniensis Lour./ 广州山柑（photo by XiangXiu SU/ 苏享修拍摄）

270 Brassicaceae 十字花科

Plant with a pungent, watery juice / 植株具辛辣、水状汁液
Cruciform corolla / 十字形花冠
Tetradynamous / 四强雄蕊
Silique or silicle / 长角果或短角果

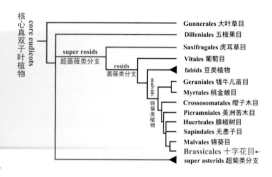

Annual or perennial **herbs**, **shrubs**, woody **climbers** or rarely **trees**; sometimes aquatic **herbs**; with a pungent, watery juice. **Leaves** often aromatic (usually mustard), simple or rarely compound, alternate (spiral) or opposite; basal leaves rosette present or absent; margin entire, toothed, lobed or pinnatifid; petioles present to absent; stipules absent; hairs, if present, simple, glandular or stellate. **Inflorescence** usually corymbs, racemes, spikes, fascicles or solitary flower on long pedicels originating from axils of rosette leaves. **Flowers** bisexual or rarely unisexual, mostly actinomorphic; rarely bracteate; nectaries often around the stamens and carpels. **Sepals** free or very rarely fused, imbricate, inner pair often with a pouch. **Petals** free or rarely absent, usually clawed, often cross-shaped (cruciform). **Stamens** usually 6, in 2 whorls, tetradynamous, rarely 2–4 or up to 24, usually opposite the sepals; anthers 2-loculed, basifixed, dehiscing by longitudinal slits, introrse. **Ovary** superior; carpels fused, usually 2 or rarely 4–6 carpels; locules (1–) 2, with false septum; ovules many per locule; placentation parietal or rarely apical; style 1, short; stigma capitate or conical, entire or 2-lobed. **Fruit** a 2-valved capsule (silique or silicle), dehiscent or indehiscent, nutlet-like, lomentaceous, or samaroid, segmented or not, terete, angled or rarely a schizocarp. **Seeds** without endosperm; cotyledons incumbent, accumbent, or conduplicate.

一年生或多年生**草本**、**灌木**、**木质藤本**或稀为**乔木**；有时为水生**草本**；具辛辣、水状汁液。**叶**常具芳香味（多为芥末味），单叶或稀复叶，互生（螺旋状）或对生；基生叶莲座状或缺；全缘、具齿、分裂或羽裂；叶柄有或无；托叶缺；如被毛，则为单毛、腺毛或星状毛。**花序**常伞房状、总状、穗状、簇生或单花生于莲座叶发出的花葶上。**花**两性或稀单性，绝大多数辐射对称；稀具苞片；蜜腺常围绕雄蕊与心皮着生。**花萼**离生或稀合生，覆瓦状，内侧一对常具囊。**花瓣**离生或稀缺，常具爪，十字状（十字形）。**雄蕊**常6枚，2轮，四强雄蕊，稀2~4枚或多达24枚，常与花萼对生；花药2室，基着，纵裂，内向。**子房**上位，心皮合生，常2枚或稀4~6枚；(1~) 2室，具假隔膜；胚珠每室多数；侧膜胎座或稀顶生胎座；花柱1个，短；柱头头状或圆锥形，光滑或2裂。**果实**为2瓣裂的蒴果（长角果或短角果），开裂或不裂，小坚果状、节荚状或翅果状，分节或不分节，圆柱形，具棱或稀为分果。**种子**无胚乳；子叶背倚、缘倚或对折。

World 321/3,660, worldwide, except Antarctica, mainly in temperate regions.
China 84/ca. 400.
This area 6/11.

Raphanus raphanistrum L./ 野萝卜

全世界共 321 属 3 660 种，除南极洲外世界广布，主要分布于温带地区。中国产 84 属约 400 种。

本地区有 6 属 11 种。

1. Stamens 2/ 雄蕊 2 枚 .... **1. *Lepidium*/ 独行菜属**
1. Stamens 6, rarely 4/ 雄蕊 6 枚，稀 4 枚
   2. The base of inner sepals saccate/ 内轮萼片基部囊状
      3. Silicles/ 短角果 .... **2. *Capsella*/ 荠菜属**
      3. Siliques/ 长角果
         4. Stigmas distinctly 2-lobed/ 柱头 2 裂 .... **3. *Orychophragmus*/ 诸葛菜属**
         4. Stigmas entire/ 柱头不裂 .... **4. *Raphanus*/ 萝卜属**
   2. The base of inner sepals not saccate / 内轮萼片基部不成囊状。
      5. Stigmas entire/ 柱头不裂 .... **5. *Rorippa*/ 蔊菜属**
      5. Stigmas distinctly 2-lobed/ 柱头 2 裂
         6. Ovules 1 or 2 per ovary/ 子房内有 1 或 2 枚胚珠 .... **1. *Lepidium*/ 独行菜属**
         6. Ovules 4 or more per ovary/ 子房内有胚珠 4 枚或更多 .... **6. *Cardamine*/ 碎米荠属**

1. *Lepidium virginicum* L./ 北美独行菜（1a. Inflorescence/ 花序；1b. Silicles/ 短角果）；
2. *Capsella bursa-pastoris* (L.) Medik./ 荠；
3. *Orychophragmus violaceus* (L.) O.E. Schulz/ 诸葛菜；
4. *Raphanus raphanistrum* L./ 野萝卜；
5. *Rorippa indica* (L.) Hiern/ 蔊菜；
6. *Cardamine impatiens* L./ 弹裂碎米荠

276 *Santalaceae* 檀香科

Hemiparasites / 半寄生植物
Stems jointed / 茎有关节
Leaves simple, margin entire / 单叶，全缘
Perianth 1-whorled / 花被片 1 轮

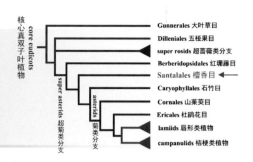

Trees, **shrubs** or **herbs**, usually root or stem hemiparasites; roots modified to haustoria; stems jointed, breaking apart easily at the constricted nodes. **Leaves** simple, alternate or opposite, often coriaceous or slightly succulent; sometimes scale-like; margin entire; stipules absent. **Inflorescence** mostly axillary; solitary flower, spikes, cymes, panicles, fascicles, corymbs or umbels; bracteate leafy. **Flowers** ± inconspicuous, bisexual or unisexual (plants dioecious or monoecious), actinomorphic, often a nectar disk. **Perianth** 1-whorled, free, fused or absent; greenish, valvate. **Stamens** equal number of tepals, opposite the tepals; anthers dorsifixed, 2-loculed, dehiscing longitudinally. **Ovary** inferior, part-inferior or rarely superior; carpels 3–4, fused; locules 1 or 5–12; ovules 1–3 (–5); placentation basal to free central. **Fruit** a drupe, nut, or viscous berry with a single seed.

乔木、灌木或草本，常根或茎半寄生；根变态为吸器；茎有关节，易在缢缩的关节处断开。**叶**为单叶，互生或对生，常革质或稍肉质；有时为鳞片状；全缘；托叶缺。**花序**大多腋生；单花、穗状、聚伞、圆锥状、簇生、伞房或伞形；苞片常叶状。**花**多少不显著，两性或单性（雌雄异株或雌雄同株），辐射对称，常具蜜腺盘。**花被片** 1 轮，离生、合生或缺；淡绿色，镊合状。**雄蕊**与花被片同数，与花被对生；花药背着，2 室，纵裂。**子房**下位、半下位或稀上位；心皮 3～4 枚，合生；1 或 5～12 室；胚珠每室 1～3 (～5) 枚；基底胎座或特立中央胎座。**果实**为核果、坚果或含 1 粒种子的黏性浆果。

World (34–)39–44/450–990, tropical and temperate regions.
China10/51.
This area 2/2.

全世界共（34~）39~44 属 450~990 种，分布于热带和温带地区。
中国产 10 属 51 种。
本地区有 2 属 2 种。

1. Herbs; leaves alternate/ 草本；叶互生 ....**1. *Thesium*/** 百蕊草属
1. Shrubs; leaves opposite/ 灌木；叶对生 ....**2. *Korthalsella*/** 栗寄生属

SANTALALES. Santalaceae. *Thesium chinense* Turcz./ 檀香目 檀香科 百蕊草

A. Whole plant/ 植株
B_1. Flowering branch/ 花枝
B_2. Lateral view of flower/ 花侧面观
B_3. Lateral view of young fruit/ 幼果侧面观
B_4. Lateral view of fruit/ 果实侧面观
C_1. Vertical section of flower (inside)/ 花纵切面 (内侧)
C_2. Lateral view of perianth lobes/ 花被裂片侧面观
C_3–C_4. Cross section of ovary/ 子房横切面
D. Ventral view of flower/ 花腹面观
E. Anther attached to perianth via viscin thread/ 花药通过黏丝贴生于花被
F. Vertical section of young fruit/ 幼果纵切面
G. Perianth tube unfold (inside)/ 花被筒展开 (内侧)
H_1. Lateral view of flower, amplified/ 花侧面观，放大
H_2. Lateral view of young fruit, amplified/ 幼果侧面观，放大

(Scale/ 标尺： A–B_4 1 cm; C_1–H_2 1 mm)

1. *Thesium chinense* Turcz./ 百蕊草；

2. *Korthalsella japonica* (Thunb.) Engl./ 栗寄生
（2a. Parasite on *Quercus*/ 寄生于栎属植物上；2b. Parasite on *Smilax*/ 寄生于菝葜属植物上；photo by XiYang YE/ 叶喜阳拍摄）

282　Plumbaginaceae 白花丹科

Salts glands / 具盐质颗粒
Flowers bisexual, actinomorphic, sessile / 两性花，辐射对称，无柄
Calyx persistent, tube 5-ribbed / 萼宿存，萼筒 5 棱
Ovary superior, 1-loculed, ovule 1 / 子房上位，1 室，1 枚胚珠

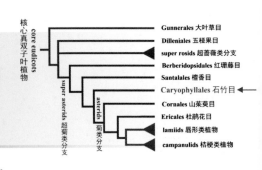

Shrublets, **shrubs**, or perennial **herbs**, less often woody climbers or annual herbs; xerophytes or halophytes. **Leaves** simple, alternate or basal; margin entire or lobed; stipules absent; often with salts glands. **Inflorescence** a spikes, panicles, heads or racemes; sometimes scapose; bracteate. **Flowers** bisexual, actinomorphic; sessile or very short pedicellate; bracteoles (1–) 2, often a 5-lobed disk. **Sepals** fused or rarely free, lobes 5, shorter than tube, 5-ribbed; persistent. **Petals** fused or almost free, contorted, often persistent. **Stamens** opposite the petals, filaments basally attached to petals; anthers 2-loculed, dehiscing longitudinally, introrse; sometimes heterostylous. **Ovary** superior; carpels fused; 1-loculed; ovule 1; placentation basal; styles 1 or 5; stigmas 5. **Fruit** an achene, capsule or nut.

小灌木、**灌木**或多年生**草本**，稀为木质藤本或一年生草本；旱生或盐生植物。**叶**为单叶，互生或基生；全缘或分裂；托叶缺；常具盐质腺体。**花序**为穗状、圆锥状、头状或总状；有时具花葶；具苞片。**花**两性，辐射对称；无柄或具极短花梗；小苞片 (1~) 2 枚，常具 5 裂花盘。**花萼**合生或稀离生，5 裂，短于萼筒，具 5 棱；宿存。**花瓣**合生或几离生，扭曲，常宿存。**雄蕊**与花瓣对生，花丝基部贴生于花瓣；花药 2 室，纵裂，内向；有时花柱异长。**子房**上位；心皮合生；1 室；胚珠 1 枚；基底胎座；花柱 1 个或 5 个；柱头 5 个。**果实**为瘦果、蒴果或坚果。

World 27/836, arid, saline or coastal habitats worldwide, especially in Mediterranean to Central Asia.
China 7/46.
This area 2/2.

全世界共 27 属 836 种，全球干旱、盐碱地或沿海生境，主要分布于地中海至中亚。
中国产 7 属 46 种。
本地区有 2 属 2 种。

1. Leaves cauline, alternate/ 叶茎生，互生 ....**1. *Plumbago*/** 白花丹属
1. Leaves basal, rosette/ 叶基生，莲座状 ....**2. *Limonium*/** 补血草属

1. *Plumbago zeylanica* L./ 蓝花丹;   2. *Limonium sinense* (Girard) Kuntze/ 补血草

283 Polygonaceae 蓼科

Usually herbs, stems often with swollen nodes / 多草本，茎节膨大
Leaves simple, alternate / 单叶，互生
Stipules often united into a sheath / 托叶常成鞘
Achene, often trigonous / 瘦果，常三棱形

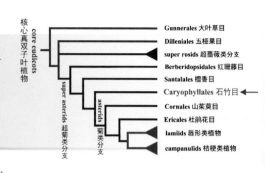

Perennial or annual **herbs**, rarely **shrubs**, or **small trees** or **woody climbers**; stems often with swollen nodes. **Leaves** simple, alternate, rarely opposite or whorled; sometimes gland-dotted; petiolate present or absent; stipules often united to a sheath (ocrea). **Inflorescence** terminal or axillary; panicles, fascicles, racemes or spikes, rarely umbel-like; pedicel occasionally articulate; bracteate. **Flowers** small, bisexual, or rarely unisexual; actinomorphic; bracteolate. **Perianth** 1–2- whorled; 5 or 3+3; persistent in fruit; sometimes enlarged in fruit or inner ones enlarged, with wings, tubercles, or spines. **Stamens** usually (3–) 6–9, rarely more; filaments free or fused at base; anthers 2-loculed, dehiscing longitudinally; disk annular (often lobed). **Ovary** superior, 1-loculed; ovule 1; styles 2 or 3, rarely 4, free or connate at lower part; placentation basal or rarely free central. **Fruit** a trigonous, biconvex, or biconcave achene.

多年生或一年生**草本**，稀**灌木**，或**小乔木**或**木质藤本**；茎节常膨大。**叶**为单叶，互生，稀对生或轮生；有时具腺点；叶柄有或无；托叶常联合成鞘（托叶鞘）。**花序**顶生或腋生；圆锥状、簇生、总状或穗状，稀伞形；花梗偶具关节；具苞片。**花**小，两性，或稀单性；辐射对称；具小苞片。**花被片** 1~2 轮；5 或 3+3 枚；果期宿存；有时增大或内轮增大，具翅、瘤突或刺。**雄蕊**常 (3~) 6~9 枚，稀更多；花丝分离或基部合生；花药 2 室，纵裂；花盘环形（常分裂）。**子房**上位，1 室；胚珠 1 枚；花柱 2 或 3 个，稀 4 个，分离或下半部合生；基底胎座或稀特立中央胎座。**果实**为三棱形、双凸镜形或双凹镜形瘦果。

World 50/1,150, widely distributed in northern temperate regions, few in tropics.
China 12/236.
This area 6/29.

全世界共 50 属 1 150 种，广布于北温带，少数至热带。
中国产 12 属 236 种。
本地区有 6 属 29 种。

1. Perianth 6/ 花被片 6 枚 ....**1. *Rumex*/ 酸模属**
1. Perianth 5/ 花被片 5 枚
   2. Stems twining or erect; perianth enlarged in fruit/ 茎缠绕或直立；花被片在果期增大
      3. Stems twining; flowers bisexual/ 茎缠绕；花两性 ....**2. *Fallopia*/ 何首乌属**
      3. Stems erect; flowers unisexual/ 茎直立；花单性 ....**4. *Reynoutria*/ 虎杖属**
   2. Stems erect; perianth not enlarged in fruit/ 茎直立；花被片在果期不增大

4. Achenes much longer than persistent perianth/ 瘦果显著长于宿存花被片 ....**5. *Fagopyrum*/ 荞麦属**
4. Achenes shorter than persistent perianth/ 瘦果短于宿存花被片
　　5. Inflorescence solitary flower or fascicles/ 花单生或数朵簇生 ....**3. *Polygonum*/ 萹蓄属**
　　5. Inflorescence racemes or panicles/ 花序总状或圆锥状 ....**6. *Persicaria*/ 蓼属**

1. *Rumex japonicus* Houtt./ 羊蹄；
2. *Fallopia multiflora* (Thunb.) Haraldson/ 何首乌；
3. *Polygonum aviculare* L./ 萹蓄；
4. *Reynoutria japonica* Houtt./ 虎杖；
5. *Fagopyrum dibotrys* (D. Don) H. Hara/ 金荞；
6. *Persicaria lapathifolia* (L.) Delarbre var. *salicifolium* (Sibth.) Miyabe/ 绵毛马蓼

284 Droseraceae 茅膏菜科

Carnivorous herbs / 食虫草本
Leaves often basal and rosette / 叶常基生，莲座状
Leaves with sticky or sensitive hair / 叶被黏性毛或感应毛
Ovary superior; loculicidal capsule / 子房上位；室背开裂蒴果

Perennial or rarely annual **herbs** to **subshrubs**, carnivorous, sometimes submerged aquatics; with or without tubers or rhizomes. **Leaves** often basal and rosette, simple, alternate or whorled; petioles present to absent; stipules present or absent; leaf blade with sticky, glandular hair, or with sensitive hairs that trigger closing of blade to trap prey. **Inflorescence** thyrses, panicles, cymes or solitary flower; scapose. **Flowers** 4 or 5-merous; bisexual, actinomorphic. **Sepals** fused basally or rarely free, imbricate, lobes longer than tube. **Petals** free, imbricate or contorted, clawed. **Stamen** filaments free or basally fused; anthers extrorse, dehiscing longitudinally. **Ovary** superior; carpels fused; 1-loculed; ovules (3–) many; placentation parietal or basal; styles free or fused. **Fruit** a loculicidal capsule, rarely indehiscent.

多年生或稀一年生**草本**至**亚灌木**，食虫植物，有时为沉水植物；具块茎或根状茎，或无。**叶**常基生，莲座状，单叶，互生或轮生；叶柄有或无；托叶有或无；叶片被黏性腺毛，或感应毛，可触发叶片捕捉猎物。**花序**聚伞圆锥状、聚伞状或为单花；具花葶。**花**4或5基数；两性，辐射对称。**花萼**基部合生或稀离生，覆瓦状，裂片长于萼筒。**花瓣**离生，覆瓦状或扭曲，具爪。**雄蕊**花丝分离或基部合生；花药外向，纵裂。**子房**上位；心皮合生；1室；胚珠3枚至多枚；侧膜胎座或基底胎座；花柱分离或合生。**果实**为室背开裂的蒴果，稀不裂。

World 3/115, mainly in tropical, subtropical and temperate regions.
China 2/7.
This area 1/3.

全世界共3属115种，主要分布于热带、亚热带和温带地区，少量在寒带。
中国产2属7种。
本地区有1属3种。

Drosera L. 茅膏菜属

Drosera peltata Sm. ex Willd./ 茅膏菜
（a. Flower/ 花；b. Cauline leaves with sticky, glandular hair/ 茎生叶，被黏性腺毛）

295 Caryophyllaceae 石竹科

Herbs, stems swollen at nodes / 草本，茎节膨大
Leaves simple, opposite, entire / 单叶，对生，全缘
Placentation often free central / 常为特立中央胎座
Embryo strongly curved, testa micro-characteristics various / 胚强烈弯曲，种皮纹饰多样

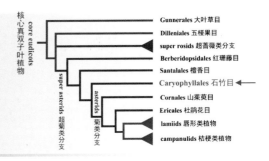

Perennial to annual **herbs**, rarely **small trees**, **subshrubs**, **shrubs** or **woody climbers**; stems and branches usually swollen at nodes. **Leaves** simple, opposite, rarely whorled or alternate; often connate at base; margin usually entire; petioles absent or less often present; stipules present or absent. **Inflorescence** cymes, umbels, panicles or rarely solitary flower. **Flowers** bisexual or rarely unisexual, actinomorphic or rarely zygomorphic; bracteoles 2, sometimes epicalyx. **Perianth** 1- or 2-whorled, sometimes many; usually 4- or 5-merous. **Sepals** free or fused, usually imbricate, or connate into a tube, leaf-like or scarious, persistent, sometimes bracteate below calyx. **Petals** rarely absent, free, often comprising claw and limb; limb entire or split, usually with coronal scales at juncture of claw and limb. **Stamens** usually opposite the sepals or alternate, (2–) 5–10, 2-whorled; filaments free or fused basally, nectary on the inside of fused stamens or external at the base; anthers usually dorsifixed. **Ovary** superior or rarely inferior; carpels fused, up to 7 or 10; locules 1(–5); ovules 1 to many; placentation basal, free central or axile. **Fruit** a capsule or rarely a nut, berry or achene. **Seeds** 1 to numerous; embryos strongly curved; testa granular, striate or tuberculate, rarely smooth or spongy.

多年生至一年生草本，稀小乔木、亚灌木、灌木或木质藤本；茎和小枝结节常膨大。叶为单叶，对生，稀轮生或互生；常基部合生；多全缘；无叶柄或稀存在；托叶有或无。花序聚伞状、伞形、圆锥状或稀为单花。花两性或稀单性，辐射对称或稀两侧对称；小苞片2枚，有时具副萼。花被片1或2轮，有时多数；常4或5基数。花萼离生或合生，常覆瓦状，或合生成管，叶状或干膜质，宿存，有时下部具苞片。花瓣稀缺，离生，常由爪和瓣片组成；瓣片全缘或撕裂状，瓣片与爪结合处常具鳞片。雄蕊常与萼片对生或互生，(2~)5~10枚，2轮；花丝分离或基部合生，蜜腺盘在合生雄蕊内侧或雄蕊基部的外侧；花药常背着。子房上位或稀下位；心皮合生，达7或10枚；1(~5)室；胚珠1枚至多数；基底胎座、特立中央胎座或中轴胎座。果实为蒴果或稀坚果、浆果或瘦果。种子1粒至多数；胚强烈弯曲；种皮颖粒状、具条纹、具瘤，稀光滑或海绵质。

World 81/ca. 2,625, worldwide, mainly in northern and warm temperate regions, Mediterranean, West Asia to West China and Himalayan regions, few in Africa, Oceania and America.
China 31/390.
This area 9/18.

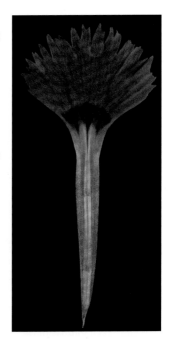

Dianthus plumarius L./ 常夏石竹（Petal comprise claw and limb/ 花瓣由爪和瓣片组成）

全世界共81属约2 625种，世界广布，主要分布于北半球温带和暖温带、地中海地区、亚洲西部至中国西部和喜马拉雅区域，少数在非洲、大洋洲和美洲。

中国产31属390种。

本地区有9属18种。

1. Stipules present/ 有托叶
    2. Leaves obovate or spatulate; abaxial sepals midvein raised into keels/ 叶倒卵形或匙形；花萼背部中脉隆起呈脊 .... **1. *Polycarpon*/ 多荚草属**
    2. Leaves linear; sepals without keel/ 叶线形；花萼不具脊 .... **2. *Spergularia*/ 拟漆姑属**
1. Stipules absent/ 无托叶
    3. Sepals free/ 萼片离生
        4. Capsule teeth as many as styles/ 蒴果裂齿与花柱同数
            5. Petals entire/ 花瓣全缘 .... **3. *Sagina*/ 漆姑草属**
            5. Petals 2-lobed/ 花瓣2裂 .... **4. *Stellaria*/ 繁缕属**
        4. Capsule teeth 2 × as many as styles/ 蒴果顶端裂齿为花柱数2倍
            6. Styles (4 or)5/ 花柱通常5枚，稀4枚
                7. Petals deeply 2-lobed; capsule ovoid, shorter than calyx, 5-valved to middle/ 花瓣2深裂；蒴果卵形，短于萼片，5瓣裂至中部 .... **5. *Myosoton*/ 鹅肠菜属**
                7. Petals 2-lobed to 1/3 their length; capsule cylindric, apex 8–10-toothed/ 花瓣2裂至1/3处；蒴果圆筒形，先端具8~10齿 .... **6. *Cerastium*/ 卷耳属**
            6. Styles (2 or)3/ 花柱通常3枚，稀2枚 .... **7. *Arenaria*/ 无心菜属**
    3. Sepals connate/ 萼片合生
        8. Styles 3 or 5/ 花柱3或5枚 .... **8. *Silene*/ 蝇子草属**
        8. Styles 2/ 花柱2枚 .... **9. *Dianthus*/ 石竹属**

1. *Polycarpon prostratum* (Forssk.) Asch. & Schweinf. ex Asch./ 多荚草（1a. Stipules present/ 具托叶；1b. Abaxial sepals midvein raised into keels/ 花萼背部中脉隆起呈脊）；

2. *Spergularia marina* (L.) Griseb./ 拟漆姑（photo by Zhen ZHANG/ 张振拍摄）；

3. *Sagina japonica* (SW.) Ohwi/ 漆姑草；

4. *Stellaria media* (L.) Vill./ 繁缕；

5. *Myosoton aquaticum* (L.) Moench/ 鹅肠菜；

6. *Cerastium glomeratum* Thuill./ 球序卷耳

7. *Arenaria serpyllifolia* L./ 无心菜；

8. *Silene aprica* Turcz./ 女娄菜；

9. *Dianthus superbus* L./ 瞿麦（photo by XinXin ZHU/ 朱鑫鑫拍摄）

CARYOPHYLLALES. Caryophyllaceae. *Stellaria media* (L.) Vill./ 石竹目 石竹科繁缕

A. Creeping plant/ 平卧植株；

B. Stems pale purplish, with 1 line of hairs/ 茎紫红色，一侧被1列短柔毛；

C. Frontal view of corolla/ 花冠正面观；

D. Vertical profile of corolla/ 花冠纵剖；

E. Flower segregated/ 花部离析；

E_1. Abaxial view of calyx/ 花萼背面观；

E_2. Styles 3/ 花柱3；

E_3. Pedicle with 1 line of hairs/ 花梗一侧被1列毛；

F. Vertical section of young fruit/ 幼果纵切

（Scale/ 标尺：A－B=1 cm；C－D&F=1 mm；E=5 mm）

297 Amaranthaceae 苋科

Usually succulent or halophytic herbs / 多肉质或盐生草本
Bracteoles often dry papery, showy / 小苞片干纸质，显著
Perianth 1-whorled, persistent / 花被片 1 轮，宿存
Fruit usually a dry utricle, or pyxidium / 果实常为胞果或盖果

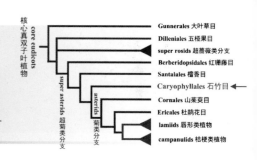

Annual or perennial **herbs**, **subshrubs** to **shrubs**, rarely **woody climbers** or **small trees**, often succulent and/or halophytic; stems sometimes succulent, often swollen nodes; indumentum of vesicular hairs (furfuraceous or farinose), ramified, stellate, rarely of glandular hairs, or glabrous. **Leaves** usually simple, alternate or opposite, reduced in some succulents or rarely absent; margin usually entire or rarely lobed, toothed or pinnatifid; stipules absent. **Inflorescence** a spikes, heads, thyrses, cymes or rarely solitary flower; often bracteate. **Flowers** usually bisexual or unisexual, actinomorphic, disk often present sometimes with appendages; bracteoles often dry papery, showy. **Perianth** 1-whorled, free or basally fused, sepaloid to petaloid, usually imbricate, dry to fleshy, often persistent. **Stamen** opposite the perianths, filaments free or fused into a cup at base; anthers 1 or 2-loculed; introrse or extroese; staminodes 0–5. **Ovary** superior or part-inferior; carpels fused; 1-loculed; ovule 1 (to several); placentation usually basal. **Fruit** usually a dry utricle, or pyxidium, or a fleshy capsule, indehiscent.

一年生或多年生**草本**、**亚灌木至灌木**，稀**木质藤本**或**小乔木**，常肉质，或盐生植物；茎有时肉质，节常膨大；毛被多样，泡状毛（鳞秕状或粉状）、分枝毛或星状毛，稀腺毛，或光滑。**叶**常单叶，互生或对生，部分多肉植物中退化，或稀缺；常全缘，或稀分裂、有锯齿或羽裂；托叶缺。**花序**穗状、头状、聚伞圆锥状、聚伞状或稀为单花；常具苞片。**花**常两性或单性，辐射对称，花盘常存在，有时具附属物；小苞片常干纸质，显著。**花被片** 1 轮，离生或基部合生，萼片状至花瓣状，常覆瓦状，干燥至肉质，常宿存。**雄蕊**与花被对生，花丝分离或基部合生成杯状；花药 1 或 2 室；内向或外向；退化雄蕊 0~5 枚。**子房**上位或半下位；心皮合生；1 室；胚珠 1 枚（至数枚）；常基底胎座。**果实**常为干燥的胞果、盖果或肉质蒴果，不裂。

World 170/2, 300, worldwide, especially in disturbed, saline or arid habitats.
China 54/234.
This area 10/23.

全世界共 170 属 2 300 种，世界广布，尤其是被干扰后的生境或干旱环境。
中国产 54 属 234 种。
本地区有 10 属 23 种。

1. Bracteoles and perianth dry membranous/ 苞片和花被片为干膜质
  2. Leaves alternate/ 叶互生
    3. Ovary with 2 to many ovules/ 子房有 2 至多枚胚珠 .... **1. *Celosia*/ 青葙属**

3. Ovary with 1 ovule/ 子房仅 1 枚胚珠 .... **2. *Amaranthus*/ 苋属**
　2. Leaves opposite/ 叶对生
　　4. Inflorescence heads/ 头状花序 .... **3. *Alternanthera*/ 莲子草属**
　　4. Inflorescence spikes/ 穗状花序 .... **4. *Achyranthes*/ 牛膝属**
1. Bracteoles and perianth not dry membranous/ 苞片和花被片非干膜质
　5. Embryo spiral/ 胚螺旋形
　　6. Bracteoles rudimentary, membranous, scale-like/ 小苞片不发达，膜质，鳞片状 .... **5. *Suaeda*/ 碱蓬属**
　　6. Bracteoles developed, herbaceous or succulent/ 小苞片发达，草质或肉质 .... **6. *Kali*/ 猪毛菜属**
　5. Embryo annular or semi-annular/ 胚环形或半环形
　　7. Flowers borne in axil of succulent bracts, appearing sunken into rachis/ 花着生于肉质的紧密排列的苞腋内，外观似花嵌入花序轴 .... **7. *Salicornia*/ 盐角草属**
　　7. Flowers free from rachis/ 花不嵌入花序轴
　　　8. Plant body usually furfuraceous or glandular/ 植物体常被糠秕状毛或具腺毛
　　　　9. Plants covered with glandular hairs/ 植株具腺毛 .... **8. *Dysphania*/ 腺毛藜属**
　　　　9. Plants covered with furfuraceous/ 植株被糠秕状毛 .... **9. *Chenopodium*/ 藜属**
　　　8. Plant body pubescent/ 植物体有柔毛 .... **10. *Bassia*/ 沙冰藜属 (*Kochia*/ 地肤属)**

1. *Celosia argentea* L./ 青葙；
2. *Amaranthus viridis* L./ 皱果苋；
3. *Alternanthera philoxeroides* (Mart.) Griseb./ 喜旱莲子草；
4. *Achyranthes bidentata* Blume/ 牛膝

5. *Suaeda glauca* (Bunge) Bunge/ 碱蓬；

6. *Kali tragus* (L.) Scop./ 刺沙蓬；

7. *Salicornia europaea* L./ 盐角草；

8. *Dysphania ambrosioides* (L.) Mosyakin & Clemants/ 土荆芥；

9. *Chenopodium album* L./ 藜；

10. *Bassia scoparia* (L.) A.J. Scott/ 地肤

304 Aizoaceae 番杏科

Usually succulent herbs / 常为肉质草本
Stamens usually many, staminodes petaloid / 雄蕊常多数, 退化雄蕊花瓣状
Fruit usually a loculicidal capsule / 多为室背开裂蒴果

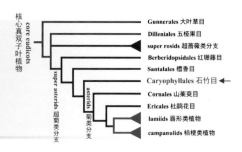

Annual to perennial succulent **herbs**, **subshrubs** or **shrubs**, rarely spiny. **Leaves** fleshy, sometimes reduced, usually simple, opposite or alternate; margin entire; petiole present or absent; stipules absent or a stipule-like sheath present at base of petiole. **Inflorescence** solitary flower, cymes, heads or umbels. **Flowers** bisexual or rarely unisexual, actinomorphic; nectaries separate or in a ring around ovary; sometimes bracteate. **Perianth** (4–) 5 (–8), petals absent but replaced by many petaloid staminodes or perianth internally petaloid and externally sepaloid. **Stamens** usually many, filaments free or fused basally; often staminodes. **Ovary** superior to inferior; carpels fused; locules (1–) 2–20, often 5; ovules 1 to many per locule; placentation usually axile or parietal to basal; stigmas as many as carpels. **Fruit** usually a loculicidal capsule, or rarely a berry or nut.

一年生至多年生肉质**草本**、**亚灌木**或**灌木**, 稀具刺。**叶**肉质, 有时退化, 常单叶, 对生或互生; 全缘; 叶柄有或无; 托叶缺, 或叶柄基部具托叶状鞘。**花序**为单花、聚伞状、头状或伞形。**花**两性, 或稀单性, 辐射对称; 蜜腺分离或成环围绕子房; 有时具苞片。**花被片** (4~) 5 (~8) 枚, 花瓣缺, 退化雄蕊花瓣状, 或内侧花被花瓣状, 外侧花被萼片状。**雄蕊**常多数, 花丝分离或基部合生; 常具退化雄蕊。**子房**上位至下位; 心皮合生; (1~) 2~20 室, 常 5 室; 胚珠每室 1 枚至多数; 常中轴胎座或侧膜胎座至基底胎座; 柱头与心皮同数。**果实**多为室背开裂蒴果, 或稀为浆果或坚果。

World 135/1,800, widely distributed in subtropical arid regions, especially in South Africa, also in Australia, Western America, pantropical regions.
China 3/3.
This area 2/2.

全世界共 135 属 1 800 种, 广布于亚热带干旱地区, 尤其是南非、澳大利亚、美洲西部, 泛热带也有分布。中国产 3 属 3 种。
本地区有 2 属 2 种。

1. Flowers epigynous; fruit a nut/ 花上位, 果为坚果 .... 1. *Tetragonia*/ 番杏属
1. Flowers perigynous; fruit a capsule/ 花周位, 果为蒴果 .... 2. *Sesuvium*/ 海马齿属

1. *Tetragonia tetragonoides* (Pall.) Kuntze/ 番杏;
2. *Sesuvium portulacastrum* (L.) L./ 海马齿
（2a. Capsule/ 蒴果；2b. Flower/ 花）

305 Phytolaccaceae 商陆科

Leaves simple, entire / 单叶，全缘
Perianth 1-whorled, persistent / 花被片 1 轮，宿存
Stamens persistent / 雄蕊宿存
Carpels usually free / 心皮常离生

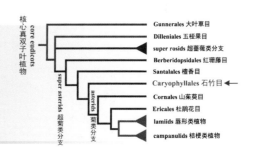

Herbs, **shrubs** or rarely **trees**, mostly glabrous. **Leaves** simple, sometimes gland-dotted, alternate; margin entire; petioles present; sometimes stipules, often spiny. **Inflorescence** terminal, axillary, or leaf-opposed; cymes, racemes, spikes, panicles, thyrses or rarely solitary flower. **Flowers** bisexual or unisexual, actinomorphic or slightly zygomorphic. **Perianths** 1-whorled, 4 or 5, small, imbricate, free or rarely fused, persistent in fruit. **Stamens** 1–2-whorled, inserted on a fleshy disk; filaments free or fused, persistent in fruit; anthers 2-loculed, dorsifixed, introrse; dehiscing longitudinally; sometimes staminodes petaloid. **Ovary** superior or part-inferior; carpels free, or fused; locules 1–16; ovule 1 per locule; placentation basal or axile; styles persistent, as many as carpels; stigmas 1–16. **Fruit** a berry, drupe, or capsule.

草本、灌木或稀乔木，多光滑。**叶**为单叶，有时具腺点，互生；全缘；具叶柄；有时具托叶，常刺状。**花序**顶生、腋生或与叶对生；聚伞状、总状、穗状、圆锥状、聚伞圆锥状或稀为单花。**花**两性或单性，辐射对称或稍两侧对称。**花被片** 1 轮，4 或 5 枚，小，覆瓦状，离生或稀合生，果期宿存。**雄蕊** 1~2 轮，嵌生于肉质花盘上；花丝分离或合生，果期宿存；花药 2 室，背着，内向；纵裂；有时退化雄蕊花瓣状。**子房**上位或半下位；心皮离生或合生；1~16 室；胚珠每室 1 枚；基底胎座或中轴胎座；花柱宿存，与心皮同数；柱头 1~16 个。**果实**为浆果、核果或蒴果。

World 5/32, tropical and subtropical regions, especially in tropical America and South Africa.
China 2/5.
This area 1/2.

全世界共 5 属 32 种，分布于热带和亚热带，尤其是热带美洲和南非，少数至亚热带边缘及温带地区。
中国产 2 属 5 种。
本地区有 1 属 2 种。

Phytolacca L. 商陆属

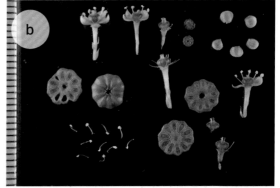

Phytolacca americana L./ 垂序商陆

（a. Inflorescence/ 花序；b. Flower segregated/ 花离析）

308 Nyctaginaceae 紫茉莉科

Stems with swollen nodes / 茎节膨大
Bracts involucral, large and colored / 总苞大而艳丽
Perianth 1-whorled, constricted beyond the ovary / 花被片 1 轮，子房以上缢缩
Achene enclosed by persistent perianth / 瘦果包在宿存花被内

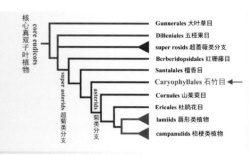

Trees, **shrubs**, **woody climbers**, or annual or perennial **herbs**; stems with swollen nodes in herbs, spines. **Leaves** simple, alternate, opposite or whorled; petioles present or absent; stipules absent. **Inflorescence** solitary flower, cymes, panicles, spikes or umbels; bracts involucral, sometimes large and brightly colored. **Flowers** bisexual or sometimes unisexual (plants dioecious), actinomorphic or zygomorphic. **Perianth** 1-whorled, constricted beyond the ovary, base persistent, closely enclosing ovary which appears inferior; limb petaloid beyond constriction, tubular, funnelform, or campanulate, apex 5–10-lobed, lobes plicate or valvate in bud, persistent or caducous. **Stamen** filaments free or attached to perianth, sometimes dimorphic; anthers 2-loculed, dehiscing longitudinally. **Ovary** superior; carpel 1; 1-loculed; ovule 1; placentation basal; style 1, long; stigma globose. **Fruit** an achene, sometimes with drupe-like anthocarp.

乔木、**灌木**、**木质藤本**，或一年生或多年生**草本**；草本茎节常膨大，具刺。**叶**为单叶，互生、对生或轮生；叶柄有或无；托叶缺。**花序**为单花、聚伞状、圆锥状、穗状或伞形；苞片呈总苞，有时大型，艳丽。花两性或有时单性（雌雄异株），辐射对称或两侧对称。**花被** 1 轮，子房以上缢缩，基部宿存，紧贴子房似下位；缢缩部分以上瓣片呈花瓣状，花被筒管状、漏斗状或钟状，先端 5~10 裂，裂片在芽中皱褶或镊合状，宿存或早落。**雄蕊**花丝分离或贴生于花被片，有时二型；花药 2 室，纵裂。**子房**上位；心皮 1 枚；1 室；胚珠 1 枚；基底胎座；花柱 1 个，长；柱头球形。**果实**为瘦果，有时为核果状掺花果。

World 30/300, tropics and subtropics.
China 6/13.
This area 3/3.

全世界共 30 属 300 种，分布于热带和亚热带地区。
中国产 6 属 13 种。
本地区有 3 属 3 种。

1. Shrubs or spiny vines; leaves often alternate/ 灌木或具刺藤本；叶常互生 ....**1. *Bougainvillea*/ 叶子花属**
1. Herbs or subshrubs; leaves always opposite/ 草本或亚灌木；叶常对生
   2. Flowers enclosed by a calyx-like involucre; fruit without sticky glands/ 花为萼状总苞包被；果实无黏腺 ....**2. *Mirabilis*/ 紫茉莉属**
   2. Flowers in thyrse; fruit with sticky glands/ 聚伞圆锥花序；果实具黏腺 ....**3. *Boerhavia*/ 黄细心属**

1. *Bougainvillea glabra* Choisy/ 光叶子花；
2. *Mirabilis jalapa* L./ 紫茉莉；
3. *Boerhavia diffusa* L./ 黄细心

309 Molluginaceae 粟米草科

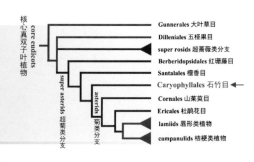

Leaves simple, often in a basal rosette / 单叶，常基生莲座状
Leaves margin entire, stipules absent / 叶全缘，托叶缺
Perianths 5, 1-whorled / 花被片 5 枚，1 轮
Curved embryo / 胚弯曲

Perennial or annual **herbs** to **subshrubs**, sometimes succulent. **Leaves** simple, often in a basal rosette or in pseudowhorls on stems; margin entire; stipules absent or membranous; hairs stellate or glandular. **Inflorescence** cymes or solitary flower. **Flowers** bisexual or unisexual, actinomorphic. **Perianth** 5, rarely 4, 1-whorled or rarely 2-whorled, free or ±fused into a tube at base, imbricate, persistent. **Stamens** 3–5 or many, filaments free or fused at base in bundles; anthers dehiscing longitudinally; staminodes 5(–8). **Ovary** superior; carpels free, basally fused or 1; locules 1–10; ovules 1 to many per locule; stigmas as many as locules; placentation axile. **Fruit** usually a loculicidal capsule. **Seeds** with curved embryo.

多年生或一年生**草本**至**亚灌木**，有时肉质。**叶**为单叶，常基生莲座状或于茎上呈假轮生；全缘；托叶缺或膜质；具星状毛或腺毛。**花序**聚伞状或为单花。**花**两性或单性，辐射对称。**花被片**5枚，稀4枚，1轮或稀2轮，离生或多少在基部合生成筒，覆瓦状，宿存。**雄蕊**3~5枚或多数，花丝分离或基部合生成束；花药纵裂；退化雄蕊5(~8)枚。**子房**上位；心皮离生，基部合生或仅1枚；1~10室；胚珠1枚至多数；柱头与子房室同数；中轴胎座。**果实**多为室背开裂的蒴果。**种子**具弯曲的胚。

World 13/120, tropical and subtropical arid regions, mainly in southern Africa.
China 2/6.
This area 1/1.

全世界共13属120种，分布于热带和亚热带干旱地区，主要分布于非洲南部。
中国产2属6种。
本地区有1属1种。

Mollugo L. 粟米草属

Mollugo stricta L./ 粟米草 (a. Plants/ 植株；b. Fruits/ 果实；c. Loculicidal capsule/ 室背开裂蒴果)

312 Basellaceae 落葵科

Fleshy twining herbs, glabrous / 肉质缠绕草本，光滑
Leaves simple, alternate, entire, estipulate / 单叶互生，全缘，无托叶
Bracteoles petaloid, persistent / 小苞片花瓣状，宿存
Utricle / 胞果

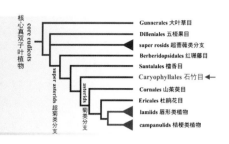

Perennial herbaceous **climbers** or **herbs**, usually fleshy, rhizomatous, glabrous. **Leaves** basally aggregated, simple, alternate; margin entire or rarely toothed; usually petiolate; stipules absent. **Inflorescence** spikes, racemes, panicles or cymes; with 3 bracts, thin. **Flowers** small, bisexual or rarely unisexual, actinomorphic; bracteoles 2 or 2+2, petaloid, persistent, ± enlarging in fruit. **Perianth** 5, free or ±fused at basally, persistent in fruit. **Stamens** 5, opposite to petals; filaments inserted on perianth; anthers dorsifixed. **Ovary** superior, carpels 3, fused; 1-loculed; ovule 1; placentation basal; style 1; stigmas 3. **Fruit** a utricle, often surrounded by persistent bracteoles and perianth, dry or fleshy.

多年生草质**藤本**或**草本**，常肉质，具根状茎，光滑。**叶**基部簇生，单叶，互生；全缘或稀具齿；常具叶柄；托叶缺。**花序**穗状、总状、圆锥状或聚伞状；苞片3枚，薄。**花**小，两性或稀单性，辐射对称；小苞片2枚或2+2枚，花瓣状，宿存，果期多少增大。**花被片**5枚，离生或至少基部合生，果期宿存。**雄蕊**5枚，与花瓣对生；花丝着生于花被；花药背着。**子房**上位，心皮3枚，合生；1室；胚珠1枚；基底胎座；花柱1个；柱头3个。**果实**为胞果，常有宿存小苞片和花被包被，干燥或肉质。

World 4/19, Asia, Africa and tropical Latin America.
China 2/3. (Introduced and naturalized)
This area 2/2.

全世界共4属19种，分布于亚洲、非洲和拉丁美洲热带地区。
中国产2属3种。（引入逸生）
本地区有2属2种。

1. Flowers sessile/ 花无柄 .... **1. *Basella*/ 落葵属**
1. Flowers pedicellate/ 花具柄 .... **2. *Anredera*/ 落葵薯属**

1. *Basella alba* L./ 落葵；
2. *Anredera cordifolia* (Ten.) Steenis / 落葵薯

314 Talinaceae 土人参科

Succulent herbs / 肉质草本
Leaves simple, entire, estipulate / 单叶，全缘，无托叶
Stem glabrous / 茎光滑
Ovary superior, placentation free central / 子房上位，特立中央胎座

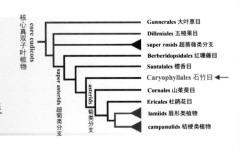

Herbs or **subshrubs**, usually succulent, rarely **small trees**; often with thick roots; stems erect, glabrous. **Leaves** simple, alternate or partly opposite; margin entire or rarely slightly toothed; sessile or shortly petiolate; stipules absent. **Inflorescence** axillary solitary flower or terminal cymes or panicles. **Flowers** small; bisexual or unisexual, actinomorphic; bracteoles 2, deciduous or persistent. **Perianth** 5 (–10), free or basally fused, imbricate, petaloid. **Stamen** filaments free, or attached to base of perianth; anthers dorsifixed, extrorse. **Ovary** superior; carpels fused; 1–5-loculed; ovules 1 to many; placentation free central. **Fruit** a loculicidal capsule or berry-like. **Seeds** black, glossy.

草本或亚灌木，常肉质，稀小乔木；根系多粗壮；茎直立，光滑。叶为单叶，互生或部分对生；全缘或稀稍具齿；无柄或具短柄；托叶缺。花序为单花腋生或顶生聚伞状或圆锥状。花小；两性或单性，辐射对称；小苞片 2 枚，早落或宿存。花被片 5(~10) 枚，离生或基部合生，覆瓦状排列，花瓣状。雄蕊花丝分离，或贴生于花被片基部；花药背着，外向。子房上位；心皮合生；1~5 室；胚珠 1 至多枚；特立中央胎座。果实为室背开裂蒴果或浆果。种子黑色，具光泽。

World 2/27, America, Africa mainland and Madagascar.
China 1/2. (Naturalized)
This area 1/1.(Naturalized)

全世界共 2 属 27 种，分布于美洲、非洲大陆和马达加斯加。
中国产 1 属 2 种。（归化）
本地区有 1 属 1 种。（归化）

***Talinum* Adans. 土人参属**

Talinum paniculatum (Jacq.) Gaerth./ 土人参（a. Flowers/ 花；b. Plants/ 植株）

315 Portulacaceae 马齿苋科

Fleshy herbs / 肉质草本
Petals early wilting / 花瓣早萎缩
Placentation usually free central / 常为特立中央胎座
Capsule circumscissile / 蒴果盖裂

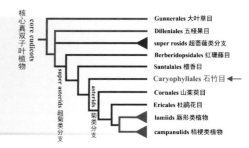

Perennial or annual **herbs**, rarely **shrub-like**, erect or creeping, usually fleshy. **Leaves** simple, alternate or opposite; margin entire; usually sessile, flat or terete; stipules absent but often scales or hairs in leaf axils. **Inflorescence** terminal, solitary or clustered flowers, with pedicel or not, subtended by an involucre of leaves. **Flowers** bisexual, actinomorphic; bracteoles 2, large or 2+2. **Sepals** usually 2. **Petals** 4 or 5, free or shortly connate at base, early wilting. **Stamens** 4 to numerous, adnate to base of petals; anthers dorsifixed, introrse. **Ovary** inferior to part-inferior; carpels fused; 1-loculed; ovules many; placentation usually free central, rarely basal; stigma 2–9-lobed. **Fruit** a circumscissile capsule (pyxidium), dehiscing with perianth remnants.

多年生或一年生**草本**，稀**灌木状**，直立或匍匐，常肉质。**叶**为单叶，互生或对生；全缘；常无柄，或扁平或圆柱形；托叶缺但叶腋常具鳞片或毛被。**花序**顶生，单花或簇生，具花梗或无，包于叶状总苞内。**花**两性，辐射对称；小苞片2枚，大型或2+2枚。**花萼**常2枚。**花瓣**4或5枚，离生或基部稍合生，早萎缩。**雄蕊**4枚至多数，贴生于花瓣基部；花药背着，内向。**子房**下位至半下位；心皮合生；1室；胚珠多数；常特立中央胎座，稀基底胎座；柱头2~9裂。**果实**为周裂蒴果（盖果），开裂时有花被残存。

World 1/116, tropical and subtropical, few in temperate regions.
China 1/5.
This area 1/1.

全世界共1属116种，分布于热带与亚热带，少量至温带地区。
中国产1属5种。
本地区有1属1种。

Portulaca L. 马齿苋属

Portulaca oleracea L./ 马齿苋（a. Pyxidium/ 盖果；b. Plants/ 植株）

317 Cactaceae 仙人掌科

Succulent plants / 多浆植物
Leaves often absent / 叶常退化至缺
Perianths free, numerous / 花被片多数，分离
Ovary inferior, 1-loculed, ovules many / 子房下位，1 室，胚珠多数

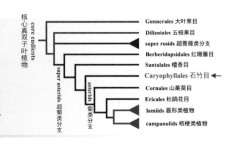

Succulent perennials, **shrubs**, **trees**, terrestrial or epiphytic; stems jointed, terete, globose, flattened, or fluted, mostly leafless and variously spiny. **Leaves** alternate, flat or subulate to terete, vestigial, or entirely absent; spines, glochids, and flowers always arising from cushion-like, axillary areoles. **Inflorescence** solitary flower or cymes; bracteate. **Flowers** bisexual, rarely unisexual; actinomorphic or occasionally zygomorphic; usually a short to elongated hypanthium. **Perianth** free, segments usually numerous, in a sepaloid to petaloid series. **Stamens** numerous; anthers 2-loculed; introrse, dehiscing longitudinally. **Ovary** inferior, rarely superior to part-inferior, 1-loculed, ovules many; placentation parietal or basal; style 1; stigmas 2 to numerous, papillate, rarely 2-lobed. **Fruit** juicy or dry berry or capsule. **Seeds** usually numerous, often arillate or strophiolate; embryo curved.

多年生多肉类植物，**灌木、乔木**，地生或附生；茎具关节，圆柱形，球形，扁平或具沟槽，大部分无叶，具各类刺。**叶**互生，扁平或钻形至圆柱形，退化或完全缺失；刺、芒刺和花常从腋生垫状小窠发出。**花序**为单花或聚伞状；具苞片。**花**两性，稀单性；辐射对称或偶两侧对称；常具短缩或伸长的被丝托。**花被片**分离，被片多数，萼片状至花瓣状连续。**雄蕊**多数；花药 2 室；内向，纵裂。**子房**下位，稀上位至半下位，1 室，胚珠多数；侧膜胎座或基底胎座；花柱 1 个；柱头 2 至多数，具乳突，稀 2 裂。**果实**为多汁或干燥的浆果或蒴果。**种子**多数，常具假种皮或种阜；胚弯曲。

World ca. 124/1, 500, tropical America, subtropical regions, sometimes in Alpine regions.
China 60+/600+ (Cultivated).
This area 1/1.

全世界共约 124 属 /1 500 种，分布于热带美洲、亚热带地区，有时高山地区也有。
中国栽培 60 余属 600 余种。
本地区有 1 属 1 种。

Opuntia Mill. 仙人掌属

Opuntia dillenii (Ker Gawl.) Haw./ 仙人掌
(a. Fruits/ 果实；b. Flowers/ 花)

318 Nyssaceae 蓝果树科

Leaves simple, estipulate / 单叶，托叶缺
Inflorescence often in heads / 花序常呈头状
Flowers 4 or 5-merous, actinomorphic / 花 4 或 5 基数，辐射对称
Ovary inferior, 1 ovule per locule / 子房下位，每室 1 胚珠

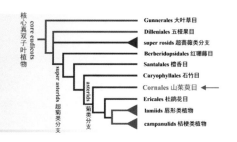

Trees or **shrubs**; sometime resinous. **Leaves** simple, alternate or opposite; margin entire or rarely undulate; petioles present; stipules absent. **In monoecious plants**: inflorescence paniculate cymes, terminal and axillary. **In dioecious or polygamo-monoecious plants**: male ones in heads, racemes, or umbels; female ones solitary or in 2–12-flowered heads. **Flowers** 4 or 5-merous, ± actinomorphic; bracteate or not. **Sepals** tube adnate to ovary; lobes small or obsolete. **Petals** usually 5–10 (or absent), free, valvate. **Stamens** in male flowers usually 10 or more in 2-whorled, ± free, around a nectariferous disk; anthers dorsifixed, dehiscing longitudinally. **Ovary** inferior, 1-loculed or 6–10-loculed, 1 ovule per locule; style 1. **Fruit** a drupe, fleshy or hard when dry.

乔木或**灌木**；有时具树脂。**叶**为单叶，互生或对生；全缘或稀波浪状；具叶柄；托叶缺。**雌雄同株植物**：花序聚伞圆锥形，顶生或腋生。**雌雄异株或杂性同株植物**：雄花序头状、总状或伞形；雌花序单花或 2~12 朵花组成头状花序。**花** 4 或 5 基数，多少辐射对称；苞片存在或缺。**花萼**萼筒贴生于子房；裂片小或退化。**花瓣**常 5~10 枚（或缺），分离，镊合状。**雄蕊**（雄花中）常 10 或更多，成 2 轮，多少分离，围绕蜜腺盘；花药背着，纵裂。**子房**下位，1 室或 6~10 室，胚珠每室 1 枚；花柱 1 个。**果实**为核果，肉质或干后坚硬。

World 5/37, mainly in East Asia, also Indo-Malaysia and eastern North America.
China 5/14.
This area 1/1.

全世界共 5 属 37 种，主要分布于东亚、印度 – 马来西亚和北美东部亦有分布。
中国产 5 属 14 种。
本地区有 1 属 1 种。

Camptotheca Decne. 喜树属

Camptotheca acuminata Decne./ 喜树

320 Hydrangeaceae 绣球科

Woody with peeling bark / 木本，具剥落状树皮
Leaves simple and opposite / 单叶对生
Flowers dimorphic / 花二型
Ovary inferior, often ribbed / 子房下位，常具棱

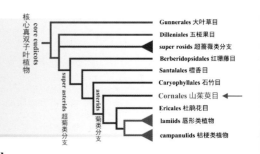

Perennial **shrubs** to **subshrubs**, **herbs** or woody **climbers**, or **small trees**; bark often peeling. **Leaves** simple, opposite or alternate; margin entire, toothed or rarely lobed; petioles present; stipules absent. **Inflorescence** cymes, heads, corymbs, panicles or racemes. **Flowers** 4- or 5-merous; bisexual or rarely unisexual, actinomorphic or sometimes zygomorphic, sometimes dimorphic fertile and sterile flowers with enlarged petaloid calyx. **Sepals** small, fused or rarely free, valvate, the lobes often reduced but enlarged in sterile flowers. **Petals** fused or less often free. **Stamens** 2-whorled to many; filaments distinct or slightly connate; anthers usually basifixed. **Ovary** inferior or rarely part-inferior or superior, often ribbed; carpels 2–5, fused; locules 1 to several; ovules 1 to many per locule; placentation parietal to axile; stigmas 2–5, usually elongated. **Fruit** a septicidal or loculicidal capsule or rarely berry.

多年生**灌木**至**亚灌木**、**草本**或木质**藤本**，或**小乔木**；树皮常剥落。**叶**为单叶，对生或互生；全缘、具齿或稀分裂；叶柄存在；托叶缺。**花序**聚伞状、头状、伞房状、圆锥状或总状。**花** 4 或 5 基数；两性或稀单性，辐射对称或有时两侧对称，有时分可育与不可育两态型，不育花具增大花瓣状萼片。**花萼**小，合生或稀离生，镊合状，裂片常退化，在不育花中增大。**花瓣**合生或偶离生。**雄蕊** 2 轮至多数；花丝分离或稍合生；花药常背着。**子房**下位或稀半下位或上位，常具棱；心皮 2~5 枚，合生；1 至数室；胚珠每室 1 至多枚；侧膜胎座至中轴胎座；柱头 2~5 个，常伸长。**果实**为室间开裂或室背开裂蒴果，或稀为浆果。

World 17/190, Northern temperate to subtropical regions.
China 11/125.
This area 2/4.

全世界共 17 属 190 种，分布于北温带至亚热带地区。
中国产 11 属 125 种。
本地区有 2 属 4 种。

1. Filaments flat; calyx lobes never enlarged and petaloid/花丝扁平；萼片绝不增大成花瓣状 .... **1.** *Deutzia*/ 溲疏属
1. Filaments linear; calyx lobes sometimes enlarged and petaloid/ 花丝丝状；萼片有时增大成花瓣状 .... **2.** *Hydrangea*/ 绣球属

CORNALES. Hydrangeaceae. *Deutzia glauca* W.C. Cheng (Cultivated)/ 山茱萸目 绣球科 黄山溲疏（栽培）

A. Flowering branch/ 花枝
B₁. Stamens/ 雄蕊
B₂. Androecium segregated/ 雄蕊群离析
C. Cross section of ovary/ 子房横切面
D. Ventral view of disk and calyx lobes/ 花盘和花萼裂片腹面观
E. Perianth segregated/ 花被片离析
F. Vertical section of flower/ 花纵切面
G. Lateral view of pistil/ 雌蕊侧面观
(Scale/ 标尺：A 1 cm; B₁–G 1 mm)

1. *Deutzia crenata* Siebold & Zucc./ 齿叶溲疏；

2. *Hydrangea robusta* Hook. f. & Thomson/ 粗枝绣球

324　Cornaceae 山茱萸科

Leaves simple, estipulate / 单叶，托叶缺
Flowers bisexual, actinomorphic / 花两性，辐射对称
Ovary inferior; ovule 1 per locule / 子房下位；胚珠每室 1 枚
Drupe with persistent calyx and disk / 核果具宿存的花萼和花盘

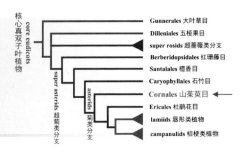

Shrubs, **trees**, or **herblike shrubs**, sometimes spiny. **Leaves** usually alternate, sometimes opposite, simple; petioles present; stipules absent. **Inflorescence** cymes, panicles, corymbs, umbels, or heads, terminal, or axillary; often bracteate. **Flowers** bisexual, actinomorphic; usually 4- or 5-merous. **Sepals** 4–10, fused, lobes toothlike or reduced to absent. **Petals** as many as sepal lobes, free or fused at base; valvate. **Stamens** as many as petals or 2–4 × as many, free, arising from an enlarged disk; filaments filiform or awn-shaped, longer than style; anthers 2-loculed, dehiscing longitudinally. **Ovary** inferior; locule 1 or 2; ovule 1 per locule; style and stigma simple, or 2- or 3-lobed. **Fruit** a drupe, crowned with persistent calyx and disk; stones surface sometimes ribbed.

灌木、乔木或草本状灌木，有时多刺。**叶**互生，偶对生，单叶；具叶柄；托叶缺。**花序**聚伞状、圆锥状、伞房状、伞形或头状，顶生或腋生；常具苞片。**花**两性，辐射对称；多 4 或 5 基数。**花萼** 4~10 枚，合生，萼裂片牙齿状或退化至缺失。**花瓣**与萼裂片同数，离生或基部合生；镊合状。**雄蕊**与花瓣同数，或为 2~4 倍，分离，自增大的花盘上伸出；花丝丝状或芒状，长于花柱；花药 2 室，纵裂。**子房**下位；1 或 2 室；胚珠每室 1 枚；花柱与柱头不分枝，或 2 或 3 裂。**果实**为核果，冠以宿存的花萼及花盘；果核有时具棱。

World 2/85, scattered worldwide.
China 2/36.
This area 2/6.

全世界共 2 属 85 种，全球散布。
中国产 2 属 36 种。
本地区有 2 属 6 种。

1. Leaves alternate, base often asymmetric/ 叶互生，叶基常不对称 .... **1. *Alangium*/ 八角枫属**
1. Leaves opposite, base often symmetric / 叶对生，叶基常对称 .... **2. *Cornus*/ 山茱萸属**

1. *Alangium platanifolium* (Siebold & Zucc.) Harms/ 瓜木； 2. *Cornus walteri* Wangerin/ 毛梾

325 Balsaminaceae 凤仙花科

Herbs, ± succulent; stems ± translucent / 草本，多少肉质；茎多少透明
Leaves simple, often gland-tipped / 单叶，叶齿常具腺状小尖
Flowers bisexual, zygomorphic, spurred / 花两性，两侧对称，有距
Fruit a fleshy explosive dehiscent capsule / 肉质弹裂蒴果

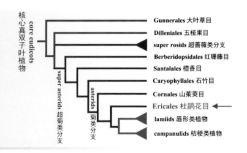

Perennial or rarely annual **herbs**; succulent to woody, rarely epiphytic or subshrubs, or very rarely semi-aquatic; stems ± translucent, often rooting at lower nodes. **Leaves** simple, alternate, opposite or whorled; margin crenate or toothed, often gland-tipped; petioles rarely absent, often with glands; stipules usually absent. **Inflorescence** axillary solitary flower, racemes or cymes; bracteate. **Flowers** bisexual, usually zygomorphic, often resupinate; sometimes bracteoles. **Sepals** free, 3 often with 2 reduced lateral sepals, 1 nectar tipped spur, often petaloid. **Petals** 5, free, upper petal (standard) flat or cucullate, small or large, often crested abaxially, lateral petals free or united in pairs (wing). **Stamens** 5, alternating with petals, connate or nearly so into a ring surrounding ovary and stigma, falling off in one piece before stigma mature; filaments short, flat with a scale-like appendage inside; anthers 2-loculed, connivent, opening by a slit or pore, introrse. **Ovary** superior; carpels fused; locules 4 or 5; ovules 2 to many per locule; placentation axile; style 1 or absent; stigmas 1 or 5. **Fruit** a fleshy explosive dehiscent capsule or drupe to berry-like.

多年生或稀一年生**草本**；肉质至木质，稀附生或半灌木，极稀半水生；茎多少透明，下部节常生根。**叶**为单叶，互生、对生或轮生；有锯齿或牙齿，常具腺状突尖；叶柄稀缺，常具腺体；托叶常缺。**花序**腋生，单花、总状或聚伞花序；具苞片。**花**两性，常两侧对称，多倒置；有时具小苞片。**花萼**离生，3 枚，侧生 2 枚退化，下方 1 枚有具蜜腺的距，常花瓣状。**花瓣** 5 枚，离生，上方花瓣（旗瓣）扁平或兜状，小或大，背面常具脊，侧方花瓣离生或两两联合（翼瓣）。**雄蕊** 5 枚，与花瓣互生，合生或几成环状围绕子房和柱头，柱头成熟前整片凋落；花丝短，扁平，内部具鳞片状附属物；花药 2 室，靠合，纵裂或孔裂，内向。**子房**上位；心皮合生；4 或 5 室；胚珠每室 2 至多枚；中轴胎座；花柱 1 个或缺；柱头 1 或 5 个。**果实**为肉质弹裂蒴果，或核果至浆果状核果。

World 2/ca. 1, 001, temperate North America, temperate to tropical Asia, Sub-Saharan Africa, Madagascar and the Seychelles.

China 2/271.

This area 1/1 (Naturalized).

全世界共 2 属约 1 001 种，分布于北美温带，亚洲温带至热带，非洲撒哈拉以南，马达加斯加和塞舌尔。

中国产 2 属 271 种。

本地区有 1 属 1 种（归化）。

Impatiens L. 凤仙花属

Impatiens balsamina L./ 凤仙花

332 Pentaphylacaceae 五列木科

Leaves simple, alternate, estipulate / 单叶互生，无托叶
Flowers 5-merous, usually < 2 cm in diam. / 花 5 基数，直径常小于 2 cm
Style persistent / 花柱宿存
Embryo U-shaped / 胚 U 形

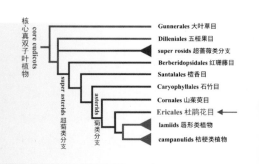

Shrubs or **trees**. **Leaves** sometimes with black dots below, simple, alternate; margin entire or toothed, sometimes inrolled; petioles present; stipules absent. **Inflorescence** axillary, usually arranged into pseudospikes or pseudoracemes along branchlets below apex, or solitary to several in a cluster. **Flowers** 5-merous; usually < 2 cm in diam., rarely up to 5 cm, bisexual or sometimes unisexual, actinomorphic; bracteoles 2, persistent. **Sepals** 5, free or fused, imbricate, persistent. **Petals** 5, basally fused or rarely free, imbricate. **Stamens** 5, opposite the sepals, filaments free, fused and/or attached to petals; anthers 2-loculed, sometimes locellate; basifixed, introrse, hairy. **Ovary** superior or less often part-inferior; carpels fused; loculed (1–) 2–5 (–10); ovules 1 to many per locule; placentation axile, rarely apical or parietal; style 1, persistent; stigmas 1–5. **Fruit** a berry, drupe or loculicidal capsule. **Seeds** with U-shaped embryo.

灌木或**乔木**。叶下部有时具黑色腺点，单叶，互生；全缘或具齿，偶内卷；叶柄存在；托叶缺。**花序**腋生，常在小枝上排列成假穗状或假总状，或单花至数朵簇生。**花** 5 基数；宽常小于 2 cm，稀可达 5 cm，两性或有时单性，辐射对称；小苞片 2 枚，宿存。**花萼** 5 枚，离生或合生，覆瓦状，宿存。**花瓣** 5 枚，基部合生或稀离生，覆瓦状。**雄蕊** 5 枚，与花萼对生，花丝分离、合生或贴生于花瓣；花药 2 室，有时具分格；背着，内向，被毛。**子房**上位或偶半下位；心皮合生；(1~) 2~5 (~10) 室；胚珠每室 1 至多枚；中轴胎座，稀顶生胎座或侧膜胎座；花柱 1 个，宿存；柱头 1~5 个。**果实**为浆果、核果或室背开裂的蒴果。种子具 U 形胚。

World 12/ca. 350, tropical and subtropical Africa, tropical America, Southeast Asia and Pacific Islands.
China 7/130.
This area 4/8.

全世界共 12 属约 350 种，分布于热带和亚热带非洲，热带美洲，东南亚和太平洋诸岛。
中国产 7 属 130 种。
本地区有 4 属 8 种。

1. Flowers larger, ca. 2 cm in diam./ 花较大，直径约 2 cm .... **1. *Ternstroemia*/ 厚皮香属**
1. Flowers less than 1 cm in diam./ 花较小，直径不超过 1 cm
   2. Flowers unisexual, plants dioecious/ 花单性，雌雄异株 .... **2. *Eurya*/ 柃属**
   2. Plants bisexual/ 花两性

3. Terminal bud pubescent/ 顶芽有毛 .... **3. *Adinandra*/ 杨桐属**
3. Terminal bud glabrescent/ 顶芽无毛 .... **4. *Cleyera*/ 红淡比属**

1. *Ternstroemia gymnanthera* (Wight & Arn.) Bedd./ 厚皮香;
2. *Eurya japonica* Thunb./ 柃木 （2a. Female flowers/ 雌花; 2b. Male flowers/ 雄花）;
3. *Adinandra millettii* (Hook. & Arn.) Benth. & Hook. f. ex Hance/ 杨桐;
4. *Cleyera japonica* Thunb./ 红淡比

334 Ebenaceae 柿科

Woody, with pseudoterminal bud/ 木本，具假顶芽
Leaves simple, entire, estipulate / 单叶全缘，托叶缺
Flowers unisexual, actinomorphic / 单性花，辐射对称
Sepals enlarged in fruit / 萼片果时增大

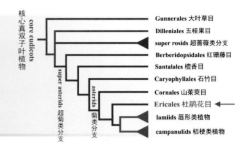

Trees or **shrubs**; usually with pseudoterminal bud; occasionally with spine-tipped branchlets. **Leaves** simple, often alternate or rarely opposite to whorled; margin entire; petioles usually present; stipules absent. **Inflorescence** usually axillary cymes, solitary flower, panicles or racemes. **Flowers** usually unisexual (plants usually dioecious, sometimes monoecious), or functionally unisexual; actinomorphic; female flowers often solitary, axillary, without stamens; male flowers often in clusters; bracteolate. **Sepals** fused, 3–7-lobed, imbricate or valvate, persistent and often enlarged in female or bisexual flowers. **Petals** fused, 3–7-lobed, urceolate, campanulate, or tubular, lobes contorted or valvate, caducous. **Stamens** 2-whorled, opposite the sepals, attached to the petals, often unequal; usually 2–4× as many as petal lobes; filaments free or united in pairs; anthers basifixed, introrse; usually staminodes in female flowers. **Ovary** superior or inferior; carpels fused, usually pistillodes in male flowers; locules 3–10(–16); ovules 1–2 per locule; placentation apical or axile; styles 2–5(–8); stigmas 2–8. **Fruit** often a fleshy berry. **Seeds** usually flat and oblong, asymmetric.

乔木或**灌木**；常具假顶芽；小枝先端偶为刺尖。**叶**为单叶，常互生或稀对生至轮生；全缘；叶柄常存在；托叶缺。**花序**常为腋生聚伞状、单花、圆锥状或总状。**花**常单性（多雌雄异株，偶雌雄同株），或为功能性单性花；辐射对称；雌花常为单花，腋生，无雄蕊；雄花常簇生；具小苞片。**花萼**合生，3~7裂，覆瓦状或镊合状，雌花或两性花中宿存并花后增大。**花瓣**合生，3~7裂，坛状、钟状或管状，裂片旋转状或镊合状，早落。**雄蕊** 2轮，与花萼对生，贴生于花瓣，常不等大；多为花瓣裂片的2~4倍；花丝分离或两两联合；花药基着，内向；雌花中常具退化雄蕊。**子房**上位或下位；心皮合生，雄花中常有退化雌蕊；3~10(~16)室；胚珠每室1~2枚；顶生胎座或中轴胎座；花柱2~5(~8)个；柱头2~8个。**果实**常为肉质浆果。**种子**多扁椭圆形，不对称。

World 4/548, pantropical, few in temperate.
China 1/60.
This area 1/3.

全世界共4属548种，泛热带分布，少数至温带。
中国产1属60种。
本地区有1属3种。

Diospyros L. 柿属

Diospyros kaki Thunb. var. *silvestris* Makino/ 野柿

1-1. *Diospyros kaki* Thunb. var. *silvestris* Makino/ 野柿;   1-2. *D. morrisiana* Hance/ 罗浮柿

335 Primulaceae 报春花科

Leaves simple, often basal, sometimes with gland / 单叶，常基生，有时具腺点
Ovary superior, 1-loculed / 子房上位，1 室
Placentation free central / 特立中央胎座
Capsule dehiscing by valves / 蒴果瓣裂

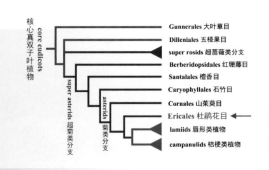

Perennial or annual **herbs** or **trees**, **shrubs**, woody **climbers**. **Leaves** sometimes gland-dotted, simple, alternate, opposite or whorled, often basal; margin entire, crenate or toothed, rarely pinnatifid; petioles present or absent; stipules absent, rarely present. **Inflorescence** solitary flower, heads, umbels, panicles or racemes. **Flowers** bisexual or unisexual, actinomorphic or rarely zygomorphic; often with bracts. **Sepals** fused or free, sometimes persistent. **Petals** fused (at least basally) or rarely free or very rarely absent, sometimes appendaged. **Stamens** attached to the petals, opposite the petals or alternate, sometimes heterostylous, filaments free or ±fused; anthers free or fused, basifixed or dorsifixed, introrse or extrorse; occasionally with scale-like staminodes. **Ovary** superior, rarely inferior to part inferior; carpels fused; 1-loculed; ovules few to many; placentation free central or very rarely basal. **Fruit** a capsule, dehiscing by valves, rarely circumscissile or indehiscent, berry or drupe.

多年生或一年生**草本**或**乔木**、**灌木**或**木质藤本**。**叶**常具腺点，单叶，互生、对生或轮生，常基生；全缘、具钝齿或牙齿，稀羽裂；叶柄有或无；托叶缺，稀存在。**花序**为单花、头状、伞形、圆锥状或总状。**花**两性或单性，辐射对称或稀两侧对称；常具苞片。**花萼**合生或离生，有时宿存。**花瓣**合生（至少基部如此）或稀离生或极稀缺失，有时具附属物。**雄蕊**贴生于花瓣，与花瓣对生或互生，有时花柱异长；花丝分离或合生；花药分离或合生，基着或背着，内向或外向；偶具鳞片状退化雄蕊。**子房**上位，稀下位至半下位；心皮合生；1 室；胚珠少数至多数；特立中央胎座或极稀为基底胎座。**果实**为蒴果，瓣裂、稀盖裂或不裂，浆果或核果。

World 58/2, 590, worldwide.
China 19/ca. 655 (Including *Kaufmannia*, *Sadiria*).
This area 9/26.

全世界共 58 属 2 590 种。
中国产 19 属约 655 种（含金钟报春属和管金牛属）。
本地区有 9 属 26 种。

Anagallis arvensis L./ 琉璃繁缕

1. Herbs/ 草本
    2. Corolla lobes contorted in bud/ 花冠裂片在花蕾中旋转状排列
        3. Capsule circumscissile; filaments pubescent/ 蒴果盖裂，花丝具毛 ....**1.** *Anagallis*/ 琉璃繁缕属
        3. Capsule valvular; filaments glabrous/ 蒴果瓣裂，花丝无毛 ....**2.** *Lysimachia*/ 珍珠菜属

2. Corolla lobes imbricate in bud/ 花冠裂片在花蕾中覆瓦状排列

    4. Flower solitary/ 花单生 ....**3. *Stimpsonia*/ 假婆婆纳属**

    4. Flowers in umbels, racemes, or spikes/ 花呈伞形、总状或穗状花序

        5. Corolla tube usually shorter than calyx, constricted at throat/ 花冠筒常短于花萼，喉部缢缩 ....**4. *Androsace*/ 点地梅属**

        5. Corolla tube longer than calyx, not constricted at throat/ 花冠筒长于花萼，喉部不缢缩 ....**5. *Primula*/ 报春花属**

1. Shrubs/ 灌木

    6. Ovary part-inferior to inferior; bracteoles 2/ 子房半下位至下位；小苞片2枚 ....**6. *Maesa*/ 杜茎山属**

    6. Ovary superior; bracteoles absent/ 子房上位；无小苞片

        7. Inflorescence on a long peduncle/ 总花梗长 ....**7. *Ardisia*/ 紫金牛属**

        7. Inflorescence without a peduncle/ 无总花梗

            8. Inflorescence racemose or paniclate/ 花呈总状或圆锥花序 ....**8. *Embelia*/ 酸藤子属**

            8. Inflorescence fascicled or umbel/ 花簇生或呈伞形花序 ....**9. *Myrsine*/ 铁仔属**

ERICALES. Primulaceae. *Androsace umbellata* (Lour.) Merr./ 杜鹃花目 报春花科 点地梅

A. Whole plant/ 植株

B_1. Corolla segregated/ 花冠离析

B_2. Vertical section of flower/ 花纵切面

B_3. Vertical section of corolla tube/ 花冠筒纵切面

C. Cross section of ovary/ 子房横切面

D_1. Dorsal view of flower/ 花背面观

D_2. Ventral view of calyx/ 花萼腹面观

D_3. Ventral view of flower/ 花腹面观

D_4. Dorsal view of calyx/ 花萼背面观

E_1. Ventral view of leaf/ 叶腹面观

E_2. Dorsal view of leaf/ 叶背面观

F_1. Single Flower/ 花

F_2. Umbel/ 伞形花序

(Scale/ 标尺：A 1 cm; B_1–D_4 1 mm; E_1–F_2 1 cm)

1. *Anagallis arvensis* L./ 琉璃繁缕（1a. Capsule circumscissile/ 蒴果盖裂；1b. Flower/ 花）；
2. *Lysimachia capillipes* Hemsl./ 细梗香草；
3. *Stimpsonia chamaedryoides* C. Wight ex A. Gray/ 假婆婆纳（Fruiting stage/ 果期）；
4. *Androsace umbellata* (Lour.) Merr./ 点地梅；
5. *Primula cicutariifolia* Pax/ 毛茛叶报春

6. *Maesa japonica* (Thunb.) Moritzi & Zoll./ 杜茎山；
7. *Ardisia crenata* Sims/ 朱砂根；
8. *Embelia vestita* Roxb./ 密齿酸藤子；
9. *Myrsine seguinii* H. Lév./ 密花树

336 Theaceae 山茶科

Usually evergreen woody plants / 常绿木本
Leaves simple, alternate / 单叶，互生
Often axillary solitary flower / 花单生叶腋
Ovary superior, loculicidal capsule / 子房上位，蒴果室背开裂

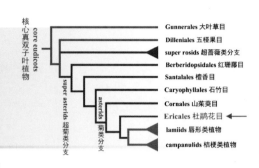

Trees or **shrubs**. **Leaves** evergreen or rarely deciduous, simple, alternate; margin toothed, rarely entire; petioles present; stipules absent. **Inflorescence** axillary or subterminal, solitary flower, or sometimes cluster or racemes. **Flowers** bisexual or unisexual, actinomorphic; bracteoles 2–8 or rarely more, persistent or caducous, sometimes undifferentiated from sepals. **Sepals** free or fused, imbricate, sometimes unequal, persistent. **Petals** free or basally fused, imbricate, lobes longer than tube. **Stamens** many, filaments free or rarely fused, attached to petal base; anthers usually dorsifixed, 2-loculed, introrse, laterally and dehiscing longitudinally. **Ovary** superior; carpels fused; locules 3–5 (–10); ovules 2–6 (–12) per locule; placentation axile; styles 1-several. **Fruit** a loculicidal capsule or ± drupe.

乔木或**灌木**。叶常绿或稀落叶，单叶，互生；具齿，稀全缘；叶柄存在；托叶缺。花序腋生或近顶生，单花，或有时簇生或总状。花两性或单性，辐射对称；小苞片 2~8 枚或稀多数，宿存或早落，有时与花萼不分化。花萼离生或合生，覆瓦状，有时不等大，宿存。花瓣离生或基部合生，覆瓦状，裂片长于冠筒。雄蕊多数，花丝分离或稀合生，贴生于花瓣基部；花药常背着，2 室，内向，侧方纵向开裂。子房上位，心皮合生；3~5 (~10) 室；胚珠每室 2~6 (~12) 枚；中轴胎座；花柱 1 至数枚。果实为室背开裂的蒴果或多少为核果。

World 9/ca. 250, Eastern, Southern and Southeast Asia, tropical America and southeast North America.
China 6/145.
This area 2/5.

全世界共 9 属约 250 种，分布于亚洲的东部、南部和东南部，热带美洲和北美洲的东南部。
中国产 6 属 145 种。
本地区有 2 属 5 种。

1. Seeds wingless/ 种子无翅 ....1. *Camellia*/ 山茶属
1. Seeds winged/ 种子具翅 ....2. *Schima*/ 木荷属

1-1. *Camellia fraterna* Hance/ 毛花连蕊茶；
1-2. *C. japonica* L./ 山茶；

2. *Schima superba* Gardner & Champ. / 木荷
（2a. Seeds/ 种子；2b. Flower/ 花）

337 **Symplocaceae** 山矾科

Leaves simple, alternate, estipulate / 单叶互生，无托叶
Flowers usually < 1 cm in diam., bisexual / 两性花，直径常小于 1 cm
Stamens many, filamentes fused / 雄蕊多数，花丝合生
Ovary inferior, calyx persistent / 子房下位，萼片宿存

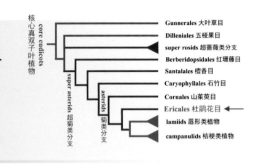

Trees or **shrubs**. **Leaves** often sweet tasting, simple, alternate; margin toothed, rarely entire; petioles present; stipules absent. **Inflorescence** racemes, spikes, glomerules, panicles or rarely solitary flower; bract 1. **Flowers** small (5–10 mm in diam.), bisexual or rarely unisexual (polygamomonoecious or dioecious), actinomorphic, hypanthium present; bracteoles 2. **Sepals** basally fused, valvate or imbricate, persistent in fruit. **Petals** fused at least basally, imbricate. **Stamens** usually many, filaments sometimes fused, attached to petals; anthers introrse; sometimes staminodes in female flowers. **Ovary** inferior, rarely part-inferior; carpels fused; locules 2–5; ovules (2–)4 per locule; placentation ± axile; style 1, filiform; stigmas small, capitate or 2–5-lobed. **Fruit** usually a berry-like drupe, with persistent calyx on the top.

乔木或**灌木**。**叶**常有甜味，单叶，互生；具齿，稀全缘；叶柄存在；托叶缺。**花序**为总状、穗状、团伞状、圆锥状或稀为单花；具 1 枚苞片。**花**小（直径 5~10 mm），两性或稀单性（杂性同株或雌雄异株），辐射对称，具被丝托；小苞片 2 枚。**花萼**基部合生，镊合状或覆瓦状，果期宿存。**花瓣**至少基部合生，覆瓦状。**雄蕊**常多数，花丝有时合生，贴生于花瓣；花药内向；有时雌花中具退化雄蕊。**子房**下位，稀半下位；心皮合生；2~5 室；胚珠每室 (2~) 4 枚；多少中轴胎座；花柱 1 个，丝状；柱头小，头状或 2~5 裂。**果实**常为浆果状核果，顶端具宿存花萼。

World 2/260, Americas, Eastern Asia and Australasia.
China 2/42.
This area 1/6.

全世界共 2 属 260 种，分布于美洲、东亚、澳大拉西亚。
中国产 2 属 42 种。
本地区有 1 属 6 种。

Symplocos Jacq. 山矾属

Symplocos paniculata Miq./ 白檀（a. Flowering branch/ 花枝；b. Fruiting branch/ 果枝）

339 *Styracaceae* 安息香科

Woody / 木本
Leaves simple, alternate, margin often entire / 单叶互生，常全缘
Indumentum of either stellate hairs or peltate scales / 植株被星状毛或盾状鳞片
Flowers 5-merous, often cernuous / 花 5 基数，常俯垂

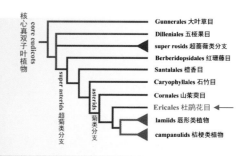

Trees or **shrubs**; bark sometimes resinous, rarely thorny. **Leaves** usually alternate, leaf blade entire or slightly toothed, indumentum of either stellate hairs or peltate scales, often brown or reddish brown; petioles present or absent; stipules absent. **Inflorescence** racemes, cymes, panicles or rarely 1–2 floweres, often cernuous. **Flowers** 5-merous, bisexual, actinomorphic; often bracteoles absent. **Sepals** fused, valvate, 5-toothed, persistent. **Petals** fused, campanulate, lobes usually 5, imbricate or valvate. **Stamens** 2-whorled, usually twice as many as corolla lobes, rarely many; opposite the sepals; filaments sometimes basally adnate to petals; anthers basifixed, introrse. **Ovary** superior to inferior; carpels fused; locules (1–) 3 or 5; ovules 1 to many per locule; placentation usually axile; style 1; stigmas capitate or 2–5-lobed. **Fruit** a fleshy drupe, with dry epicarp, loculicidal capsule, or samara.

乔木或**灌木**；有时具树脂，稀具刺。**叶**常互生，全缘或稍具齿，被星状毛或盾状鳞片，常为褐色或红褐色；叶柄有或无；托叶缺。**花序**总状、聚伞状、圆锥状或稀 1~2 朵花，常俯垂。**花** 5 基数，两性，辐射对称；小苞片常缺。**花萼**合生，镊合状，5 齿，宿存。**花瓣**合生，钟状，裂片常 5 枚，镊合状或覆瓦状。**雄蕊** 2 轮，常为花冠裂片的 2 倍，稀多数；与花萼对生；花丝常基部与花瓣贴生；花药背着，内向。**子房**上位至下位；心皮合生；(1~) 3 或 5 室；胚珠每室 1 至多数；常中轴胎座；花柱 1 个；柱头头状或 2~5 裂。**果实**为具干燥外果皮的肉质核果、室背开裂蒴果或翅果。

World 11/ca. 160, tropical Asia and America, few in Mediterranean, and temperate North America and East Asia.
China 11/ca. 55.
This area 1/5.

全世界共 11 属约 160 种，分布于热带亚洲和美洲，少量至地中海，北美温带和东亚。
中国产 11 属约 55 种。
本地区有 1 属 5 种。

Styrax L. 安息香属

Styrax confusus Hemsl./ 赛山梅

ERICALES. Styracaceae. *Melliodendron xylocarpum* Hand.–Mazz./ 杜鹃花目 安息香科 陀螺果

A. Dorsal view of flowering branch/ 花枝背面观
B. Ventral view of flowering branch/ 花枝腹面观
C_1. Cross section of ovary/ 子房横切面
C_2. Ventral view of stamen/ 雄蕊腹面观
C_3. Lateral view of stamen/ 雄蕊侧面观
C_4. Androecium/ 雄蕊群
C_5. Pistil/ 雌蕊
C_6. Androecium unfold (inside)/ 雄蕊群展开（内侧）
D. Flower segregated/ 花离析
E. Dorsal view of corolla/ 花冠背面观

(Scale/ 标尺： A–B none/ 无；C_1 1 mm; C_2–E 5 mm)

342 Actinidiaceae 猕猴桃科

Woody climbers / 木质藤本
Hairs common / 常被毛
Stamens numerous, in bundles / 雄蕊多数, 成束
Ovary superior, berry with stiff hairs / 子房上位, 浆果被刚毛

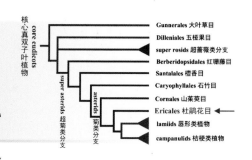

Woody climbers, trees or **shrubs**. **Leaves** simple, alternate, rarely opposite; margin entire or toothed; petioles present; stipules absent; hairs common, especially on young branches. **Inflorescence** axillary dense panicles, cymes or solitary flower; rarely cauliflorous. **Flowers** bisexual or unisexual (polygamous or functionally dioecious), actinomorphic, showy; often bracteolate, small. **Sepals** (2 or 3) 5, free or basally fused, imbricate, rarely valvate; persistent. **Petals** (4 or) 5, sometimes more, imbricate, deciduous. **Stamens** 10 to numerous, in bundles; distinct or adnate to base of petals; anthers 2-loculed, versatile, dehiscing porcidally or longitudinally. **Ovary** superior, disk absent; carpels fused, locules 3–5 or more; placentation axile; ovules many per locule; styles as many as carpels, distinct or connate (then only one style), generally persistent in fruit. **Fruit** a berry sometimes with stiff hairs or leathery capsule.

木质**藤本**、**乔木**或**灌木**。**叶**为单叶, 互生, 稀对生 ; 全缘或具齿 ; 叶柄存在 ; 托叶缺 ; 常被毛, 特别是嫩枝。**花序**腋生密集圆锥状、聚伞状或为单花 ; 稀老茎生花。**花**两性或单性（杂性或功能性雌雄异株）, 辐射对称, 显著 ; 常具小苞片, 小。**花萼** (2 或 3 或) 5 枚, 离生或基部合生, 覆瓦状, 稀镊合状 ; 宿存。**花瓣** (4 或) 5 枚, 有时多数, 覆瓦状, 脱落。**雄蕊** 10 枚至多数, 成束 ; 离生或基部与花瓣合生 ; 花药 2 室, 丁字状着药, 顶孔开裂或纵裂。**子房**上位, 花盘缺 ; 心皮合生, 3~5 室或更多 ; 中轴胎座 ; 胚珠每室多数 ; 花柱与心皮同数, 分离或合生（此时花柱仅 1 枚）, 常于果期宿存。**果实**为浆果, 有时被刚毛, 或为革质蒴果。

World 3/357, widely distributed in tropical regions, especially in Southeast Asia.
China 3/66.
This area 1/3.

全世界共 3 属 357 种, 广布于热带, 尤其是东南亚。
中国产 3 属 66 种。
本地区有 1 属 3 种。

***Actinidia* Lindl. 猕猴桃属**

1-1. *Actinidia chinensis* Planch./ 中华猕猴桃 ;

1-2. *A. chinensis* Planch. var. *deliciosa* (A. Chev.) A. Chev./ 美味猕猴桃（Cultivated/ 栽培）

345 Ericaceae 杜鹃花科

Evergreen woody plants, often with scale / 常绿木本，常具鳞片
Leaves simple, alternate and coriaceous / 单叶互生，常革质
Petals fused, with persistent calyx / 合瓣花，花萼宿存
Stamens with spurs or awns; anther dehiscing porcidally / 雄蕊有距或芒；花药孔裂

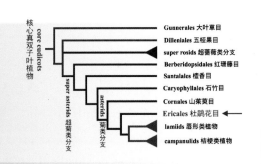

Shrubs, or rarely **woody climbers** or **small trees**, or **herbs**, buds often with scale. **Leaves** simple, sometimes gland-dotted; usually alternate, opposite or whorled; margin entire or toothed; petiole present; stipules absent. **Inflorescence** racemes, spikes, heads, corymbs, panicles or rarely solitary flower; bracts paired, basal. **Flowers** bisexual or rarely unisexual (dioecious); (4 or) 5-merous; actinomorphic or slighty zygomorphic; usually bracteoles 2–3. **Sepals** free or basally fused; persistent. **Petals** fused or sometimes free or absent; urceolate, campanulate, tubular or funnel-shaped, imbricate. **Stamens** 10, filaments sometimes fused at base; connective with spurs or awns; anthers dehiscing porcidally. **Ovary** superior or inferior; carpels fused; locules (1–) 4–5(–14); ovules 1 to many per locule; placentation axile; style 1, slender, stigmas capitate. **Fruit** a capsule or berry, rarely a drupe, with persistent calyx.

灌木，稀木质藤本或小乔木，或草本，芽常具鳞片。叶为单叶，有时具腺点；常互生、对生或轮生；全缘或具齿；具叶柄；托叶缺。花序总状、穗状、头状、伞房状、圆锥状或稀为单花；苞片成对，基部着生。花两性或稀单性（雌雄异株）；(4 或) 5 基数；辐射对称或稍两侧对称；常具 2~3 枚小苞片。花萼离生或基部合生；宿存。花瓣合生或有时离生或缺；坛状、钟状、管状或漏斗状，覆瓦状。雄蕊 10 枚，花丝有时基部合生；药隔有距或芒；花药，孔裂。子房上位或下位；心皮合生；(1~) 4~5(~14) 室；胚珠每室 1 枚至多数；中轴胎座；花柱 1 个，细长，柱头头状。果实为蒴果或浆果，稀为核果，具宿存萼片。

World ca. 124/4,100, worldwide.
China 23/837.
This area 2/6.

全世界共约 124 属 4 100 种，世界广布。
中国产 23 属 837 种。
本地区有 2 属 6 种。

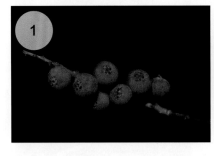

1. *Vaccinium bracteatum* Thunb./ 南烛；
2. *Rhododendron ovatum* (Lindl.) Planch. ex Maxim./ 马银花

1. Ovary inferior; berry/ 子房下位；浆果 .... **1. *Vaccinium*/ 越橘属**
1. Ovary superior; capsule/ 子房上位；蒴果 .... **2. *Rhododendron*/ 杜鹃花属**

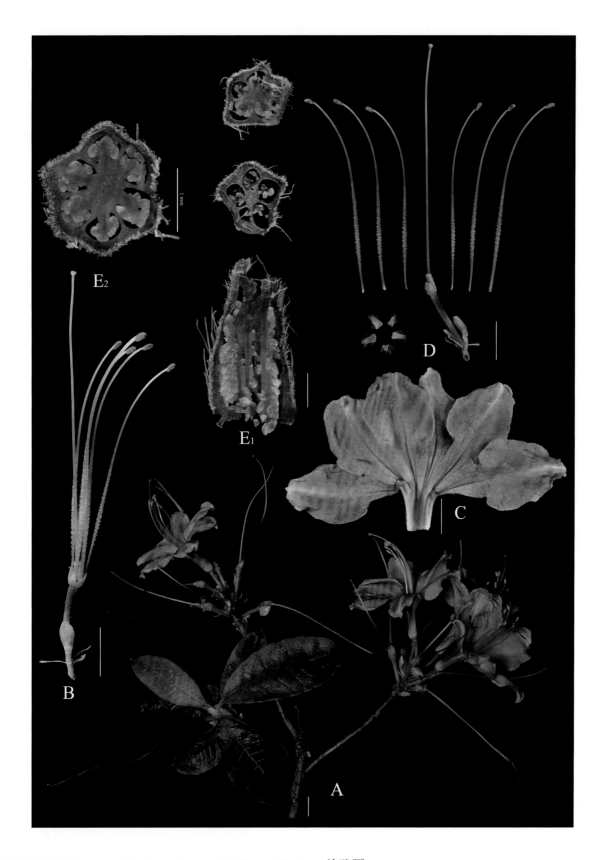

ERICALES. Ericaceae. *Rhododendron molle* (Blume) G. Don/ 羊踯躅

A. Flowering branch/ 花枝
B. Flower, corolla removed/ 花，除花冠
C–D. Flower segregated/ 花离析
E_1. Vertical section of ovary
E_2. Cross section of ovary
(Scale/ 标尺：A–D 1cm; E 1mm)

351 Garryaceae 丝缨花科

Evergreen woody plant / 常绿木本
Branches green, opposite / 小枝绿色，对生
Leaves simple, opposite / 单叶对生
Flowers unisexual, 4-merous / 花单性，4 基数

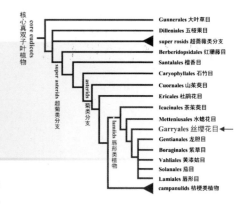

Trees or **shrubs**, evergreen; branches opposite, green and cylindrical, often with cone-shaped teminal bud. **Leaves** simple, opposite; margin entire or toothed; petioles present; stipules absent. **Inflorescence** terminal catkins or panicles; often bracts 1–3. **Flowers** inconspicuous, unisexual (dioecious); actinomorphic; 4-merous, subtended by 1 or 2 bracteoles. **Male flowers**: stamens 4, alternate petals; filaments free from perianth; anthers basifixed, introrse. **Female flowers**: perianth 2 or reduced to absent or 4-merous; carpels fused or 1; 1-loculed; ovules 1–3; placentation apical, style short, thick, stigmas capitate. **Fruit** a dry berry or fleshy drupe.

乔木或灌木，常绿；小枝对生，绿色，圆柱形，顶芽常松果状。叶为单叶，对生；全缘或具齿；具叶柄；托叶缺。花序为顶生柔荑花序或圆锥花序；常有 1~3 枚苞片。花不显，单性（雌雄异株）；辐射对称；4 基数，为 1 或 2 枚小苞片包围。雄花：雄蕊 4 枚，与花瓣互生；花丝与花瓣分离；花药基着，内向。雌花：花被片 2 枚或退化至缺，或 4 基数；心皮合生或 1 枚；1 室；胚珠 1~3 枚；顶生胎座，花柱短厚，柱头头状。果实为干燥的浆果或肉质核果。

World 2/17, North America, Central America, Caribbean and East Asia.
China 1/10.
This area 1/1.

全世界共 2 属 17 种，分布于北美洲、中美洲、加勒比海地区和东亚。
中国产 1 属 10 种。
本地区有 1 属 1 种。

***Aucuba* Thunb. 桃叶珊瑚属**

Aucuba japonica Thunb. / 青木（a. Female inflorescence / 雌花序；b. Male inflorescence / 雄花序）

352 Rubiaceae 茜草科

Leaves opposite or whorled, margin entire / 叶全缘，对生或轮生
Stipules interpetiolar, conspicuous / 叶柄间托叶明显
Corolla tubular at base / 花冠基部筒状
Ovary inferior / 子房下位

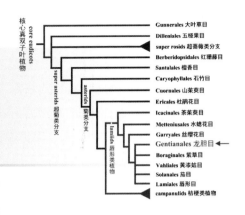

Trees, **shrubs**, subshrubs, annual or perennial **herbs**, **woody climbers**; sometimes creeping and rooting at nodes, succulent, aquatic or epiphytic. **Leaves** simple, opposite, rarely whorled; margin entire, very rarely toothed; petioles present or absent; stipules usually interpetiolar, less often intrapetiolar, sometimes united around the stem into a sheath. **Inflorescence** panicles, cymes, racemes, fascicles or solitary flower; rarely cauliflorous. **Flowers** bisexual or rarely unisexual, actinomorphic, rarely zygomorphic; distylous, or rarely tristylous. **Sepals** usually fused to inferior ovary in hypanthium or ovary portion. **Petals** usually fused, large and often showy to reduced; variously funnel, salverform, tubular, campanulate, or occasionally rotate to urceolate. **Stamens** alternating with petals, filaments usually attached to petals, filaments fused or free; anthers 2-loculed, dorsifixed to basifixed, usually pubescent; dehiscencing longitudinal, introrse; connective often prolonged into an apical and sometimes basal appendage. **Ovary** inferior, rarely superior; carpels fused; locules (1–) 2 (–many); ovules 1 to many per locule; placentation axile or less often parietal; often heterostylous. **Fruit** a berry, drupe, less often a schizocarp or capsule.

乔木、灌木、亚灌木、一年生或多年生草本或木质藤本；有时匍匐，节上生根、肉质、水生或附生植物。叶为单叶，对生，稀轮生；全缘，稀有锯齿；叶柄有或无；托叶常为叶柄内托叶，稀为叶柄间托叶，有时联合成鞘，抱茎。花序为圆锥状、聚伞状、总状、簇生或为单花；稀老茎生花。花两性或稀单性，辐射对称，稀两侧对称；两型花柱，稀三型花柱。花萼与下位子房或部分子房合生成被丝托。花瓣常合生，大而显著，或退化；呈多种漏斗状、高脚碟状、管状、钟状或偶为圆盘状至坛状。雄蕊与花瓣互生，花丝常贴生于花瓣，花丝合生或离生；花药2室，背着至基着，常被毛；纵裂，内向；药隔常于先端伸长，基部有时具附属物。子房下位，稀上位；心皮合生；(1~)2(至多)室；胚珠每室1至多枚；中轴胎座或偶侧膜胎座；花柱异长。果实为浆果或核果，稀为分果或蒴果。

World ca. 614/13, 150, tropical and subtropical regions, few in northern temperate regions.

China 103/ ca. 743.

This area 16/28.

全世界共约614属13 150种，广布于热带至亚热带地区，少数至北温带。

中国产103属约743种。

本地区有16属28种。

Gardenia jasminoides J. Ellis / 栀子

1. Inflorescence heads/ 头状花序
   2. Shrubs or trees/ 灌木或乔木 ....**1.** *Adina*/ 水团花属
   2. Climber/ 藤本 ....**2.** *Uncaria*/ 钩藤属
1. Inflorescence not a head/ 非头状花序
   3. Plants often with petaloid calyx/ 常有花瓣状萼片 ....**3.** *Mussaenda*/ 玉叶金花属
   3. Plants without petaloid calyx/ 无花瓣状萼片
      4. Ovules 2 or more per locule/ 子房每室有胚珠 2 至多数
         5. Fruit a capsule/ 果为蒴果 ....**4.** *Hedyotis*/ 耳草属
         5. Fruit fleshy, indehiscent/ 果肉质，不裂
            6. Ovary 1-loculed/ 子房 1 室 ....**5.** *Gardenia*/ 栀子属
            6. Ovary 2-loculed/ 子房 2 室
               7. Flowers unisexual or polygamodioecious/ 花单性或杂性异株 ....**6.** *Diplospora*/ 狗骨柴属
               7. Flowers bisexual/ 花两性
                  8. Cymes axillary, or opposite with leaves, or borne at leafless nodes; seeds embedded in pulp/ 聚伞花序腋生，或与叶对生，或生于无叶的节上；种子沉没于果肉内 ....**7.** *Aidia*/ 茜树属
                  8. Cymes terminal or axillary; seeds exposed from pulp/ 聚伞花序顶生或腋生；种子暴露 ....**8.** *Tarenna*/ 乌口树属
      4. Ovule 1 per locule/ 子房每室只有 1 枚胚珠
         9. Trees, shrubs or woody climbers/ 木本、灌木或木质藤本
            10. Inflorescence heads/ 头状花序 ....**9.** *Morinda*/ 巴戟天属
            10. Inflorescence not a head/ 非头状花序
               11. Shrub or trees, erect; rarely climber/ 直立灌木或乔木；稀为藤本
                  12. Inflorescence terminal/ 花序顶生 ....**10.** *Psychotria*/ 九节属
                  12. Flowers axillary, rarely terminal/ 花腋生，稀顶生
                     13. Shrubs armed/ 有刺灌木 ....**11.** *Damnacanthus*/ 虎刺属
                     13. Shrubs unarmed/ 无刺灌木 ....**12.** *Serissa*/ 六月雪属
               11. Climber/ 藤本 ....**13.** *Paederia*/ 鸡矢藤属
         9. Herbs or herbaceous climbers/ 草本或草质藤本
            14. Stipules large, leaf-like, in whorls/ 托叶大型，叶状，数枚轮生
               15. Flowers 5-merous; fruit fleshy/ 花 5 基数；果肉质 ....**14.** *Rubia*/ 茜草属
               15. Flowers 4-merous; fruit dry/ 花 4 基数；果干燥 ....**15.** *Galium*/ 拉拉藤属
            14. Stipules small, not like above/ 托叶小，非上述特征
               16. Stems twining/ 茎缠绕 ....**13.** *Paederia*/ 鸡矢藤属
               16. Stems erect, not twining/ 茎直立，非缠绕状 ....**16.** *Spermacoce*/ 纽扣草属

1. *Adina pilulifera* (Lam.) Franch. ex Drake/ 水团花；
2. *Uncaria rhynchophylla* (Miq.) Miq. ex Havil./ 钩藤；
3. *Mussaenda pubescens* W.T. Aiton/ 玉叶金花；
4. *Hedyotis strigulosa* (Bartl. ex DC.) Fosberg/ 肉叶耳草；
5. *Gardenia jasminoides* J. Ellis/ 栀子（5a. Flower/ 花；
5b. Cross section of ovary/ 子房横切面）；
6. *Diplospora dubia* (Lindl.) Masam./ 狗骨柴；
7. *Aidia cochinchinensis* Lour./ 茜树

8. *Tarenna mollissima* (Hook. & Arn.) B.L. Rob./ 白花苦灯笼；

9. *Morinda umbellata* L. subsp. *obovata* Y.Z. Ruan/ 羊角藤；

10. *Psychotria asiatica* L./ 九节；

11. *Damnacanthus indicus* C.F. Gaertn./ 虎刺；

12. *Serissa serissoides* (DC.) Druce/ 白马骨；

13. *Paederia foetida* L./ 鸡矢藤；

14. *Rubia argyi* (H. Lév. & Vaniot) H. Hara ex Lauener & D.K. Ferguson/ 东南茜草；

15. *Galium spurium* L./ 猪殃殃；

16. *Spermacoce remota* Lam./ 光叶丰花草（photo by XinXin ZHU/ 朱鑫鑫拍摄）

353　Gentianaceae 龙胆科

Usually rhizomatous herbs / 多具根状茎草本
Leaves simple, entire, opposite / 单叶全缘，对生
Tepals fused, 4 or 5-merous / 花被片合生，4 或 5 基数
Ovary superior, fruit a 2-valved capsule / 子房上位，2 瓣裂蒴果

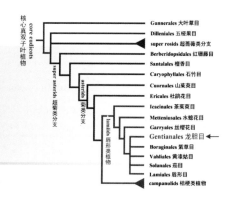

Annual or perennial, rhizomatous **herbs**, **shrubs**, **small trees** or **woody climbers**; sometimes mycoheterotrophic. **Leaves** simple, opposite, less often alternate or whorled, rarely reduced to scales; margin entire, very rarely toothed; petioles present or absent; stipules absent, sometimes an interpetiolar. **Inflorescence** cymes, racemes, spikes, heads, sometimes reduced to sessile clusters or solitary flower; sometimes bracteate. **Flowers** bisexual or rarely unisexual (dioecious), actinomorphic or sometimes zygomorphic; often 4 or 5-merous. **Sepals** fused, tubular, obconic, campanulate, or rotate, lobes joined at least basally, usually imbricate, often large, persistent. **Petals** fused, tubular, obconic, salverform, funnelform, campanulate, or rotate, rarely with basal spurs; usually contorted or rarely imbricate, sometimes nectariferous, sometimes appendages or plicae. **Stamens** alternating with petals, filaments free, attached to petals; anthers 2-loculed, dorsifixed, rarely basifixed, usually introrse; sometimes staminodes 1–4. **Ovary** superior; carpels fused, often glands at base; locules 1–2; ovules 3 or more per locule; placentation parietal or rarely free central, or axile; styles 2-lobed. **Fruit** a 2-valved capsule or rarely a berry.

一年生或多年生，具根状茎**草本**、**灌木**、**小乔木**或**木质藤本**；有时为真菌异养型。**叶**为单叶，对生，偶互生或轮生，稀退化为鳞片；全缘，罕有锯齿；叶柄有或无；托叶缺，偶具叶柄内托叶。**花序**为聚伞状、总状、穗状、头状，有时简化为无柄，簇生或为单花；有时具苞片。**花**两性或稀单性（雌雄异株），辐射对称或有时两侧对称；常 4 或 5 基数。**花萼**合生，管状、倒圆锥形、钟状或圆盘状，裂片至少在基部合生，常覆瓦状，多大型，宿存。**花瓣**合生，花冠管状、倒圆锥形、高脚碟状、漏斗状、钟状或圆盘状，基部稀有距；常扭曲或稀覆瓦状，有时具蜜汁，偶有附属物或褶皱。**雄蕊**与花瓣互生，花丝分离，着生于花瓣；花药 2 室，背着，稀基着，常内向；有时具 1~4 枚退化雄蕊。**子房**上位；心皮合生，基部常具腺体；1~2 室；胚珠每室 3 至多枚；侧膜胎座或稀为特立中央胎座，或中轴胎座；花柱 2 裂。**果实**为 2 瓣裂的蒴果或稀为浆果。

World 80/700+, worldwide, except Antarctica, mainly in northern temperate and cold temperate regions.
China 22/ca. 420.
This area 3/3.

全世界共 80 属 700 余种，世界广布，除南极洲，主要分布于北半球温带和寒温带。

Centaurium pulchellum (Sw.) Druce var. *altaicum* (Griseb.) Kitag. & H. Hara / 百金花

中国产 22 属约 420 种。

本地区有 3 属 3 种。

1. Corolla with plicae extending between lobes/ 花冠裂片间具褶 .... **1. *Gentiana*/ 龙胆属**
1. Corolla without plicae between lobes/ 花冠裂片间无褶
    2. Anthers helically coiled after flowering/ 花药在花后螺旋状扭转 .... **2. *Centaurium*/ 百金花属**
    2. Anthers not coiled/ 花药不扭转 .... **3. *Swertia*/ 獐牙菜属**

1. *Gentiana zollingeri* Fawc./ 笔龙胆；
2. *Centaurium pulchellum* (Sw.) Druce var. *altaicum* (Griseb.) Kitag. & H. Hara/ 百金花；
3. *Swertia hickinii* Burkill/ 江浙獐牙菜

354 Loganiaceae 马钱科

Leaves simple, opposite / 单叶对生
Stipules interpetiolar, encircling the stem / 叶柄间托叶，抱茎
Sepals and petals fused, 4 or 5-merous / 合瓣花，4 或 5 基数
Stamens inserted to petals; filaments unequal in length / 雄蕊着生于花瓣上；花丝不等长

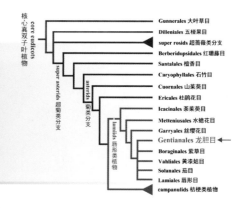

Trees, **shrubs**, **woody climbers**, or **herbs**, sometimes epiphytes, sometimes with axillary spines or tendrils. **Leaves** simple, opposite, occasionally alternate, rarely whorled, fascicled; stipules interpetiolar to encircling the stem, or reduced to lines connecting petiole base; margin entire or toothed. **Inflorescence** cymes, often grouped into thyrses, sometimes umbel-like, scorpioid, or reduced to a single flower; bracts usually small. **Flowers** bisexual or unisexual (plants dioecious or gynodioecious); actinomorphic or zygomorphic; nectar disk present. **Sepals** fused, 4- or 5-lobed, mostly persistent, imbricate or valvate. **Petals** fused; lobes 4 or 5(–16), valvate to imbricate, or contorted in bud. **Stamens** inserted to petals, equal in number to corolla lobes and alternating with them or sometimes fewer; filaments free, unequal in length; anthers basifixed, 2–4-loculed, dehiscing longitudinally, introrsely or extrorsely. **Ovary** superior or rarely part-inferior, (1 or)2(–4)-loculed; ovules 1 to many per locule; placentation axile or parietal; style simple, terminal, persistent or deciduous; stigmas usually capitate, entire or shortly 2–4-lobed. **Fruit** a capsule, berry, or drupe, 1 to many seeded. **Seeds** sometimes winged.

乔木、灌木、木质藤本或**草本**，有时附生，有时叶腋具刺或卷须。**叶**为单叶，对生，偶互生，稀轮生或簇生；叶柄内托叶，抱茎，或退化为线形，与叶基联合；全缘或具齿。**花序**为聚伞状，常组成聚伞圆锥状，有时伞形、蝎尾状，或退化为单花；苞片常小。**花**两性或单性（雌雄异株或雄全异株）；辐射对称或两侧对称；具蜜腺盘。**花萼**合生，4 或 5 裂，多宿存，覆瓦状或镊合状。**花瓣**合生；4 或 5 裂（稀达 16 裂），镊合状至覆瓦状，或在芽内扭转。**雄蕊**着生于花瓣，与花冠裂片同数目互生，有时较少；花丝分离，不等长；花药基着，2~4 室，纵裂，内向或外向。**子房**上位或稀半下位，(1 或) 2 (~4) 室；胚珠每室 1 至多枚；中轴胎座或侧膜胎座；花柱简单，顶生，宿存或脱落；花柱常头状，全缘或稍 2~4 裂。**果实**为蒴果、浆果或核果，含 1 至多粒种子。**种子**有时具翅。

World ca. 15/400, tropics to temperate regions.
China 5/28.
This area 2/3.

全世界共约 15 属 400 种，分布于热带至温带地区。
中国产 5 属 28 种。
本地区有 2 属 3 种。

Gardneria multiflora Makino / 蓬莱葛

1. Herbs, erect/ 直立草本 .... **1. *Mitrasacme*/ 尖帽草属**
1. Woody climber/ 木质藤本 .... **2. *Gardneria*/ 蓬莱葛属**

1. *Mitrasacme pygmaea* R. Br./ 水田白；   2. *Gardneria multiflora* Makino/ 蓬莱葛

356 Apocynaceae 夹竹桃科

Sap usually milky / 常有乳汁
Leaves simple and opposite, margin entire / 单叶对生，全缘
Flowers with corona or scales / 花具副花冠或鳞片
Seeds often with coma / 种子常具种缨

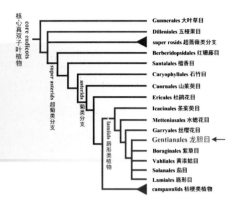

Trees, **shrubs**, **woody climbers**, or less often **herbs**; sometimes succulent; with milky sap. **Leaves** simple, usually opposite or whorled, rarely alternate; sometimes reduced to absent; margin entire; petioles usually present; stipules absent or sometimes present. **Inflorescence** panicles or umbels, cymes, spikes, head-like, corymbs, solitary flower or racemes, sometimes cauliflorous. **Flowers** bisexual or rarely unisexual (dioecious), actinomorphic; fragrant or foetid. **Sepals** fused, usually 5-lobed, imbricate, often with glandular hairs inside. **Petals** usually 5, at least basally fused, imbricate or rarely valvate; often with coronal corona or scales. **Stamens** usually 5, often attached to the base of corolla tube; filaments short, sometimes fused; anthers free or fused to top of style; dorsifixed or basifixed, pollen sometimes aggregated into pollinia; connective often with apical appendages. **Ovary** superior, inferior or part-inferior; carpels 2 usually fused or free; locules (1–) 2; ovules (1–) 2 to many per locule; placentation axile to apical or parietal; styles 1–2. **Fruit** often a berry, drupe, capsule, or paired follicle. **Seeds** often with coma.

乔木、灌木、木质藤本或稀为**草本**；有时肉质；具乳汁。**叶**为单叶，常对生或轮生，稀互生；有时退化至缺；全缘；常具叶柄；托叶缺或有时存在。**花序**为圆锥状或伞形、聚伞状、穗状、头状、伞房状或为单花或总状，有时老茎生花。**花**两性或稀单性（雌雄异株），辐射对称；芳香或恶臭。**花萼**合生，常5裂，覆瓦状，内侧常具腺毛。**花瓣**常5枚，至少在基部合生，覆瓦状或稀镊合状；常具花冠状副花冠或鳞片。**雄蕊**常5枚，多贴生于花冠筒基部；花丝短，有时合生；花药离生或合生于花柱顶端；背着或基着，花粉有时黏合成花粉块；药隔顶端常具附属物。**子房**上位、下位或半下位；心皮2枚，常合生或离生；(1~) 2室；胚珠每室 (1~) 2至多数；中轴胎座、顶生胎座或侧膜胎座；花柱1~2个。**果实**常为浆果、核果、蒴果或双生的蓇葖果。**种子**常具种缨。

World ca. 366/5,100, tropical and subtropical regions, few in temperate regions. China 87/423.
This area 8/12.

全世界共约366属5 100种，主要分布于热带和亚热带，少数至温带地区。中国产87属423种。
本地区有8属12种。

Alyxia sinensis Champ. ex Benth./ 链珠藤

1. Pollen united into pollinia/ 花粉粒黏合成花粉块
   2. Leaves mostly fleshy/ 叶多肉质 ....**1.** *Hoya*/ 球兰属

2. Leaves not fleshy/ 叶非肉质
　　3. Corona lobes absent or reduced to scales/ 副花冠裂片缺或退化为小鳞片 .... **2. *Gymnema*/ 匙羹藤属**
　　3. Corona lobes well developed or corona forming a continuous ring/ 副花冠裂片显著，或副花冠环状
　　　　4. Stigmas elongated or beaked/ 柱头伸长或喙状 .... **3. *Metaplexis*/ 萝藦属**
　　　　4. Stigmas disclike/ 柱头盘状 .... **4. *Cynanchum*/ 鹅绒藤属**
1. Pollen not united into pollinia/ 花粉粒不形成花粉块
　　8. Leaves whorled/ 叶轮生 .... **5. *Alyxia*/ 链珠藤属**
　　8. All leaves opposite/ 叶对生
　　　　9. Anther apex with long soft hairs/ 花药先端具长柔毛 .... **6. *Sindechites*/ 毛药藤属**
　　　　9. Anther apex glabrous/ 花药先端光滑
　　　　　　10. Follicles divergent or parallel, linear or fusiform/ 蓇葖果略叉开或平行，线形或纺锤形 ....
　　　　　　.. **7. *Trachelospermum*/ 络石属**
　　　　　　10. Follicles divaricate, narrowly ovoid/ 蓇葖果叉开，狭卵形 .... **8. *Anodendron*/ 鳝藤属**

GENTIANALES. Apocynaceae. *Metaplexis japonica* (Thunb.) Makino/ 龙胆目 夹竹桃科 萝藦
A. Flowering branch/ 花枝
B. Part of gynostemium/ 部分合蕊柱
C. Corolla segregated/ 花冠离析
D. Cross section of corolla tube/ 花冠筒横切面
E. Corolla unfold (inside)/ 花冠展开（内侧）
F. Cymes/ 聚伞花序
G. Gynostemium segregated/ 合蕊柱离析
H. Translator/ 载粉器
I. Vertical profile of flower/ 花纵剖面
(Scale/标尺：A 1 cm; B–C 1 mm; D–F 1 cm; G–I 1 mm)

1. *Hoya carnosa* (L. f.) R. Br./ 球兰；
2. *Gymnema sylvestre* (Retz.) R. Br. ex Schult./ 匙羹藤；
3. *Metaplexis japonica* (Thunb.) Makino/ 萝藦；
4. *Cynanchum fordii* Hemsl./ 山白前；
5. *Alyxia sinensis* Champ. ex Benth./ 链珠藤；
6. *Sindechites henryi* Oliv./ 毛药藤；
7. *Trachelospermum jasminoides* (Lindl.) Lem./ 络石；
8. *Anodendron affine* (Hook. & Arn.) Druce/ 鳝藤

357 Boraginaceae 紫草科

Leaves simple, alternate / 单叶，互生
Herbs with bristle to prickle-like hairs / 草本被刚毛至刺状毛
Inflorescences often scorpioid / 常为蝎尾状花序
Corolla throat often with scale-like corona / 花冠喉部常具鳞片状副花冠
Nutlets / 小坚果

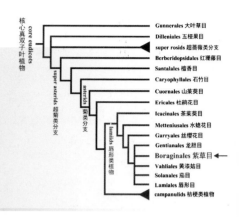

Herbs, less often woody **climbers**, **shrubs**, or **trees**, rarely root-parasites. **Leaves** simple, alternate, rarely opposite; margin entire or serrate; petioles present or absent; stipules absent; hairs usually bristle to prickle-like in herbs. **Inflorescence** often scorpioid or panicles, cymes, head-like, spikes, corymbs or solitary flower. **Flowers** bisexual, actinomorphic. **Sepals** usually 5-lobed, mostly persistent. **Petals** fused, usually imbricate or contorted; tubular, campanulate, rotate, funnelform, or salverform; often throat with scale-like corona. **Stamens** 5, inserted on corolla tube or rarely at throat, included or rarely exserted; anthers introrse, 2-loculed, usually dorsifixed, dehiscing longitudinally. **Ovary** superior, carpels 2, fused; 2–4-loculed; ovule 1-2 per locule; placentation axile to basal; style terminal or gynobasic, branched or not. **Fruit** 1–4-seeded drupes or nutlets, or capsule.

草本，偶为木质**藤本**、**灌木**或**乔木**，稀根寄生性。**叶**为单叶，互生，稀对生；全缘或具齿；叶柄有或无；托叶缺；草本类群常被刚毛至刺状毛。**花序**常为蝎尾状或圆锥状、聚伞状、头状、穗状、伞房状或为单花。**花**两性，辐射对称。**花萼**常5裂，多宿存。**花瓣**合生，常覆瓦状或旋卷状；管状、钟状、圆盘状、漏斗状或高脚碟状；喉部常具鳞片状副花冠。**雄蕊** 5枚，嵌生于花冠筒或稀喉部，内藏或稀伸出；花药内向，2室，常背着，纵裂。**子房**上位，心皮2枚，合生；2~4室；胚珠每室1~2枚；中轴胎座至基底胎座；花柱顶生或着生于子房基部，分枝或不分枝。**果实**为1~4粒种子的核果或小坚果，或为蒴果。

World ca. 143/2,785, mainly in temperate regions, few in tropical mountane habitats.
China 44/ca. 300.
This area 7/9.

全世界共约143属2 785种，主要分布于温带，少数至热带山区。
中国产44属约300种。
本地区有7属9种。

1. Ovary (2 or) 4-lobed/ 子房4裂或2裂
   2. Nutlets attachment scar concave/小坚果着生面内凹 .... **1. *Symphytum*/ 聚合草属**
   2. Nutlets attachment scar not concave/ 小坚果着生面不内凹
      3. Anthers mucronulate at apex/ 花药顶端有小尖头 ....**2. *Lithospermum*/ 紫草属**
      3. Anthers without mucronulate/ 花药顶端无小尖头

4. Nutlets with glochids or wings/ 小坚果有锚状刺或翅 ....**3. *Cynoglossum*/ 琉璃草属**
4. Nutlets without glochids or wings/ 小坚果无锚状刺或翅
  5. Nutlets tetrahedral or lenticular/ 小坚果四面体形或透镜状 ....**4. *Trigonotis*/ 附地菜属**
  5. Nutlets neither tetrahedral nor lenticular/ 小坚果非四面体形或透镜状
   6. Cupular emergence with 2 layers/ 小坚果具 2 层碗状突起 ....**5. *Thyrocarpus*/ 盾果草属**
   6. Cupular emergence with 1 layer/ 小坚果具 1 层碗状突起 ....**6. *Bothriospermum*/ 斑种草属**
1. Ovary undivided/ 子房不分裂 ....**7. *Ehretia*/ 厚壳树属**

BORAGINALES. Boraginaceae. *Trigonotis peduncularis* (Trevir.) Benth. ex Baker & S.Moore/ 紫草目 紫草科 附地菜
A. Whole plant/ 植株
B. Corolla unfold (inside)/ 花冠展开 (内侧)
C. Stamens/ 雄蕊
D. Flowering branch/ 花枝
E. Nutlets/ 小坚果
(Scale/ 标尺： A 5 cm; B 1 mm; C 0.2 mm; D 1 cm; E 1 mm)

1. *Symphytum officinale* L./ 聚合草（photo by WeiLiang MA/ 马炜梁拍摄）;
2. *Lithospermum zollingeri* A. DC./ 梓木草;
3. *Cynoglossum lanceolatum* Forssk./ 小花琉璃草;
4. *Trigonotis peduncularis* (Trevir) Benth. ex Baker & S. Moore/ 附地菜;
5. *Thyrocarpus sampsonii* Hance/ 盾果草;
6. *Bothriospermum zeylanicum* (J. Jacq.) Druce/ 柔弱斑种草;
7. *Ehretia acuminata* R. Br./ 厚壳树

359 Convolvulaceae 旋花科

Twining vines with milky sap / 缠绕藤本，有乳汁
Leaf bases cordate or hastate / 叶基心形或戟形
Flowers 5-merous, funnelform / 花 5 基数，花冠漏斗状
Capsule dehiscing by valves / 蒴果，瓣裂

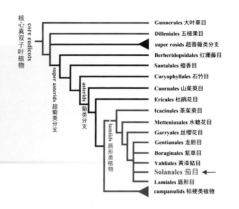

Herbs to shrubs, usually with twining or creeping stems or erect, rarely **trees**, or **twining** parasitic herbs; often with slightly milky juice. **Leaves** simple, or sometimes compound or reduced to scale-like; alternate; margin entire or rarely lobed, bases cordate or hastate; petioles present; stipules absent. **Inflorescence** solitary, cymes, heads, racemes, thyrses or panicles. **Flowers** bisexual or rarely unisexual (plants dioecious); usually 5-merous; actinomorphic, often showy. **Sepals** free or sometimes fused, imbricate, equal or not, often persistent, sometimes enlarged in fruit. **Petals** fused, often showy, funnelform, campanulate, salverform or urceolate, contorted. **Stamens** alternating with corolla lobes; filaments attached to petals, sometimes hairy. **Ovary** superior; carpels 2, fused; 1–2 (–4)-loculed; ovules 2 or many per locule; placentation basal; style 1 or 2; stigmas entire or 2–3-lobed. **Fruit** a capsule, usually dehiscing by valves, circumscissile, rarely a berry or nut-like. **Seeds** usually trigonous.

草本至灌木，常具缠绕茎或匍匐茎，或直立，稀**乔木**，或为**寄生**缠绕草本；常稍具乳汁。**叶**为单叶，或有时为复叶或退化至鳞片状；互生；全缘或稀分裂，叶基心形或戟形；具叶柄；托叶缺。**花序**为单花、聚伞状、头状、总状、聚伞圆锥状或圆锥状。**花**两性或稀单性（雌雄异株）；常 5 基数；辐射对称，常显著。**花萼**离生或有时合生，覆瓦状，等大或不等大，常宿存，有时果期增大。**花瓣**合生，常显著，漏斗形、钟状、高脚碟状或坛状，旋转状排列。**雄蕊**与花冠裂片互生；花丝贴生于花瓣，有时被毛。**子房**上位；心皮 2 枚，合生；1~2 (~4) 室；胚珠每室 2 枚或多数；基底胎座；花柱 1 个或 2 个；柱头全缘或 2~3 裂。**果实**为蒴果，常瓣裂、周裂，稀浆果或坚果状。**种子**常为三棱形。

World 58/1,650, tropics to temperate regions, especially in America, tropical and subtropical Asia.
China 20/128.
This area 6/20.

全世界共 58 属 1 650 种，广布于热带至温带，尤其是美洲、亚洲的热带和亚热带。
中国产 20 属 128 种。
本地区有 6 属 20 种。

1. Plants parasitic/ 寄生植物 .... **1. *Cuscuta*/ 菟丝子属**
1. Plants not parasitic/ 非寄生植物
  2. Pollen finely spiny/ 花粉粒具刺 .... **2. *Ipomoea*/ 番薯属**
  2. Pollen never spiny/ 花粉粒无刺

3. Ovary deeply 2-lobed/子房 2 深裂 .... **3. *Dichondra*/ 马蹄金属**
3. Ovary not lobed/ 子房不分裂
    4. Styles 2/花柱2个 .... **4. *Evolvulus*/ 土丁桂属**
    4. Style 1/ 花柱 1 个
        5. Calyx not enclosed in bracts; flowers usually yellow/花萼不为苞片所包；花常为黄色 .... **5. *Merremia*/ 鱼黄草属**
        5. Calyx enclosed in bracts; flower red, pink/ 花萼包于苞片中；花常为红色、粉色 .... **6. *Calystegia*/ 打碗花属**

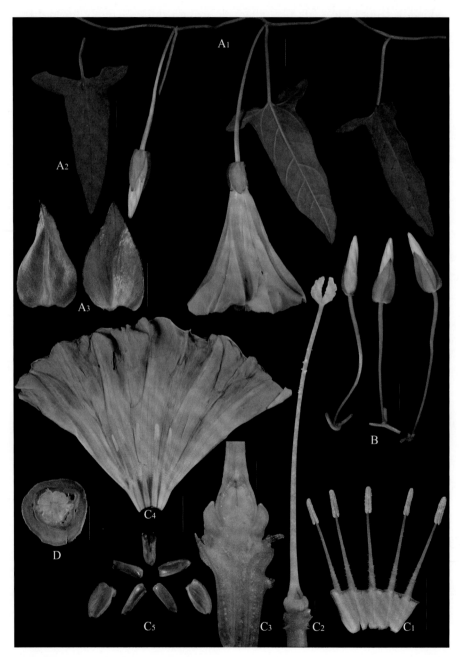

SOLANALES. Convolvulaceae. *Calystegia hederacea* Wall. / 茄目 旋花科 打碗花

A_1. Flowering branch/ 花枝

A_2. Leaf/ 叶

A_3. Bracts/ 苞片

B. Flower, bud stage/ 花 (蕾期)

C_1. Androeceum unfold/ 雄蕊群展开

C_2. Pistil/ 雌蕊

C_3. Vertical section of ovary/ 子房纵切面

C_4. Corolla unfold (inside)/ 花冠展开 (内侧)

C_5. Calyx segregated, with two bracts/ 花萼离析，具 2 枚苞片

D. Cross section of ovary/ 子房横切面

(Scale/ 标尺： A_1–A_2 1 cm; A_3 5 mm; B–C_1 1 cm; C_2–C_3 1 mm; C_4–C_5 1 cm; D 1mm)

1. *Cuscuta campestris* Yunck./ 原野菟丝子；
2. *Ipomoea nil* (L.) Roth/ 牵牛；
3. *Dichondra micrantha* Urb./ 马蹄金；
4. *Evolvulus alsinoides* (L.) L./ 土丁桂；
5. *Merremia sibirica* (L.) Hallier f./ 北鱼黄草；
6. *Calystegia soldanella* (L.) R. Br./ 肾叶打碗花

360 Solanaceae 茄科

Usually simple leaves, estipulate / 常单叶，托叶缺
Calyx enlarged in fruit / 花萼果期增大
Flowers bisexual, 5-merous / 花两性，5 基数
Fruit a berry or capsule / 浆果或蒴果

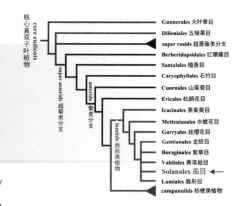

Herbs, **shrubs**, small **trees**, or woody **climbers**. **Leaves** simple, rarely compound; alternate; margin entire to deep lobed; petioles present; stipules absent. **Inflorescence** cymes or solitary flower; bracteate. **Flowers** mostly bisexual, usually actinomorphic or sometimes zygomorphic; 5-merous; bracteolate. **Sepals** fused, 5-lobed and persistent, often enlarged in fruit. **Petals** fused, usually 5-lobed, campanulate, tubular, funnelform or salverform, rarely 2-lipped. **Stamens** as many as corolla lobes and alternate with them; filaments attached to the petals, sometimes unequal; anthers dorsifixed or basifixed, sometimes connivent, dehiscing by longitudinal slits or terminal porcidally; staminodes 0–3. **Ovary** superior or part-inferior; carpels 2, fused; often 2-loculed, ovules 1 to many per locule; placentation mostly axile. **Fruit** a berry or less often drupe or capsule.

草本、灌木、小乔木或木质藤本。**叶**为单叶，稀复叶；互生；全缘至深裂；具叶柄；托叶缺。**花序**聚伞状或为单花；具苞片。**花**多为两性，常辐射对称或有时两侧对称；5 基数；具小苞片。**花萼**合生，5 裂且宿存，常果期增大。**花瓣**合生，常 5 裂，冠筒钟状、管状、漏斗形或高脚碟状，稀二唇形。**雄蕊**与花冠裂片同数且互生；花丝贴生于花瓣，有时不等长；花药背着或基着，有时靠合，纵裂或顶孔开裂；退化雄蕊 0~3 枚。**子房**上位或半下位；心皮 2 枚，合生；常 2 室，胚珠每室 1 至多枚；多中轴胎座。**果实**为浆果或偶为核果或蒴果。

World ca. 102/2,460, widely distributed in temperate regions and tropical regions, especially in tropical America.
China 20/102.
This area 7/16.

全世界共约 102 属 2 460 种，分布于温带和热带，尤其是美洲热带地区。
中国产 20 属 102 种。
本地区有 7 属 16 种。

1. Flowers in several- to many-flowered inflorescence/ 花多朵生于花序上 ....**1. *Solanum*/ 茄属**
1. Flowers 1–3 per axil/ 花每腋 1~3 朵
  2. Fruit enclosed in fruiting calyx/ 果实完全被果萼包围
    3. Fruiting calyx lobes free more than halfway; fruit a dry berry/ 果期萼裂片分离过半；浆果干燥 ....
    ....**2. *Nicandra*/ 假酸浆属**
    3. Fruiting calyx lobes united to near apex; fruit a juicy berry/ 果期萼裂片联合至近顶端；浆果多汁 ....

....**3. *Physalis*/** 酸浆属

2\. Fruit fully or mostly exposed, free from calyx/ 果实不被萼片包围，暴露在外

    4. Fruit a capsule/ 果为蒴果 ....**4. *Datura*/** 曼陀罗属

    4. Fruit a berry/ 果为浆果

        5. Stamens mostly exserted; leaves usually fasciculate on short shoots/ 雄蕊多伸出；叶常簇生于短枝顶端 ....**5. *Lycium*/** 枸杞属

        5. Stamens included; leaves arising along stem or forming a basal rosette/ 雄蕊内藏；叶沿茎生长或基生莲座状

            6. Anthers dehiscing by apical pores; calyx often with 10 subapical teeth/ 花药顶孔开裂；花萼常 10 齿 ....**6. *Lycianthes*/** 红丝线属

            6. Anthers dehiscing longitudinally; calyx 5-toothed or toothless/ 花药纵裂开裂；花萼 5 齿或无齿 ....**7. *Tubocapsicum*/** 龙珠属

1. *Solanum lyratum* Thunb. ex Murray/ 白英；
2. *Nicandra physalodes* (L.) Gaertn./ 假酸浆；
3. *Physalis angulata* L./ 苦蘵；
4. *Datura stramonium* L./ 曼陀罗；
5. *Lycium chinense* Mill./ 枸杞；
6. *Lycianthes biflora* (Lour.) Bitter/ 红丝线（6a. Fruiting branch/ 果枝；6b. Calyx with 10 teeth/ 花萼具 10 齿）；
7. *Tubocapsicum anomalum* (Franch. & Sav.) Makino/ 龙珠

366 Oleaceae 木樨科

Branches often 4-angled / 枝常四棱形
Leaves opposite / 叶对生
Flowers 2-merous, actinomorphic / 花 2 基数，辐射对称
Ovary superior, 2-loculed / 子房上位，2 室

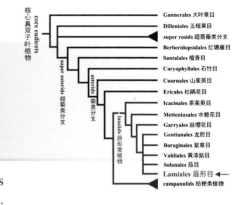

Trees, erect or scandent **shrubs**, woody **climbers**, or **herbs**; stems often 4-angled. **Leaves** simple, or pinnately compound; often opposite; margin entire to toothed; petioles present; stipules absent. **Inflorescence** panicles, cymes, or fascicles, rarely solitary flower; bracteate. **Flowers** bisexual or rarely unisexual, actinomorphic; nectar disk present. **Sepals** usually 4-lobed, valvate. **Petals** fused, rarely free to base, 4-lobed. **Stamens** 2, rarely 4, alternate with petals; filaments attached to the petals. **Ovary** superior, carpels fused; 2-loculed; ovules often 2 per locule; placentation axile; style 1 or absent; stigma 2-lobed or capitate. **Fruit** a drupe, berry, capsule, or samara.

乔木、直立或攀缘**灌木**、木质**藤本**或**草本**；茎常 4 棱形。**叶**为单叶，或羽状复叶；常对生；全缘至具齿；具叶柄；托叶缺。**花序**为圆锥状、聚伞状或簇生，稀为单花；具苞片。**花**两性或稀单性，辐射对称；具蜜腺盘。**花萼**常 4 裂，镊合状。**花瓣**合生，稀分离至基部；4 裂。**雄蕊** 2 枚，稀 4 枚，与花瓣互生；花丝贴生于花瓣。**子房**上位，心皮合生；2 室；胚珠每室常 2 枚；中轴胎座；花柱 1 个或缺；柱头 2 裂或头状。**果实**为核果、浆果、蒴果或翅果。

World 24/615, widely distributed in temperate and tropical regions, especially in Asia.
China 10/160.
This area 6/13.

全世界共 24 属 615 种，广布于温带和热带地区，尤其是亚洲。
中国产 10 属 160 种。
本地区有 6 属 13 种。

1. Fruit a samara/ 果为翅果
    2. Leaves simple/ 单叶 ....**1. *Fontanesia*/ 雪柳属**
    2. Leaves pinnately compound/ 羽状复叶 ....**2. *Fraxinus*/ 梣属**
1. Fruit a drupe or berry/ 果为核果或浆果
    3. Fruit a drupe/ 果为核果
        4. Inflorescence axillary/ 花序腋生 ....**3. *Osmanthus*/ 木樨属**
        4. Inflorescence terminal/ 花序顶生 ....**4. *Chionanthus*/ 流苏树属**
    3. Fruit a berry or berrylike/ 果为浆果或浆果状
        5. Leaves simple/ 单叶 ....**5. *Ligustrum*/ 女贞属**
        5. Leaves pinnately compound, 3-foliolate / 羽状复叶或 3 出复叶 ....**6. *Jasminum*/ 素馨属**

1. *Fontanesia phillyreoides* Labill subsp. *fortunei* (Carrière) Yalt./ 雪柳；
2. *Fraxinus chinensis* Roxb./ 白蜡树；
3. *Osmanthus fragrans* (Thunb.) Lour./ 木樨；
4. *Chionanthus retusus* Lindl. & Paxton/ 流苏树；
5. *Ligustrum obtusifolium* Siebold & Zucc. subsp. *microphyllum* (Nakai) P.S. Green/ 东亚女贞；
6. *Jasminum lanceolarium* Roxb./ 清香藤

369 Gesneriaceae 苦苣苔科

Herbs hairy and slighty fleshy / 多毛且稍肉质
Leaves basal or opposite / 叶基生或对生
Corolla 5-lobed, 2-lipped / 花冠 5 裂，2 唇形
Carpels 2, 1-loculed / 子房 2 心皮，1 室

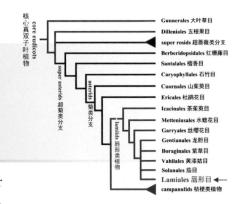

Perennial or rarely annual **herbs** or **small shrubs**, **woody climbers** or rarely **trees**, terrestrial or epiphytic. **Leaves** simple, opposite, sometimes alternate or whorled, or basal, rosette forming; ± succulent; often softy hairy; margin entire, toothed or rarely pinnatifid; petioles present or rarely absent; stipules absent. **Inflorescence** cymes, panicles or solitary flower; usually pedunculate; bracts in pairs. **Flowers** bisexual or rarely unisexual (monoecious), usually zygomorphic or actinomorphic; often bracteolate. **Sepals** fused, 5-lobed; tubular or campanulate, 2-lipped. **Petals** fused, imbricate, usually 2-lipped, sometimes with a spur, nectariferous. **Stamens** opposite the sepals, unequal; filaments attached to petals; anthers basifixed or rarely dorsifixed; 2-loculed. **Ovary** superior, part-inferior or inferior; carpels 2, fused; 1-loculed, placentation parietal or rarely axile; style 1; stigmas 1 or 2. **Fruit** a loculicidal or septicidal capsule. **Seeds** numerous and minute.

多年生或稀一年生**草本**或**小灌木**、**木质藤本**或稀为**乔木**，地生或附生。**叶**为单叶，对生，有时互生或轮生，或基生，成莲座状；多少肉质；常被柔毛；全缘、具齿或稀羽裂；具叶柄，稀缺；托叶缺。**花序**聚伞状、圆锥状或为单花；常具花梗；苞片成对。**花**两性或稀单性（雌雄同株），常两侧对称或辐射对称；多具小苞片。**花萼**合生，5 裂；萼筒管状或钟状，二唇形。**花瓣**合生，覆瓦状，常二唇形，有时有距，具蜜汁。**雄蕊**与花萼对生，不等大；花丝贴生于花瓣；花药基着或稀背着；2 室。**子房**上位、半下位或下位；心皮 2 枚，合生；1 室，侧膜胎座或稀中轴胎座；花柱 1 个；柱头 1 或 2 个。**果实**为室背开裂或室间开裂的蒴果。**种子**多而细小。

Lysionotus pauciflorus Maxim./ 吊石苣苔

World ca. 150/3,500, worldwide.
China 60/421.
This area 2/2.

全世界共约 150 属 3 500 种，世界分布。
中国产 60 属 421 种。
本地区有 2 属 2 种。

1. Stemless; all leaves basal/ 无地上茎；叶全基生 ....**1. *Oreocharis*/ 马铃苣苔属**
1. Stems branched or simple; leaves along stem/ 茎分枝或单一；叶茎生 ....**2. *Lysionotus*/ 吊石苣苔属**

1. *Oreocharis maximowiczii* C.B. Clarke/ 大花石上莲
(1a. Flowering plants/ 花期植株; 1b. Capsule/ 蒴果);

2. *Lysionotus pauciflorus* Maxim./ 吊石苣苔
(2a. Flowering plants/ 花期植株; 2b. Seeds/ 种子)

370 Plantaginaceae 车前科

Usually herbs / 常为草本
Inflorescence terminal and unbranched / 花序顶生，不分枝
Corolla 2-lipped ± equals in size / 上下唇瓣近等大
Stamens 4, somewhat didynamous / 雄蕊 4 枚，多少为二强雄蕊

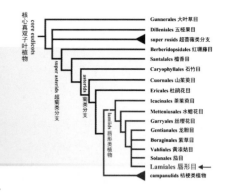

Herbs, or shrubs, rarely climbers; terrestrial, xerophytic or aquatic. **Leaves** simple or compound, alternate, opposite, or occasionally whorled; margin entire to variously toothed; petioles present or absent; stipules absent. **Inflorescence** often a racemes or spikes, sometimes flowers solitary or paired; terminal and unbranched. **Flowers** usually bisexual or unisexual; usually zygomorphic or actinomorphic. **Sepals** fused, imbricate. **Petals** fused, imbricate, often 2-lipped, two lips ± equals in size; sometimes with a nectar spur. **Stamens** 4 or 2, somewhat didynamous; filaments adnate to petals; anthers dorsifixed. **Ovary** superior or inferior; carpels 2, fused; locules 1–2; ovules 1 to many per locule; placentation axile or rarely basal; stigma 2-lobed. **Fruit** a capsule, septicidal, poricidal or circumscissile, rarely a nut, berry or drupe.

草本或灌木，稀藤本；地生、旱生或水生。**叶**为单叶或复叶，互生、对生或偶轮生；全缘至各种锯齿；叶柄有或无；托叶缺。**花序**常为总状或穗状，有时单花或双花；顶生，不分枝。**花**常两性或单性；常两侧对称或辐射对称。**花萼**合生，覆瓦状。**花瓣**合生，覆瓦状，常二唇形，上下唇瓣多少等大；有时具蜜腺距。**雄蕊** 4 或 2 枚，多少为二强雄蕊；花丝与花瓣合生；花药背着。**子房**上位或下位，心皮 2 枚，合生；1~2 室；胚珠每室 1 至多数；中轴胎座或稀基底胎座；柱头 2 裂。**果实**为蒴果，室间开裂、孔裂或周裂，稀为坚果、浆果或核果。

World ca. 90/1, 900, worldwide, mainly in temperate regions.
China 21/165.
This area 6/16.

全世界共约 90 属 1 900 种，世界广布，主要分布于温带地区。
中国产 21 属 165 种。
本地区有 6 属 16 种。

1. Petals absent/ 无花瓣 .... **1. *Callitriche*/ 水马齿属**
1. Petals present/ 有花瓣
  2. Corolla actinomorphic or nearly so/ 花冠辐射对称或近辐射对称 .... **2. *Plantago*/ 车前属**
  2. Corolla zygomorphic/ 花冠两侧对称
    3. Leaves glandular punctate/ 叶具腺点 .... **3. *Scoparia*/ 野甘草属**
    3. Leaves not glandular punctate/ 叶不具腺点
      4. Stamens 4 or 5/ 雄蕊 4 或 5 枚 .... **4. *Bacopa*/ 假马齿苋属**

4. Stamens 2/ 雄蕊 2 枚

    5. Calyx lobes 5, subequal in length/ 花萼裂片 5 枚，近等长 ....**5. *Veronicastrum*/ 腹水草属**

    5. Calyx lobes 4, if 5 then upper one much smaller than other lobes/ 花萼裂片 4 枚，如 5 枚则上方 1 枚小得多 ....**6. *Veronica*/ 婆婆纳属**

1. *Callitriche palustris* L./ 水马齿；
2. *Plantago virginica* L./ 北美车前；
3. *Scoparia dulcis* L./ 野甘草；
4. *Bacopa monnieri* (L.) Wettst./ 假马齿苋；
5. *Veronicastrum axillare* (Siebold & Zucc.) T. Yamaz./ 爬岩红；
6. *Veronica polita* Fr./ 婆婆纳

371　Scrophulariaceae 玄参科

With stellate hairs or pellicid dots / 具星状毛或透明油点
Leaves opposite / 叶对生
Inflorescence often branched / 花序常分枝
Corolla 2-lipped / 花冠二唇形

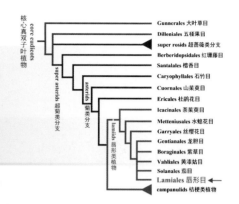

Herbs, **subshrubs**, **shrubs** or **trees**. **Leaves** opposite or rarely upper ones alternate; margin entire or toothed; petioles present or absent; stipules absent or present; usually with stellate hairs or pellicid dots. **Inflorescence** usually solitary flower, racemes, cymes, spikes or thyrses, often branched; sometimes bracteate. **Flowers** bisexual or rarely unisexual (functionally dioecious), usually zygomorphic. **Sepals** fused, 4- or 5-lobed, or free, ± unequal. **Petals** fused, 4- or 5-lobed, usually 2-lipped, tubular. **Stamens** 4, somewhat didynamous, sometime equal, rarely 5 or 2; filaments attached to petals; anthers introrse. **Ovary** superior, often nectary disk at base; carpels 2, fused, sometimes 4; usually 2-loculed; ovules 1 to many per locule; placentation axile or less often apical or parietal; stigmas often small. **Fruit** a septicidal or loculicidal capsule, rarely a berry or drupe.

草本、**亚灌木**、**灌木**或**乔木**。**叶**对生或稀上部互生；全缘或具齿；叶柄有或无；托叶缺或存在；常被星状毛或透明腺点。**花序**常为单花、总状、聚伞状、穗状或聚伞圆锥状，常分枝；有时具苞片。**花**两性或稀单性（功能性雌雄异株），常两侧对称。**花萼**合生，4 或 5 裂，或离生，多少不等大。**花瓣**合生，4 或 5 裂，常二唇形，管状。**雄蕊** 4 枚，多少为二强雄蕊，有时等长，稀 5 或 2 枚；花丝贴生于花瓣；花药内向。**子房**上位，基部常具蜜腺盘；心皮 2 枚，合生，偶为 4 枚；常 2 室；胚珠每室 1 至多枚；中轴胎座或稀为顶生胎座或侧膜胎座；柱头常细小。**果实**为室间开裂或室背开裂蒴果，稀为浆果或核果。

World 9/510, worldwide, mainly in Asia temperate regions and tropical montane habitats.
China 8/68.
This area 2/3.

全世界共 9 属 510 种，世界广布，尤其是亚洲温带和热带山地。
中国产 8 属 68 种。
本地区有 2 属 3 种。

1. Shrubs to trees, rarely herbs; flowers actinomorphic/ 灌木至乔木，稀草本；花辐射对称 .... **1. *Buddleja*/ 醉鱼草属**
1. Herbs, woody herbs to shrubs; flowers zygomorphic/ 草本、木质草本至灌木；花两侧对称 .... **2. *Scrophularia*/ 玄参属**

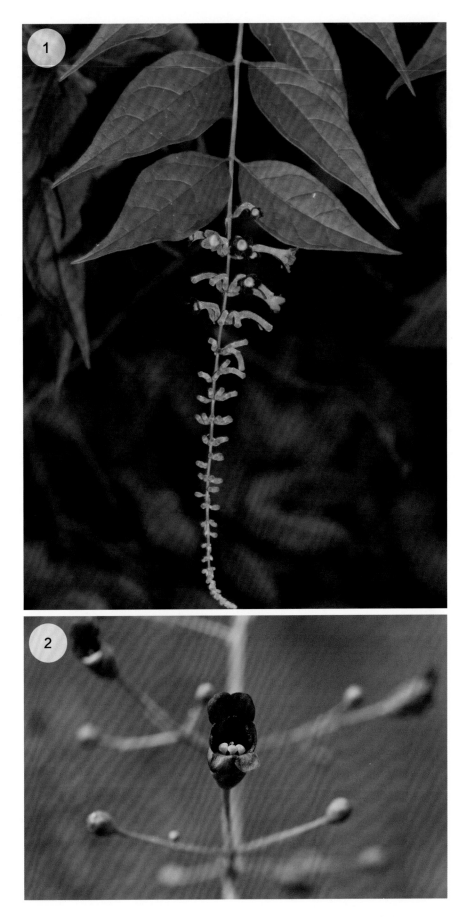

1. *Buddleja lindleyana* Fortune/ 醉鱼草；
2. *Scrophularia ningpoensis* Hemsl./ 玄参

373 Linderniaceae 母草科

Stem 4-angled / 茎四棱

Corolla lower lip larger than upper / 花冠下唇大于上唇

Stamens 4, filaments appendaged / 雄蕊 4 枚，花丝具附属物

Capsule septicidal or poricidal / 蒴果室间开裂或孔裂

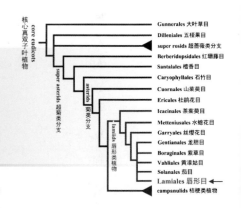

Herbs to **subshrubs**, sometimes **aquatic**; erect, prostrate, or creeping, 4-angled. **Leaves** usually simple, decussate, or basally rosette; margin entire or toothed; petioles present or absent; stipules absent. **Inflorescence** usually racemes, head, or sometimes solitary flower. **Flowers** bisexual, zygomorphic. **Sepals** fused, 5-lobed. **Petals** fused, glandular hairs inside; 2-lipped, lower lip larger than upper, extended; upper lip erect. **Stamens** 4, all fertile or 2 abaxial reduced, filaments attached to petals, abaxial pair Z-shaped or with conspicuous appendage. **Ovary** superior; carpels fused; 2-loculed; ovules usually many per locule; placentation axile to basal; style apex often enlarged. **Fruit** septicidal or poricidal capsule. **Seeds** small, numerous, with ruminate endosperm.

草本至**亚灌木**，有时**水生**；茎直立、平卧或匍匐，四棱形。**叶**常为单叶，交互对生，或基生莲座状；全缘或具齿；叶柄有或无；托叶缺。**花序**常为总状、头状，或有时为单花。**花**两性，两侧对称。**花萼**合生，5 裂。**花瓣**合生，内侧具腺毛；二唇形，下唇大于上唇，平展；上唇直立。**雄蕊** 4 枚，全部发育或前对雄蕊退化，花丝贴生于花瓣，前对雄蕊花丝 Z 形，或具明显附属物。**子房**上位；心皮合生；2 室；胚珠常每室多数；中轴胎座至基底胎座；花柱先端增大。**果实**为室间开裂或孔裂的蒴果。**种子**小，多数，具嚼烂状胚乳。

World ca. 17/253, warm temperate regions and tropics, especially in Americas.
China 8/41.
This area 4/4 (genera here following Fischer et al., 2013).

全世界共约 17 属 253 种，分布于暖温带和热带尤其是美洲。
中国产 8 属 41 种。
本地区有 4 属 4 种（属的界定参考了 Fischer 等 2013 年的工作）。

1. Seeds without alveolate endosperm, seed surface ± smooth or only weakly furrowed, endosperm weakly polygonal or undulate in transverse section/ 种子不具蜂窝状胚乳，种子表面多少光滑或仅有浅沟槽，胚乳横切面稍呈多边形或波状 ·· ***1. Lindernia*/ 陌上菜属**
1. Seeds with alveolate endosperm, seed surface with rounded pits or longitudinal furrows, endosperm star-shaped in transverse section/ 种子具蜂窝状胚乳，种子表面具圆孔或纵沟槽，胚乳横切面呈星形
    2. Anterior stamens 2, reduced/ 前对雄蕊退化 ·· ***2. Bonnaya*/ 泥花草属**
    2. All 4 stamens fertile/ 4 枚雄蕊全部可育

3. Fruit usually not exceeding length of calyx/ 果实通常不超过萼筒 ....**3. *Torenia*/ 蝴蝶草属**
3. Fruit usually distinctly exceeding length of calyx/ 果实常明显超出萼筒 ....**4. *Vandellia*/ 长蒴母草属**

1. *Lindernia procumbens* (Krock.) Philcox, seeds/ 陌上菜，种子；
2. *Bonnaya antipoda* (L.) Druce, seeds/ 泥花草，种子；
3. *Torenia violacea* (Azaola ex Blanco) Pennell/ 紫萼蝴蝶草；
4. *Vandellia anagallis* (Burm. f.) T. Yamaz./ 长蒴母草

376 Pedaliaceae 芝麻科

Often xerophytic / 多旱生
Leaves simple, nearly opposite / 单叶，近对生
Flower solitary, axillary / 花单生，叶腋
False septum present / 具假隔膜
Fruit often winged, spiny or horned / 果实常具翅、刺或角

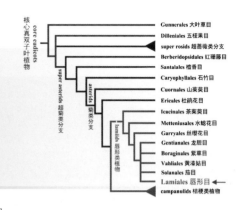

Herbs, erect or creeping, **shrubs** or **trees**, often xerophytic; stems sometimes swollen. **Leaves** simple, opposite or nearly opposite; margin entire or lobbed to pinnatifid; petioles usually present; stipules absent; conspicuous glandular hairs. **Inflorescence** axillary, solitary flower, cymes, or rarely raceme-like; pedicels with 2 (to many) nectaries at base. **Flowers** bisexual, ±zygomorphic. **Sepals** fused, 5-lobed. **Petals** fused, imbricate, spurred; obliquely campanulate; lowest lobe longest; glandular disk. **Stamens** 4, didynamous, anthers 2-loculed; filaments attached to petals; often staminode 1. **Ovary** superior or rarely inferior, 2 or rarely 3–4 carpels, fused; 2-loculed, ovules 1 to many per locule; placentation axile. **Fruit** a capsule, divided by a false septum, drupe or nut; often winged, spiny or horned.

草本，直立或匍匐，**灌木**或**乔木**，常旱生；茎有时膨大。**叶**为单叶，对生或近对生；全缘或分裂至羽裂；常具叶柄；托叶缺；腺毛明显。**花序**腋生，单花、聚伞状或稀总状式；花梗基部具 2（至多数）个蜜腺。**花**两性，多少两侧对称。**花萼**合生，5 裂。**花瓣**合生，覆瓦状，有距；钟状，偏斜；下唇最大；具蜜腺盘。**雄蕊** 4 枚，二强雄蕊，花药 2 室；花丝贴生于花瓣；常具 1 枚退化雄蕊。**子房**上位或稀下位，2 枚或稀 3~4 枚心皮，合生；2 室，胚珠每室 1 至多枚；中轴胎座。**果实**为蒴果，被假隔膜分开，核果或坚果；常具翅、刺或角。

World 15/70, tropical and subtropical Asia and Africa.
China 1/1.
This area 1/1. (Naturalized)

全世界共 15 属 70 种，分布于亚洲的热带和亚热带及非洲。
中国产 1 属 1 种。
本地区有 1 属 1 种。（归化）

Sesamum L. 芝麻属

Sesamum indicum L./ 芝麻

377 Acanthaceae 爵床科

Leaves opposite, veins prominent / 叶对生，叶脉显著
Inflorescence terminal racemes / 顶生总状花序
Bracts petaloid, prominent / 苞片花瓣状，显著
Seeds often borne on retinacula / 种子常生于珠柄钩

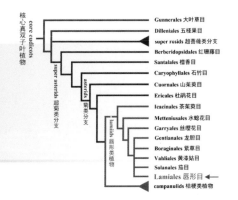

Herbs, sometimes **shrubs**, or rarely woody **climbers** or **trees**, sometimes aquatic or epiphytic; stems often swollen above the nodes. **Leaves** opposite, veins prominent; margin entire to toothed; petioles present or absent; stipules absent. **Inflorescence** terminal cymes, racemes, spikes, or solitary flower; bracts usually petaloid, prominent. **Flowers** bisexual, or rarely unisexual; zygomorphic; bracteolate. **Sepals** fused, deeply 4–5-lobed, or free. **Petals** fused, 2-lipped, sometimes upper lip absent. **Stamens** 2 or 4, rarely 5, didynamous; filaments attached to the petals; anthers dorsifixed; 2-loculed, dehiscing longitudinally. **Ovary** superior, basal nectary disk; carpels fused, 1–2-loculed; ovules 2 to many per locule; placentation axile. **Fruit** a capsule, loculicidal, often explosively dehiscent, rarely a drupe. **Seeds** often borne on retinacula.

草本，有时为**灌木**，或稀为木质**藤本**或**乔木**，有时水生或附生；茎常在节上部膨大。**叶**对生，叶脉明显；全缘至具齿；叶柄有或无；托叶缺。**花序**为顶生聚伞状、总状、穗状或单花；苞片常花瓣状，显著。**花**两性，或稀单性；两侧对称；具小苞片。**花萼**合生，4~5 深裂，或离生。**花瓣**合生，二唇形，有时上唇缺。**雄蕊** 2 或 4 枚，稀 5 枚，二强雄蕊；花丝贴生于花瓣；花药背着；2 室，纵裂。**子房**上位，基部有蜜腺盘；心皮合生，1~2 室；胚珠每室 2 至多枚；中轴胎座。**果实**为蒴果，纵裂，常为爆炸性开裂，稀为核果。**种子**常生于珠柄钩上。

World ca. 250/4,000, widely distributed in tropical and subtropical regions, few in warm temperate zones.
China 41/ca. 310.
This area 2/4.

全世界共约 250 属 4 000 种，广泛分布于热带和亚热带地区，少量至温带地区。
中国产 41 属约 310 种。
本地区有 2 属 4 种。

1. Inflorescence with 2–4 involucre/ 花序具 2~4 枚总苞片 ....**1. *Peristrophe*/ 观音草属**
1. Inflorescence without involucres/ 花序不具总苞片 ....**2. *Justicia*/ 爵床属**

1. *Peristrophe japonica* (Thunb.) Bremek./ 九头狮子草（1a. Flower/ 花；1b. Fruit/ 果实，photo by Xinxin ZHU/ 朱鑫鑫拍摄）；

2. *Justicia hayatae* Yamam./ 早田氏爵床（2a. Inflorescence/ 花序；2b. Infructescence/ 果序）

379 Lentibulariaceae 狸藻科

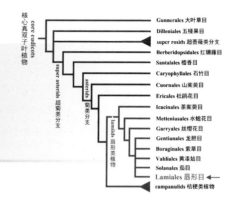

Insectivorous herbs, usually aquatic / 食虫草本，常水生
With prey-catching bladders or sticky hair / 具捕虫囊或黏毛
Leaves usually heterophylly / 叶常异型
Ovary superior, 1-loculed / 子房上位，1 室

Carnivorous herbs, aquatic or of wetlands, without true roots; stems modified into rhizoids or stolons. **Traps**: leaves **highly dissected**, bearing prey-catching bladders, small, bladder-like, each with 2 sensitive valves forming a trapdoor entrance, or leaves with sticky glandular hair. **Leaves** basally aggregated, simple or compound, alternate, usually heterophylly; margin entire or dissected; stipules absent. **Inflorescence** racemes, spikes or solitary flower; bracts and bracteole often present. **Flowers** bisexual, zygomorphic. **Sepals** 4 or 5, fused; often 2-lipped. **Petals** 5, fused, 2-lipped, lower lip larger than upper lip, spurred. **Stamens** 2, opposite with sepals; filaments adnate to petals; anther 2-loculed; dorsifixed. **Ovary** superior; carpels 2, fused; 1-loculed; ovules 2 to many; placentation free central or basal; style 1 or absent; stigma 2-lipped, upper lip often reduced. **Fruit** a capsule.

食虫草本，水生或湿生，无真正的根；茎变态为假根或匍匐枝。**捕虫器**：叶**高度分裂**，生有捕虫囊，小而囊状，每囊具有 2 个触敏活性瓣，或叶具黏性腺毛。**叶**基部簇生，单叶或复叶，互生，常异型；全缘或有缺刻；托叶缺。**花序**为总状、穗状或单花；常具苞片与小苞片。花两性，两侧对称。**花萼** 4 或 5 枚，合生；常二唇形，**花瓣** 5 枚，合生，二唇形，下唇大于上唇，有距。**雄蕊** 2 枚，与花萼对生；花丝贴生于花瓣；花药 2 室；背着。**子房上位**；心皮 2 枚，合生；1 室；胚珠每室 2 至多数；特立中央胎座或基底胎座；花柱 1 个或缺；柱头二唇形，上唇常退化。**果实**为蒴果。

World 3/ca. 290, worldwide, mainly in tropics.
China 2/ca. 27.
This area 1/4.

全世界共 3 属约 290 种，世界广布，主要分布于热带地区。
中国产 2 属约 27 种。
本地区有 1 属 4 种。

Utricularia L. 狸藻属

Utricularia aurea Lour./ 黄花狸藻
（a. Flowers/ 花；b. Prey-catching bladder/ 捕虫囊）

382 Verbenaceae 马鞭草科

Stems square, leaves opposite / 茎四棱，叶对生
Inflorescence terminal, dense cymes / 顶生密伞花序
Corolla weakly 2-lipped / 花冠略二唇形
Style terminal, stigma 2-lobed / 花柱顶生，柱头 2 裂

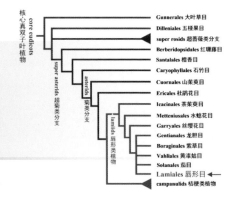

Herbs, **woody climbers**, **shrubs**, or **trees**; stems usually square in cross-section; sometimes with prickles or thorns. **Leaves** simple or compound, opposite, or occasionally whorled; sometimes aromatic; margin entire to serrate; petioles present; stipules absent. **Inflorescence** terminal, racemes, spikes, heads or cymes; bracteate, sometimes involucral or petaloid. **Flower** bisexual or unisexual (dioecious), slightly zygomorphic. **Sepals** tubular, persistent. **Petals** 4–5, fused, weakly 2-lipped, tubular or rarely campanulate. **Stamens** opposite the sepals, filaments attached to petals; 4, didynamous; anthers dorsifixed, 1- or 2-loculed, dehiscing longitudinally or sometimes a circular pore. **Ovary** superior, entire or 4-grooved; 2–4-loculed; stigma usually 2-lobed; ovules 1 or 2 per locule. **Fruit** a drupe with 1–4 pyrenes or 2(4) nutlets, or capsule.

草本、**木质藤本**、**灌木**或**乔木**；茎横切面常为方形；有时具皮刺或枝刺。**叶**为单叶或复叶，对生，或偶轮生；有时具芳香味；全缘至具齿；具叶柄；托叶缺。**花序**顶生，总状、穗状、头状或聚伞状；具苞片，有时为总苞状或花瓣状。花两性或单性（雌雄异株），稍两侧对称。**花萼**管状，宿存。**花瓣** 4~5 枚，合生，稍二唇形，管状或稀为钟状。**雄蕊**与花萼对生，花丝贴生于花瓣；4 枚，二强雄蕊；花药背着，1 或 2 室，纵裂或有时圆形孔裂。**子房**上位，全缘或具 4 沟槽；2~4 室；柱头常 2 裂；胚珠每室 1 或 2 枚。**果实**为具 1~4 核的核果或 2(4) 小坚果，或为蒴果。

World 32/ca. 840, widely distributed in tropical and temperate regions.
China 5/8.
This area 4/4.

Verbena officinalis L./ 马鞭草

全世界共 32 属约 840 种，广布于热带至温带地区。
中国产 5 属 8 种。
本地区有 4 属 4 种。

1. Inflorescence dense capitula or short spikes, with overlapping bracts/花序为密集的头状花序或短穗状花序，有重叠的苞片
    2. Shrubs, erect/ 直立灌木 ....**1.** *Lantana*/ 马缨丹属
    2. Herbs, creeping/ 匍匐草本 ....**2.** *Phyla*/ 过江藤属
1. Inflorescence elongated spikes or racemes, without overlapping bracts/ 花序为伸长的穗状花序或总状花序，

无重叠的苞片

3. Shrubs, often climbing/ 攀缘灌木 .... **3. *Duranta*/ 假连翘属**

3. Herbs/ 草本 .... **4. *Verbena*/ 马鞭草属**

1. *Lantana camara* L./ 马缨丹（1a. Plants/ 植株；1b. Inflorescence/ 花序）；

2. *Phyla nodiflora* (L.) Greene/ 过江藤；

3. *Duranta erecta* L./ 假连翘（Naturalized/ 归化）；

4. *Verbena officinalis* L./ 马鞭草

383 Lamiaceae 唇形科

Leaves opposite, stems 4-angled / 叶对生，茎四棱
Often verticillasters / 常为轮伞花序
Corolla 2-lipped, stamens didynamous / 唇形花冠，二强雄蕊
Fruit 4 nutlets / 4 个小坚果

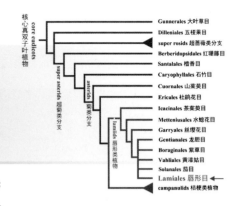

Herbs to shrubs, or trees, or rarely **woody climbers**, with aromatic oils; stems usually square in cross-section. **Leaves** simple or rarely compound, usually opposite, occasionally whorled or alternate; margin entire to serrate; petioles present; stipules absent. **Inflorescence** usually 2 dichasium per axil, often subtended by leaves or bracts, sometimes verticillasters. **Flowers** bisexual or rarely unisexual, zygomorphic or rarely actinomorphic. **Sepals** persistent, 5-toothed, sometimes 2-lipped, enlarged. **Petals** usually 2-lipped, upper lip 2-lobed and lower 3-lobed, or upper lip entire and lower 4-lobed, rarely limb 5-lobed. **Stamens** 4, didynamous to somewhat equal, sometimes reduced to 2; filaments attached to petals, filaments free or fused. **Ovary** superior, often a nectar disk below ovary; carpels 2, fused, sometimes deeply 4-lobed; placentation usually basal or basal-axile; style terminal to gynobasic. **Fruit** often a schizocarp splitting into 4 nutlets or 4 drupelets, or a drupe with 1–4 pyrenes.

草本至灌木、乔木或稀为**木质藤本**，具芳香油；茎常四棱。**叶**为单叶或稀为复叶，常对生，偶轮生或互生；全缘至有锯齿；具叶柄；托叶缺。**花序**常为腋生成对的二歧聚伞花序，常包于叶或苞片内，有时为轮伞花序。**花**两性或稀单性，两侧对称或稀辐射对称。**花萼**宿存，5 齿，有时二唇形，果期增大。**花瓣**常二唇形，上唇 2 裂，下唇 3 裂；或上唇全缘，下唇 4 裂，稀檐部 5 裂。**雄蕊** 4 枚，二强雄蕊至多少等长，有时退化为 2 枚；花丝贴生于花瓣，花丝分离或合生。**子房**上位，基部具蜜腺盘；心皮 2 枚，合生，有时 4 深裂；常为基底胎座或基生中轴胎座；花柱顶生或基生。**果实**多为分果，为 4 个小坚果或 4 个小核果，或为具 1~4 个果核的核果。

World 236/7,173, worldwide, especially in Mediterranean and Central Asia.
China 96/970.
This area 23/55.

全世界共 236 属 7 173 种，世界广布，尤其是地中海、中亚地区。
中国产 96 属 970 种。
本地区有 23 属 55 种。

Vitex rotundifolia L. f./ 单叶蔓荆

1. Style terminal/ 花柱顶生
   2. Fruit dry, usually a schizocarp/ 果实常为干燥的分果 .... **1.** ***Caryopteris*/** 莸属
   2. Fruit a fleshy drupe/ 果实为肉质核果

3. Corolla actinomorphic/ 花冠辐射对称
　　4. Stamens fewer than corolla lobes/ 雄蕊数量少于花冠裂片 .... *2. Clerodendrum*/ 大青属
　　4. Stamens as many as corolla lobes/ 雄蕊数量与花冠裂片相等 .... *3. Callicarpa*/ 紫珠属
3. Corolla zygomorphic/ 花冠两侧对称
　　5. Lower lip of corolla with middle lobe greatly elongated/ 花冠下唇中裂片显著增大
　　　... *4. Vitex*/ 牡荆属
　　5. Lower lip of corolla with middle lobe not elongated/ 花冠下唇中裂片不增大 ....
　　　... *5. Premna*/ 豆腐柴属
1. Style from bases of ovary lobes/ 花柱自子房裂瓣间伸出
　6. Style arising above base of ovary/ 花柱着生点高于子房基部
　　7. Corolla 1-lipped (lower lip only) / 花冠单唇形（仅具下唇） .... *6. Teucrium*/ 香科科属
　　7. Corolla 2-lipped / 花冠二唇形 .... *7. Ajuga*/ 筋骨草属
　6. Style inserted at base of ovary/ 花柱着生于子房基部
　　8. Ovary stipitate; seeds ± transverse; radicle curved/ 子房有柄；种子多少横生；胚根弯曲 ....
　　　... *8. Scutellaria*/ 黄芩属
　　8. Ovary generally not stipitate; seeds erect; radicle short, straight/ 子房通常无柄；种子直立；胚根短而直伸
　　　9. Stamens ascending under upper corolla lip or spreading or projected/ 雄蕊上升于花冠上唇之下或平展或凸出
　　　　10. Anthers not globose, cell not or rarely confluent at apex/ 花药非球状；药室先端不贯通或稀贯通
　　　　　11. Corolla upper lip convex, arcuate, falcate, or galeate/ 花冠上唇外凸，弧状、镰状或盔状
　　　　　　12. Stamens 4/ 雄蕊 4 枚
　　　　　　　13. Posterior stamens longer than anterior ones/ 后对雄蕊长于前对雄蕊 ....
　　　　　　　　.. *9. Glechoma*/ 活血丹属
　　　　　　　13. Posterior stamens shorter than anterior ones/ 后对雄蕊短于前对雄蕊
　　　　　　　　14. Calyx throat closed in fruit/ 花萼喉部在果实成熟时闭合 .... *10. Prunella*/ 夏枯草属
　　　　　　　　14. Calyx throat open in fruit/ 花萼喉部在果实成熟时张开
　　　　　　　　　15. Style lobes unequal in length/ 花柱裂片不等长 .... *11. Leucas*/ 绣球防风属
　　　　　　　　　15. Style lobes subequal or equal in length/ 花柱裂片近等长或等长
　　　　　　　　　　16. Nutlets ovoid, rounded at apex/ 小坚果卵球形，顶端钝圆 .... *12. Stachys*/ 水苏属
　　　　　　　　　　16. Nutlets ± acutely 3-angled, not truncate at apex/ 小坚果多少锐三棱形，顶端不平截
　　　　　　　　　　　17. Throat of corolla dilated; calyx teeth not spinescent/ 花萼喉部膨大；萼齿非针刺状
　　　　　　　　　　　　18. Lateral lobes of lower corolla lip not developed; anther cells divaricate, hairy/ 下唇侧裂片不发达，药室极叉开，被毛 .... *13. Lamium*/ 野芝麻属
　　　　　　　　　　　　18. Lateral lobes of lower corolla lip well developed, anther cells divergent, glabrous/ 下唇侧裂片发达，药室略叉开，光滑 .... *14. Matsumurella*/ 小野芝麻属
　　　　　　　　　　　17. Throat of corolla barely dilated; calyx teeth spinescent/ 花萼喉部几不膨大；萼齿针刺状 .... *15. Leonurus*/ 益母草属
　　　　　　12. Stamens 2; anther attached to filaments by joints / 雄蕊 2 枚；花药与花丝以关节相连 ....

....**16.** *Salvia*/ 鼠尾草属

11. Corolla upper lip flat or convex or corolla nearly actinomorphic/ 花冠上唇扁平或外凸或花冠近辐射对称

    19. Stamens ascending to underneath upper corolla lip/ 雄蕊上升于花冠上唇之下 ....

....**17.** *Clinopodium*/ 风轮菜属

    19. Stamens ascending from base, if spreading then erect/ 雄蕊自基部上升，如平展则直伸向前

        20. Stamens 2 or 4 (didynamous)/ 雄蕊 2 枚，如 4 枚，则二强雄蕊

            21. Stamens 4/ 雄蕊 4 枚 ....**18.** *Perilla*/ 紫苏属

            21. Stamens 2/ 雄蕊 2 枚

                22. Flowers distinctly pedicellate/ 花明显具梗 ....**19.** *Mosla*/ 石荠苎属

                22. Flowers sessile/ 花无梗 ....**20.** *Lycopus*/ 地笋属

        20. Stamens 4, equal/ 雄蕊 4 枚，等长 ....**21.** *Mentha*/ 薄荷属

10. Anthers globose; cell confluent at apex / 花药球状，药室先端贯通 ....**22.** *Elsholtzia*/ 香薷属

9. Stamens declinate, lying along or included in lower lip of corolla/ 雄蕊下倾，平卧于花冠下唇上或包于其内 ....**23.** *Isodon*/ 香茶菜属

LAMIALES. Lamiaceae. *Vitex rotundifolia* L. f./ 唇形目 唇形科 单叶蔓荆

A. Flowering branch/ 花枝

B. Lateral view of flower/ 花侧面观

C_1. Cymes/ 聚伞花序

C_2. Cymes, bud stage/ 聚伞花序，蕾期

D_1. Stigma and style/ 柱头与花柱

D_2. Cross section of ovary/ 子房横切面

D_3. Stamen/ 雄蕊

D_4. Lateral view of stamen/ 雄蕊侧面观

E. Cross section of fruit/ 果实横切面

F. Cross profile of flower/ 花纵剖面

(Scale/ 标尺：A 无；B–D_4 1 mm; E 5 mm; F 1 mm)

1. *Caryopteris incana* (Thunb. ex Houtt.) Miq./ 兰香草；
2. *Clerodendrum kaichianum* P.S. Hsu/ 浙江大青；
3. *Callicarpa cathayana* H.T. Chang/ 华紫珠；
4. *Vitex rotundifolia* L.f./ 单叶蔓荆；
5. *Premna microphylla* Turcz./ 豆腐柴；
6. *Teucrium viscidum* Blume/ 血见愁；
7. *Ajuga nipponensis* Makino/ 紫背金盘；
8. *Scutellaria indica* L./ 韩信草；
9. *Glechoma longituba* (Nakai) Kuprian./ 活血丹（9a. Vertical section of flower/ 花纵切面；9b. Flowering branch/ 花枝）

10. *Prunella vulgaris* L./ 夏枯草；
11. *Leucas chinensis* (Retz.) R. Br./ 滨海白绒草；
12. *Stachys* sp./ 水苏属；
13. *Lamium amplexicaule* L./ 宝盖草；
14. *Matsumurella chinensis* (Benth.) Bendiksby/ 小野芝麻；
15. *Leonurus japonicus* Houtt./ 益母草；
16. *Salvia japonica* Thunb./ 鼠尾草；
17. *Clinopodium chinense* (Benth.) Kuntze/ 风轮菜；
18. *Perilla frutescens* (L.) Britton/ 紫苏；
19. *Mosla scabra* (Thunb.) C.Y. Wu & H.W. Li/ 石荠苎；
20. *Lycopus lucidus* Turcz. ex Benth./ 地笋；
21. *Mentha crispata* Schrad. ex Willd./ 皱叶留兰香；
22. *Elsholtzia splendens* Nakai ex F. Maekawa/ 海州香薷；
23. *Isodon inflexus* (Thunb.) Kudô/ 内折香茶菜

384 Mazaceae 通泉草科

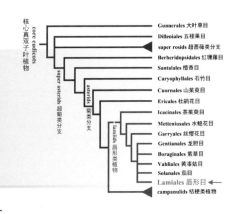

Terminal racemes, secund / 顶生总状花序，偏向一侧
Lower lip with 2 longitudinal plaits / 下唇具 2 纵向褶皱
Stamens 4, didynamous, apically connivent / 雄蕊 4 枚，二强，先端靠合
Capsule loculicidal / 蒴果室背开裂

Perennial or annual **herbs**, erect or procumbent and rooting from lower nodes. **Leaves** in a rosette or opposite, often upper leaves alternate, sometimes few or reduced to scale-like; margin entire to serrate; petiole present, winged; stipule absent. **Inflorescence** terminal racemes, secund, less often few or rarely solitary from bract axiles; bracts small or absent; bracteoles present or not. **Flowers** bisexual, zygomorphic. **Sepals** 5-lobed; funnelform or campanulate; persistent. **Petals** 2-lipped, tube tubular or cylindric; upper lip 2-lobed, small, erect; lower lip longer and broader than upper lip, with 2 longitudinal plaits. **Stamens** 4, didynamous, inserted on corolla tube, filaments glabrous or hairy; anther locules divergent, apically connivent. **Ovary** superior, hairy or glabrous; carpels 2, fused; 2-loculed; placentation axile; stigmas 2. **Fruit** a loculicidal capsule, included in cupular persistent sepals, or rarely berry-like, indehiscent.

多年生或一年生**草本**，直立或匍匐，下部茎节处常生不定根。**叶**基生莲座状或对生，上部叶常互生，有时少数或退化成鳞片状；全缘至有锯齿；具叶柄，有翅；托叶缺。**花序**顶生，总状，偏向一侧，偶少花或稀为自苞腋伸出的单花；苞片小或缺；小苞片有或无。**花**两性，两侧对称。**花萼** 5 裂；漏斗状或钟状；宿存。**花瓣**二唇形，冠筒管状或圆筒状；上唇 2 裂，小，直伸；下唇长宽均大于上唇，具 2 道纵向褶皱。**雄蕊** 4 枚，二强，着生于花冠筒上，花丝光滑或被毛；花药药室略叉开，先端靠合。**子房**上位，被毛或光滑；心皮 2 枚，合生；2 室；中轴胎座；柱头 2 个。**果实**为室背开裂的蒴果，包裹于宿存杯状萼筒内，或稀浆果状，不开裂。

World 3/33, Central and Eastern Asia, extend into Oceania.
China 3/28.
This area 1/2.

全世界共 3 属 33 种，分布于亚洲中部和东部，延伸至大洋洲。
中国产 3 属 28 种。
本地区有 1 属 2 种。

Mazus Lour. 通泉草属

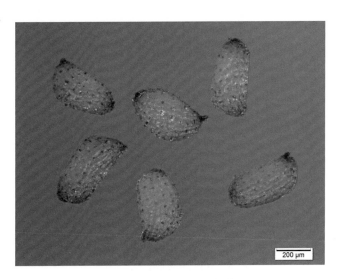

Mazus pumilus (Burm. f.) Steenis / 通泉草

LAMIALES. Mazaceae. *Mazus pumilus* (Burm. f.) Steenis/ 唇形目 通泉草科 通泉草

A. Whole plant/ 植株

B₁. Pistil/ 雌蕊

B₂. Vertical section of corolla (lower)/ 花冠纵切面 (下唇)

B₃. Vertical section of corolla (upper)/ 花冠纵切面 (上唇)

B₄. Vertical section of corolla (through symmetry axis)/ 花冠纵切面 (沿对称轴)

B₅. Dorsal view of corolla/ 花冠背面观

C₁. Ventral view of corolla/ 花冠腹面观

C₂. Calyx unfold (inside)/ 花萼展开 (内侧)

C₃. Raceme/ 总状花序

C₄. Vertical profile of flower, corolla removed/ 花纵剖面，除花冠

D. Cross section of ovary/ 子房横切面

(Scale/ 标尺： A 1 cm; B₁–D 1 mm)

386 Paulowniaceae 泡桐科

Woody plants, with conspicuous lenticels when young / 木本植物，幼时皮孔明显
Branches and leaves opposite / 小枝和叶对生
Stamens 4, didynamous / 雄蕊 4 枚，二强
Capsule, 2-valved / 蒴果，2 瓣裂

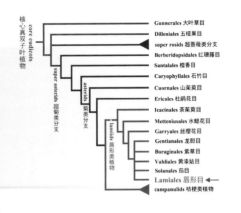

Trees, semiepiphytic pseudovines, or rarely parasitic shrubs; bark smooth with conspicuous lenticels when young, longitudinally splitting with age; branches opposite, without terminal buds. **Leaves** opposite, occasionally 3 in a whorl; glandular or stellate hairy; margin entire or lobed; petiole present; stipules absent. **Inflorescence** a large thyrses, panicle or racemes, often terminal. **Flowers** large, bisexual, zygomorphic. **Sepals** campanulate, thick, hairy; 3–5-lobbed. **Petals** tube usually curved; 2-lipped; lower lip 3-lobed, spreading; upper lip 2-lobed, erect. **Stamens** 4, didynamous; filaments inserted near tube base; anther divergent, apical connivent. **Ovary** superior; carpels fused, 2-loculed; placentation axile. **Fruit** loculicidal capsule, 2-valved or incompletely 4-valved. **Seeds** small, numerous, membranous winged.

乔木、半附生假藤本或稀为寄生灌木；树皮光滑，幼时皮孔明显，随之纵向开裂；小枝对生，顶芽缺。**叶**对生，偶 3 枚轮生；具腺毛或星状毛；全缘或分裂；具叶柄；托叶缺。**花序**为大型聚伞花序、圆锥花序或总状花序，常顶生。**花**大，两性，两侧对称。**花萼**钟状，厚，被毛；3~5 裂。**花瓣**合生，冠筒常扭曲；二唇形；下唇 3 裂，平展；上唇 2 裂，直伸。**雄蕊** 4 枚，二强；花丝着生于冠筒基部；花药略开叉，先端靠合。**子房**上位；心皮合生，2 室；中轴胎座。**果实**为室背开裂蒴果，裂为 2 瓣或不完全 4 瓣。**种子**小，多数，具膜质翅。

World 1/8, temperate East Asia, Himalaya, Myanmar to Malesia.
China 1/7.
This area 1/2.

全世界共 1 属 8 种，分布于东亚温带、喜马拉雅、缅甸至马来西亚。
中国产 1 属 7 种。
本地区有 1 属 2 种。

Paulownia Siebold & Zucc. 泡桐属

Paulownia tomentosa (Thunb.) Steud./ 毛泡桐（a. Flowers/ 花；b. Seeds/ 种子）

387 Orobanchaceae 列当科

Parasitic or hemiparasitic herbs / 寄生或半寄生草本
Leaves simple, often opposite / 单叶常对生
Inflorescences racemose or spikes / 总状或穗状花序
Seeds minute, alveolate / 种子小，表面呈蜂窝

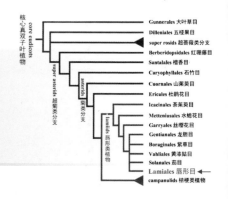

Herbs, **subshrubs** or **shrubs**, usually hemiparasitic to holoparasitic, with haustoria connections, few autotrophic. **Leaves** simple, alternate and spiral, or opposite; scale-like in holoparasites; leaves turn black in drying; margin entire, toothed to deeply lobed or pinnatifid; petioles present to absent; stipules absent. **Inflorescence** racemes or spikes, rarely solitary flower; bracts leaf-like. **Flowers** bisexual, zygomorphic. **Sepals** fused, 5-lobbed, valvate. **Petals** fused, imbricate, 2-lipped, lower lip usually larger than upper lip, upper lip 2-lobed and lower lip 3-lobed. **Stamens** alternate the petals, 4, didynamous; filaments attached to petals, sometimes with hair at base; anther with 2 unequal or equal cell or only 1 cell present. **Ovary** superior, carpels 2, fused; 1–2-loculed; ovules usually many per locule; placentation parietal or axile. **Fruit** a loculidcidal or septicidal capsule, rarely a drupe, berry or schizocarp. **Seeds** minute, alveolate surface.

草本、**亚灌木**或**灌木**，常半寄生至全寄生，具吸器，少数自养。**叶**为单叶，螺旋状互生，或对生；全寄生类型中叶片呈鳞片状；叶失活后变黑；全缘、具齿至深裂或羽裂；叶柄存在至缺；托叶缺。**花序**总状或穗状，稀单花；苞片叶状。花两性，两侧对称。**花萼**合生，5裂，镊合状。**花瓣**合生，覆瓦状，二唇形，下唇常大于上唇，上唇2裂，下唇3裂。**雄蕊**与花瓣互生，4枚，二强；花丝贴生于花瓣，有时基部被毛；花药具2个不等大或等大药室，或仅1个药室。**子房**上位，心皮2枚，合生；1~2室；胚珠每室常多数；侧膜胎座或中轴胎座。**果实**为室背开裂或室间开裂蒴果，稀核果、浆果或分果。**种子**小，表面呈蜂窝状。

World 99/2,060, worldwide, northern temperate regions to Africa and Madagascar.
China 35/ca. 471.
This area 6/7.

全世界共99属2 060种，世界广布，分布于北温带地区至非洲和马达加斯加。
中国产35属约471种。
本地区有6属7种。

1. Plants with green leaves/ 植株具绿色的叶
   2. Bracts dentate/ 苞片具齿 .... **1. *Melampyrum*/ 山罗花属**
   2. Bracts often entire/ 苞片常全缘
      3. Bracteoles 2/ 小苞片2枚
         4. Basal leaves well developed/ 基生叶正常 .... **2. *Siphonostegia*/ 阴行草属**
         4. Basal leaves scale-like/ 基生叶呈鳞片状 .... **3. *Monochasma*/ 鹿茸草属**

3. Bracteoles absent/ 无小苞片 ··· **4. *Phtheirospermum*/ 松蒿属**
1. Plants without green leaves/ 植株无绿色的叶
 5. Inflorescence racemes or spikes/ 花序总状或穗状 ··· **5. *Orobanche*/ 列当属**
 5. Inflorescence solitary flower or few/ 花序为单花或少花 ·· **6. *Aeginetia*/ 野菰属**

LAMIALES. Orobanchaceae. *Monochasma savatieri* Franch. ex Maxim./ 唇形目 列当科 白毛鹿茸草

A. Ventral view of flower/ 花腹面观
B. Cross section of ovary/ 子房横切面
C. Stamen/ 雄蕊
D. Young fruit with perianth/ 幼果，带花被
E_1. Vertical profile of corolla (upper)/ 花冠纵剖面（上唇）
E_2. Flower segregated without corolla/ 花离析，除花冠
E_3. Vertical profile of corolla (lower)/ 花冠纵剖面（下唇）

(Scale/ 标尺：A 5 mm; B–D 1 mm; E_1–E_3 5 mm)

1. *Melampyrum roseum* Maxim./ 山罗花；
2. *Siphonostegia chinensis* Benth./ 阴行草；
3. *Monochasma savatieri* Franch. ex Maxim./ 白毛鹿茸草；
4. *Phtheirospermum japonicum* (Thunb.) Kanitz/ 松蒿；
5. *Orobanche brassicae* (Novopokr.) Novopokr./ 光药列当；
6. *Aeginetia indica* L./ 野菰

392 Aquifoliaceae 冬青科

Leaves simple, alternate / 单叶，互生
Flowers small, 4-merous, unisexual / 单性花小，4 基数
Berry-like drupe / 核果浆果状
Endocarp stony, sulcate / 内果皮石质，具沟槽

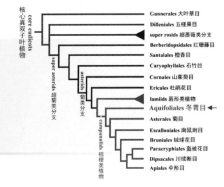

Trees or **shrubs**, usually evergreen. **Leaves** simple, usually alternate and spiral, rarely opposite or whorled; margin entire, serrate, or spinose; petioles present; stipules minute. **Inflorescence** racemes, cymes, fascicles, panicles or rarely solitary flower, axillary. **Flowers** small, functionally unisexual (plants dioecious), actinomorphic. **Sepals** 4–8-lobed, persistent. **Petals** 4–8-lobbed, lobes longer than the tube, imbricate. **Stamens** alternating with petals; anther introrse, dehiscing longitudinally; staminodes in female flowers. **Ovary** superior, carpels fused; 4–8(–10)-loculed; ovules 1(–2) per locule; placentation axile; style very short or lacking, stigma capitate. **Fruit** a berry-like drupe, pyrenes (1–) 4–6(–23); endocarp smooth, leathery, woody, or stony, striate, sulcate, or rugose, and/or pitted.

乔木或**灌木**，多常绿。**叶**为单叶，常螺旋状互生，稀对生或轮生；全缘、具齿或刺状；具叶柄；托叶细小。**花序**总状、聚伞状、簇生、圆锥状或稀为单花，腋生。**花**小，功能性单性花（雌雄异株），辐射对称。**花萼** 4~8 裂，宿存。**花瓣** 4~8 裂，裂片长于冠筒，覆瓦状。**雄蕊**与花瓣互生；花药内向，纵裂；雌花中具退化雄蕊。**子房**上位，心皮合生；4~8(~10) 室；胚珠每室 1(~2) 枚；中轴胎座；花柱极短或缺，柱头头状。**果实**为浆果状核果，分核 (1~)4~6(~23)；内果皮光滑、革质、木质或石质，具条纹、沟槽，或多皱纹，具凹点。

World 1/420, tropical to temperate regions, mainly in Central America, South America and tropical Asia. China 1/ca. 204. This area 1/8.

全世界共 1 属 420 种，分布于热带至温带地区，尤其是中、南美洲，热带亚洲。中国产 1 属约 204 种。本地区有 1 属 8 种。

Ilex L. 冬青属

1-1. *Ilex integra* Thunb./ 全缘冬青
（1-1a. Female flower/ 雌花；1-1b. Young fruits/ 幼果）；
1-2. *I. crenata* Thunb./ 齿叶冬青；
1-3. *I. rotunda* Thunb./ 铁冬青（Male flower/ 雄花）

AQUIFOLIALES. Aquifoliaceae. *Ilex latifolia* Thunb./ 冬青目 冬青科 大叶冬青

A_1. Flowering branch/ 花枝
A_2. Fruiting branch/ 果枝
B_1. Flower segregated/ 花离析
B_2. Disk (left) and stigma (right)/ 花盘（左）和柱头（右）
B_3. Cross section of ovary/ 子房横切面
C_1. Ventral view of flower/ 花腹面观
C_2. Dorsal view of flower/ 花背面观
C_3. Lateral view of flower/ 花侧面观
C_4. Pyrenes/ 果核
C_5. Cymes/ 聚伞花序

(Scale/ 标尺： A_1-A_2 5 cm; B_1-C_5 1 mm)

394 Campanulaceae 桔梗科

Herbs, lactiferous / 草本，有乳汁
Leaves simple, alternate / 单叶，互生
Corolla 5-lobed, often campanulate / 花冠 5 裂，多钟形
Placentation axile, capsule or berry / 中轴胎座，蒴果或浆果

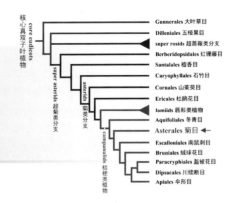

Herbs, less often **trees** or **woody climbers**; often with rhizomes, lactiferous. **Leaves** simple, alternate, rarely opposite or whorled; margin entire, toothed or pinnatifid; petiole present; stipule absent. **Inflorescence** solitary flower, cymes, racemes, spikes, umbels or heads. **Flowers** bisexual or rarely unisexual, actinomorphic or zygomorphic; usually bracteolate. **Sepals** 5-lobed, adnate to ovary and forming a hypanthium, persistent. **Petals** fused, rarely free; valvate; usually 5-lobed; campanulate. **Stamens** opposite with sepals, often 5, filaments usually attached to petals, sometimes fused forming tube; anthers basifixed or rarely dorsifixed; 2-loculed; dehiscing longitudinally; introrse. **Ovary** part-inferior or rarely superior; carpels 2–5, fused, 1–9-loculed; ovules few to many per locule; placentation axile, rarely basal or apical; style solitary, pubescent with pollen-collecting hairs below apex. **Fruit** a capsule, commonly apically loculicidal or laterally poricidal, or a berry.

草本、稀**乔木**或**木质藤本**；常具根状茎，有乳汁。**叶**为单叶，互生，稀对生或轮生；全缘，具齿或羽裂；具叶柄；托叶缺。**花序**为单花、聚伞状、总状、穗状、伞形或头状。**花**两性或稀单性，辐射对称或两侧对称；常具小苞片。**花萼** 5 裂，与子房合生形成托杯，宿存。**花瓣**合生，稀分离；镊合状；常 5 裂；钟形。**雄蕊**与花萼对生，常 5 枚，花丝常与花瓣合生，有时合生成管状；花药基着或稀背着；2 室；纵裂，内向。**子房**半下位或稀上位；心皮 2~5 枚，合生，1~9 室；胚珠每室少数至多数；中轴胎座，稀基底胎座或顶生胎座；花柱单生，近顶部有收集花粉的毛。**果实**为蒴果，常于顶端室背开裂或侧方孔裂，或为浆果。

World 84/2,380, worldwide, mainly in temperate regions and subtropics.
China 14/159.
This area 5/5.

全世界共 84 属 2 380 种，世界广布，主要分布于温带和亚热带。
中国产 14 属 159 种。
本地区有 5 属 5 种。

Lobelia chinensis Lour./ 半边莲

1. Corolla zygomorphic/ 花冠两侧对称 .... **1. *Lobelia*/ 半边莲属**
1. Corolla actinomorphic/ 花冠辐射对称
　　2. Fruit dehiscing apically/ 果顶部开裂

3. Stigma 5-fid/ 柱头 5 裂 ....**2. *Platycodon*/ 桔梗属**
3. Stigma 3-fid/ 柱头 3 裂 ....**3. *Wahlenbergia*/ 蓝花参属**
2. Fruit dehiscing laterally/ 果侧面开裂
  4. Corolla lobbed nearly to base, rotate/ 花冠深裂达基部，辐状 ....**4. *Triodanis*/ 异檐花属**
  4. Corolla lobbed to middle, campanulate/ 花冠分裂至中部，钟状 ....**5. *Adenophora*/ 沙参属**

1. *Lobelia chinensis* Lour./ 半边莲；
2. *Platycodon grandiflorus* (Jacq.) A. DC./ 桔梗；
3. *Wahlenbergia marginata* (Thunb.) A. DC./ 蓝花参；
4. *Triodanis perfoliata* (L.) Nieuwl. subsp. *biflora* (Ruiz & Pav.) Lammers/ 异檐花（photo by ChengChun PAN/ 潘成椿拍摄）；
5. *Adenophora petiolata* Pax. & K. Hoffm. subsp. *hunanensis* (Nannf.) D.Y. Hong & S. Ge/ 杏叶沙参

401 Goodeniaceae 草海桐科

Leaves simple, alternate / 单叶互生
Petal valvate, 2-lipped, lobes winged / 花瓣镊合状，二唇形，裂片具翅
Style with pollen-collecting indusium / 花柱顶端具集药杯
Drupe with persistent calyx / 核果具宿萼

Herbs, shrubs, trees, or rarely creeper. **Leaves** simple, often alternate, less often opposite or whorled; margin entire, toothed or lobed; petioles present or absent; stipules absent. **Inflorescence** a cymes, heads or solitary flower, axillary; bracts present, sometimes involucral. **Flowers** bisexual, somewhat zygomorphic or actinomorphic; often bracteolate. **Sepals** fused, tubular, 5-lobed, adnate to ovary, persistent. **Petals** fused or rarely free, valvate, 2-lipped with 2 upper petals and 3 lower or all 5 lower, lobes winged. **Stamens** 5, alternate with petals, filaments free from petals; anthers sometimes fused into a tube round the style; 2-loculed, dehiscing longitudinally, introrse. **Ovary** part-inferior to inferior or superior; 1–2 (–4)-loculed; ovules 1 to many per locule; placentation basal or axile; style with apical hairy pollen-collecting indusium, curved, stigma bilobed. **Fruit** a drupe, with persistent calyx.

草本、灌木、乔木或稀匍匐生长。叶为单叶，常互生，偶对生或轮生；全缘、具齿或分裂；叶柄有或无；托叶缺。花序为聚伞状、头状或单花，腋生；具苞片，有时总苞状。花两性，多少两侧对称或辐射对称；常具小苞片。花萼合生，管状，5裂，与子房合生，宿存。花瓣合生或稀分离，镊合状，二唇形，上唇2枚下唇3枚，或5枚全部位于下唇，裂片具翅。雄蕊5枚，与花瓣互生，花丝与花瓣分离；花药有时合生成筒，包围花柱；2室，纵裂，内向。子房半下位至下位，或上位；1~2 (~4)室；胚珠每室1至多枚；基底胎座或中轴胎座；花柱弯曲，顶端具多毛的集药杯，柱头2裂。果实为核果，具宿存花萼。

World 12/400, tropical and subtropical regions, and temperate southern hemisphere.
China 2/3.
This area 1/1.

全世界共12属400种，分布于热带与亚热带地区，以及南半球温带。
中国产2属3种。
本地区有1属1种。

Scaevola L. 草海桐属

Scaevola taccada (Gaertn.) Roxb./ 草海桐
（a. Flowers/ 花；b. Infructescence/ 果序；c. Plants/ 植株）

403 Asteraceae 菊科

- Often herbs, somtimes laticiferous / 常草本，部分有乳汁
- Capitula with phyllaries / 头状花序，有总苞
- Corolla tubular or ligulate / 花冠筒状或舌状
- Synantherous stamen / 聚药雄蕊
- Fruit an inferior achene / 下位瘦果

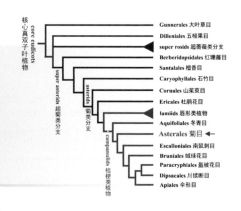

Often **herbs**, rarely subshrubs to trees; cichorioideae laticiferous, occasionally with resinous ducts. **Leaves** often in a basal rosette; cauline leaves usually alternate, without stipules. **Capitula** solitary or arranged in various inflorescence, with 1 to several series of phyllaries, receptacle usually flattened, **florets** sessile, bisexual or unisexual, 5-merous, calyx reduced; corolla gamopetalous, tubular, bilabiate, or ligulate; stamens 4–5, synantherous; ovary inferior, capels 2, 1-loculed; ovule 1, anatropous. **Fruit** an inferior achene (cypsela); pappus consisting of 1 to many rows of scales or bristles, or absent.

多**草本**，稀亚灌木至乔木；舌状花类有乳汁管，偶有树脂道。**叶**常基生莲座状；茎生叶常互生，无托叶。**头状花序**单生，或再排成各式花序，具1至多层总苞，花托常平展。小花无柄，两性或单性，5基数，花萼退化；花冠合生，管状、二唇形或舌状；雄蕊4~5枚，聚药雄蕊；子房下位，心皮2枚，1室；胚珠1枚，倒生。**果实**为下位瘦果；顶部常具1至多层鳞片状或刚毛状冠毛或无。

World 1,600–1,700/24,000–30,000, worldwide.
China 253/ca. 2,350.
This area 66/133(belonging to 13 tribes).

全世界共1 600~1 700属24 000~30 000种，世界广布。
中国产253属约2 350种。
本地区有66属133种（隶属13族）。

Aster arenarius (Kitam.) Nemoto/ 普陀狗娃花

1. Capitula homogamous and all florets ligulate; latex present/ 头状花序花同型，所有小花为舌状花；具乳汁 ....**I. Cichorieae/ 菊苣族**
1. Capitula heterogamous, or if homogamous then corolla tubular and 5-lobed or zygomorphic and pseudoligulate; plants without or rarely with latex/ 头状花序异型，如同型则为管状花，5裂；或为两侧对称的假舌状花；植株不具或少有乳汁
    2. Capitula homogamous; corolla pseudoligulate, 5-lobed and zygomorphic/ 头状花序同型；花冠假舌状，5裂，两侧对称
        3. Capitula in bracteate glomerules; phyllaries 8, decussate; style branches long/ 头状花序组成具苞片的团伞花序；总苞8枚，交互对生；花柱分枝长 ....**II. Vernonieae/ 斑鸠菊族**
        3. Capitula not in bracteate glomerules; phyllaries not decussate; style branches very short or scarcely bilobed/

头状花序不组成具苞片的团伞花序；总苞片非交互对生；花柱分枝极短或几不 2 裂 ....**III. Mutisieae/ 帚菊木族**

2. Capitula heterogamous or homogamous and corolla 3–5-lobed, actinomorphic/ 头状花序异型，或同型，花冠 3~5 裂，辐射对称
    4. Capitula each with only 1 floret, aggregated into a terminal globose pseudocephalium; leaves spiny/ 头状花序仅含 1 小花，多数头状花序在茎顶排列为球形假头状花序；叶多刺 .... **IV. Echinopeae/ 蓝刺头族**
    4. Capitula with more than 1 floret, or if with only 1 floret then leaves not spiny/ 头状花序含多朵小花，如仅 1 朵，则叶无刺
        5. Leaves at least below capitula opposite or mostly so/ 叶在花序下部对生或大多对生
            6. Style branch tips longer than stigmatic lines, prominent; capitula discoid; corolla never yellow/ 花柱分枝先端长于柱头线，显著；头状花序仅具两性盘花；花冠不为黄色 ....**V. Eupatorieae/ 泽兰族**
            6. Style branch tips shorter than stigmatic lines, or absent; capitula radiate, disciform, or discoid; corolla often yellow/ 花柱分枝先端短于柱头线，或缺；头状花序辐射状、具两性或杂性盘花；花冠常为黄色 ....**VI. Heliantheae/ 向日葵族**
        5. Leaves all alternate/ 叶全为互生
            7. Disk corolla 3- or 4-merous/ 盘花 3 或 4 基数 .... **VII. Anthemideae/ 春黄菊族**
            7. Disk corolla or all corollas 5-merous/ 盘花或全部花均为 5 基数
                8. Style shaft with a papillose-pilose thickening below branches; leaves spiny, or not spiny/ 花柱分枝下部具增厚毛环；叶多刺或无 .... **VIII. Cardueae/ 飞廉族**
                8. Style shaft without a papillose-pilose thickening below branches; leaves not spiny/ 花柱分枝下部不具增厚毛环；叶无刺
                    9. Phyllaries rather dry, never herbaceous and green throughout/ 总苞干燥，绝不为草质或绿色
                        10. Anthers rounded at base; phyllaries with distinct pale or brownish scarious margin/ 花药基部圆钝；总苞边缘明显干膜质，灰白或淡褐色 .... **VII. Anthemideae/ 春黄菊族**
                        10. Anthers tailed at base; phyllaries papery, whitish, brownish, or yellowish/ 雄蕊基部有尾；总苞纸质，白色，淡褐色或淡黄色 .... **IX. Gnaphalieae/ 鼠曲草族**
                    9. Phyllaries herbaceous, all or at least outer ones green throughout, except at very apex/ 总苞草质，除先端外，全部或至少外层全为绿色
                        11. Style branches long, slender, subulate, without an apical appendage, hairy abaxially and with stigmatic papillae over entire adaxial surface; capitula homogamous/ 花柱分枝细长，钻形，无附属物，背面被毛，内侧为柱头面，具乳头状突起；头状花序同型 ....**II. Vernonieae/ 斑鸠菊族**
                        11. Style branches not with above combination of characters; capitula homogamous or heterogamous/ 花柱分枝与上述不同；头状花序同型或异型
                            12. Receptacle paleate/ 花托有托片
                                13. Achenes with a carbonized layer in pericarp, thereby black or streaked with black/ 菊果果皮具碳化层，黑色或有黑色条纹 .... **VI. Heliantheae/ 向日葵族**

13\. Achenes without a carbonized layer in pericarp, thereby usually not black/ 菊果果皮不具碳化层,通常不为黑色 .... **X. Athroismeae/ 山黄菊族**

12\. Receptacle epaleate/ 花托无托片

 14. Style branches terminating in a triangular to subulate appendage/ 花柱分枝顶端有三角形至钻形附属物

  15. Phyllaries uniseriate, but sometimes with an outer series of much shorter bracts (calyculus); involucre cylindric/ 总苞片 1 层, 有时外层具数枚短得多的苞片(副萼); 总苞圆柱形 .... **XI. Senecioneae/ 千里光族**

  15. Phyllaries 2 to several seriate, if subuniseriate then involucre saucer-shaped to subglobose/ 总苞片 2 至数层, 如近单层则总苞碟形至半球形 .... **XII. Astereae/ 紫菀族**

 14. Style branches without an apical appendage/ 花柱分枝顶端无附属物

  16. Phyllaries uniseriate, but sometimes with an outer series of much smaller bracts (calyculus); capitula never aggregated into a compact synflorescence/ 总苞片 1 层, 有时外层具数枚小得多的苞片(副萼); 头状花序绝不排列成密集的复合花序 .... **XI. Senecioneae/ 千里光族**

  16. Phyllaries 2- or 3-seriate or imbricate in several series; capitula sometimes aggregated into a compact synflorescence/ 总苞片 2 或 3 层, 或多层覆瓦状排列; 头状花序有时形成密集的复合花序 .... **XIII. Inuleae/ 旋覆花族**

ASTERALES. Asteraceae. *Aster indicus* L./ 菊目 菊科 马兰

A. Flowering branch/ 花枝
B_1. Lateral view of disk floret tube/ 盘花花冠筒侧面观
B_2. Pistil/ 雌蕊
B_3. Synantherous unfold (inside)/ 聚药雄蕊展开（内侧）
C_1. Vertical section of capitula/ 头状花序纵切面
C_2. Lateral view of disk floret/ 盘花侧面观
C_3. Lateral view of ray floret/ 缘花侧面观
D. Dorsal view of capitula/ 头状花序背面观
(Scale/ 标尺： A 5mm; B_1–B_3 1 mm; C_1–C_3 1 mm; D 5 mm)

I. Cichorieae/ 菊苣族

1. Pappus absent in all achenes/ 菊果无冠毛 ....**1. *Lapsanastrum*/ 稻槎菜属**
1. Pappus well developed in all achenes/ 菊果冠毛发达
   2. Pappus of numerous fine cottony outer bristles intermixed with some thicker inner ones/ 冠毛外层刚毛柔软纤细，内层刚毛粗 ....**2. *Sonchus*/ 苦苣菜属**
   2. Pappus of bristles ± equal in diam. and stiffness/ 冠毛刚毛多少同质
      3. Capitulum solitary on a hollow scape single or few from a leaf rosette; achene beak usually longer than achene body/ 头状花序单一，生于莲座叶发出的中空花葶上；菊果的喙多长于菊果主体 ....**3. *Taraxacum*/ 蒲公英属**
      3. Capitula usually few to numerous, if solitary then not on a hollow scape and achene not as above/ 头状花序常多枚，如仅1枚，则花葶不为中空且菊果特征与上述不同
         4. Rosulate herbs, capitula several to many on unbranched axe/ 莲座状草本，头状花序多个，着生于不分枝花序轴上 ....**4. *Youngia*/ 黄鹌菜属**
         4. Capitula few to numerous on a branched axe/ 头状花序少数至多数，着生于分枝的花序轴上
            5. Achene isodiametric and with ribs of ± equal shape and size/ 菊果圆柱状，肋近相等 ....**5. *Ixeris*/ 苦荬菜属**
            5. Achene somewhat to distinctly compressed and/or with ± unequal ribs/ 菊果多少压扁，肋多少不等大 ....**6. *Ixeridium*/ 小苦荬属**

1. *Lapsanastrum apogonoides* (Maxim.) Pak & K. Bremer/ 稻槎菜；
2. *Sonchus oleraceus* L./ 苦苣菜（photo by WeiLing MA/ 马炜梁拍摄）；
3. *Taraxacum mongolicum* Hand.–Mazz./ 蒙古蒲公英；
4. *Youngia japonica* (L.) DC./ 黄鹌菜；
5. *Ixeris repens* (L.) A. Gray/ 沙苦荬菜；
6. *Ixeridium laevigatum* (Blume) Pak & Kawano/ 褐冠小苦荬

II. Vernonieae/ 斑鸠菊族

1. Capitula densely clustered into compound synflorescence subtended by (1–)3 leaf-like bracts, each involucre with 1–4 florets and ca. 8 phyllaries/ 头状花序密集形成复合花序，包于 (1~)3 枚叶状苞片内，每个总苞含 1~4 朵小花，约 8 枚总苞片 ....**1. *Elephantopus*/ 地胆草属**
1. Capitula in lax panicles, each involucre with more than 4 florets and always with many phyllaries/ 头状花序排列为疏散的圆锥花序，每个总苞小花多于 4 朵，总苞片多数 ....**2. *Cyanthillium*/ 夜香牛属**

1. *Elephantopus scaber* L./ 地胆草；　　　　　　2. *Cyanthillium cinereum* (L.) H. Rob./ 夜香牛

III. Mutisieae/ 帚菊木族

Only *Ainsliaea* DC.

仅兔儿风属 1 属

Ainsliaea kawakamii Hayata/ 灯台兔儿风

IV. Echinopeae/ 蓝刺头族

Only *Echinops* L.

仅蓝刺头属 1 属

Echinops grijsii Hance/ 华东蓝刺头

V. Eupatorieae/ 泽兰族

1. Phyllaries all deciduous leaving a naked receptacle, phyllaries remaining appressed until lost and not spreading with age/ 总苞片早落，花托裸露，总苞片贴伏直至掉落，后期不会开展 ....**1. *Praxelis*/ 假臭草属**
1. At least some basal phyllaries persistent, phyllaries usually spreading with age/ 至少基部部分总苞片宿存，总苞片后期常开展
    2. Pappus of scales or awns, rarely absent; receptacle paleaceous/ 冠毛鳞片状或芒状，稀缺；花托有托片 ....**2. *Ageratum*/ 藿香蓟属**
    2. Pappus of capillary setae; receptacle epaleate/ 冠毛细刚毛状；花托无托片 ....**3. *Eupatorium*/ 泽兰属**

1. *Praxelis clematidea* R.M. King & H. Rob./ 假臭草；   3. *Eupatorium cannabinum* L./ 大麻叶泽兰
2. *Ageratum conyzoides* L./ 藿香蓟；

VI. Heliantheae/ 向日葵族

1. Florets all unisexual; ray florets absent/ 全部小花单性；舌状花缺
   2. Phyllaries in male capitula free to base; inner phyllaries in female becominga hard bur/ 雄花序的总苞片分离；雌花序内层总苞片形成坚硬的刺果 .... **1. *Xanthium*/ 苍耳属**
   2. Phyllaries in male capitula connate; phyllaries in female usually with free tips forming tubercles, spines, or wings/ 雄花序的总苞片结合；雌花序总苞片先端常分离，形成瘤、刺或翅 .... **2. *Ambrosia*/ 豚草属**
1. Some or all florets bisexual; ray florets present, rarely absent/ 部分或全部小花为两性；有舌状花，稀缺失
   3. Only ray florets fertile, ray florets much longer than those of sterile disk florets/ 仅舌状花可育，舌状花远长于不育盘花 .... **3. *Parthenium*/ 银胶菊属**
   3. Disk florets fertile; ray florets present and fertile or sterile or absent/ 盘花可育；舌状花存在，可育或不可育，或缺失
      4. Ray florets with or without short tubes, persistent with corollas fused to apex of achene/ 舌状花有或无花冠筒，花冠宿存并与菊果先端合生 .... **4. *Zinnia*/ 百日菊属**
      4. Ray florets deciduous, or if ray florets absent then corollas not fused to apex of achene/ 舌状花早落，如舌状花缺，则盘花花冠不与菊果先端合生
         5. Pappus of subulate to acerose scales, or spatulate, entire to erose, fimbriate, or laciniate, sometimes aristate/ 冠毛有钻形至针形鳞片，或匙形，全缘或蚀刻，流苏状或条裂，有时具芒
            6. Pappus of 6–12 aristate, rarely linear scales with erose margin; capitula > 10 mm in diam./ 冠毛具 6~12 条芒，稀具边缘蚀刻的线状鳞片；头状花序直径大于 10mm .... **5. *Gaillardia*/ 天人菊属**
            6. Pappus absent or of fimbriate, sometimes aristate scales; capitula 3 - 5 mm in diam./ 冠毛缺或具流苏状，有时芒状鳞片；头状花序直径3~5 mm .... **6. *Galinsoga*/ 牛膝菊属**
         5. Pappus absent, or awned/ 冠毛缺，或具芒
            7. Achenes compressed/ 菊果压扁
               8. Pappus of retrorsely barbed awns/ 冠毛为具倒刺的芒
                  9. Style branches with long hairs; pappus of 2 scabrid awns/ 花柱分枝具长柔毛；冠毛为 2 枚粗糙的芒 .... **7. *Glossocardia*/ 鹿角草属**
                  9. Style branches with short minute papillae; pappus of 2–4 scabrid awns/ 花柱分枝具短细毛；冠毛为 2~4 枚粗糙的芒
                     10. Filaments pubescent; achene apex beaked/ 花丝被毛；菊果顶端有喙 .... **8. *Cosmos*/ 秋英属**
                     10. Filaments glabrous; achene apex narrow, not beaked/ 花丝无毛；菊果先端狭，无喙 .... **9. *Bidens*/ 鬼针草属**
               8. Pappus absent, or persistent, of 2 bristly cusps or scales/ 冠毛缺，或宿存，有 2 枚刚毛状尖突或鳞片 .... **10. *Coreopsis*/ 金鸡菊属**
            7. Achenes all relatively plump, or 3–5-angled in ray florets and compressed in disk florets/ 舌状花菊果鼓起，或具 3~5 棱；盘花菊果压扁
               11. Achenes enclosed by inner phyllaries or outer paleae/ 菊果包于内层总苞片或外层托苞 .... **11. *Sigesbeckia*/ 豨莶属**

11. Achenes not enclosed by inner phyllaries/ 菊果不包于内层总苞片
  12. Paleae narrow, long, flat/ 托苞狭长，扁平 ....**12.** *Eclipta*/ 鳢肠属
  12. Paleae concave or folded, ± enclosing florets/ 托苞凹或折叠，多少包藏小花
    13. Achenes in ray florets broadly ovate or elliptic, 3-angled; disk achenes ellipsoid, strongly compressed/ 舌状花菊果阔卵形或椭圆形，3棱；盘花菊果椭圆形，强烈压扁 ....**13.** *Acmella*/ 金钮扣属
    13. Achenes in bisexual florets 4 or 5-angled, or compressed/ 两性花菊果4或5棱，或压扁
      14. Ray florets sterile/ 舌状花不可育
        15. Peduncles usually distally dilated, always fistulose/ 花梗先端常膨大，空管状 ....**14.** *Tithonia*/ 肿柄菊属
        15. Peduncles never fistulose/ 花梗不为空管状 ....**15.** *Helianthus*/ 向日葵属
      14. Ray florets fertile/ 舌状花可育
        16. Outer phyllaries herbaceous and larger than inner; leaves sessile or very shortly petiolate; capitula always solitary, terminal/ 外层总苞片草质并大于内层；叶无柄或柄极短；头状花序单个顶生 ....**16.** *Sphagneticola*/ 蟛蜞菊属
        16. Outer phyllaries ± equal in size to inner; leaves usually conspicuously petiolate; synflorescence of 1–3(–6) capitula, terminal or axillary/ 外层总苞片与内层多少近等大；叶常具明显叶柄；1~3（~6）个头状花序，顶生或腋生 ....**17.** *Melanthera*/ 卤地菊属

1. *Xanthium strumarium* L./ 苍耳；
2. *Ambrosia artemisiifolia* L./ 豚草；
3. *Parthenium hysterophorus* L./ 银胶菊；
4. *Zinnia elegans* Jacq./ 百日菊（Naturalized/ 归化）；
5. *Gaillardia pulchella* Foug./ 天人菊（Naturalized, photo by WeiLiang MA/ 归化，马炜梁拍摄）

6. *Galinsoga quadriradiata* Ruiz & Pav./ 粗毛牛膝菊；
7. *Glossocardia bidens* (Retz.) Veldkamp/ 鹿角草；
8. *Cosmos bipinnatus* Cav./ 秋英（Naturalized/ 归化）；
9. *Bidens pilosa* L./ 鬼针草；
10. *Coreopsis lanceolata* L./ 线叶金鸡菊（Naturalized, photo by WeiLiang MA/ 归化，马炜梁拍摄）；
11. *Sigesbeckia orientalis* L./ 豨莶（photo by WeiLiang MA/ 马炜梁拍摄）；
12. *Eclipta prostrata* (L.) L./ 鳢肠；
13. *Acmella paniculata* (Wall. ex DC.) R.K. Jansen/ 金钮扣（photo by WeiLiang MA/ 马炜梁拍摄）；
14. *Tithonia diversifolia* (Hemsl.) A. Gray/ 肿柄菊（Naturalized/ 归化）；
15. *Helianthus annuus* L./ 向日葵（Naturalized/ 归化）；
16. *Sphagneticola calendulacea* (L.) Pruski/ 蟛蜞菊；
17. *Melanthera prostrata* (Hemsl.) W.L. Wagner & H. Rob./ 卤地菊

VII. Anthemideae/ 春黄菊族

1. Capitula radiate; ray florets often conspicuous/ 头状花序辐射状；舌状花常明显
   2. Receptacle scales present at least near margin of receptacle/ 有托片，至少花托边缘存在 .... **1. *Achillea*/ 蓍属**
   2. Receptacle scales absent, but receptacle sometimes hairy/ 无托片，但花托有时被毛
      3. Achene winged/ 菊果有翅 .... **2. *Glebionis*/ 茼蒿属**
      3. Achene not winged/ 菊果无翅 .... **3. *Chrysanthemum*/ 菊属**
1. Capitula discoid with all florets bisexual, tubular, or capitula disciform with marginal florets/ 头状花序仅有两性盘花，管状，或头状花序为杂性盘花，具缘花
   4. Marginal female florets in many rows/ 边缘有多层雌花
      5. Capitula pedunculate, terminal; achenes without persistent style/ 头状花序具柄，顶生；菊果无宿存柱头 .... **4. *Cotula*/ 山芫荽属**
      5. Capitula sessile, axillary; achenes with persistent style/ 头状花序无柄，腋生；菊果具宿存柱头 .... **5. *Soliva*/ 裸柱菊属**
   4. Marginal female florets in 1 row/ 边缘有 1 层雌花
      6. Capitula in spikes or racemes, often secund, often grouped into panicles/ 头状花序穗状或总状，常偏一侧，常组合成圆锥状
         7. Pappuslike corona present/ 具冠毛状副花冠 .... **6. *Crossostephium*/ 芙蓉菊属**
         7. Corona absent/ 无副花冠 .... **7. *Artemisia*/ 蒿属**
      6. Capitula in terminal, rounded to flat-topped panicles, clusters, or solitary/ 头状花序组成顶生圆锥花序，圆顶或平顶，或簇生，或单生 .... **8. *Ajania*/ 亚菊属**

1. *Achillea millefolium* L./ 蓍；
2. *Glebionis coronaria* (L.) Cass. ex Spach/ 茼蒿；
3. *Chrysanthemum indicum* L./ 野菊；
4. *Cotula anthemoides* L./ 芫荽菊；
5. *Soliva anthemifolia* (Juss.) R. Br./ 裸柱菊

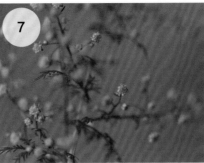

6. *Crossostephium chinense* (L.) Makino/ 芙蓉菊；

7. *Artemisia annua* L./ 黄花蒿；

8. *Ajania pallasiana* (Fisch. ex Besser) Poljakov/ 亚菊（Naturalized/ 归化）

VIII. Cardueae/ 飞廉族

1. Leaf margin unarmed/ 叶边缘无刺 .... **1. *Hemisteptia*/ 泥胡菜属**
1. Leaf margin spiny/ 叶边缘具刺 .... **2. *Cirsium*/ 蓟属**

1. *Hemisteptia lyrata* (Bunge) Fisch. & C.A. Mey./ 泥胡菜； 2. *Cirsium japonicum* DC./ 蓟

IX. Gnaphalieae/ 鼠曲草族

1. Pappus bristles connate into a ring/ 冠毛为刚毛，连合成环 .... **1. *Gamochaeta*/ 合冠鼠曲草属**
1. Pappus bristles free or coherent by patent cilia/ 冠毛为刚毛，分离，或由开展的纤毛相连
    2. Phyllaries brown or hyaline, inconspicuous/ 总苞片褐色或透明，不明显 .... **2. *Gnaphalium*/ 湿鼠曲草属**
    2. Phyllaries white, yellow, pink, or red, conspicuous/ 总苞片白色、黄色、粉色或红色，明显 ....
    .... **3. *Pseudognaphalium*/ 鼠曲草属**

1. *Gamochaeta pensylvanica* (Willd.) Cabrera/ 匙叶合冠鼠曲草；

2. *Gnaphalium japonicum* Thunb./ 细叶湿鼠曲草；

3. *Pseudognaphalium affine* (D. Don) Anderb./ 鼠曲草

X. Athroismeae/ 山黄菊族

Only *Centipeda* Lour.

仅石胡荽属 1 属

Centipeda minima (L.) A. Braun & Asch./ 石胡荽

（a. Plants/ 植株；b. Capitula/ 头状花序）

XI. Senecioneae/ 千里光族

1. Style branches terminating in an apiculate appendage/ 花柱分枝顶端尖，有附器

    2. Involucre calyculate/ 总苞有小外苞片 ....**1. *Gynura*/ 菊三七属**

    2. Involucre not calyculate/ 总苞无小外苞片 ....**2. *Emilia*/ 一点红属**

1. Style branches apically rounded or truncate/ 花柱分枝顶端圆钝或截形

    3. Capitula all bisexual, tubular/ 头状花序全为两性的管状花 ....**3. *Erechtites*/ 菊芹属**

    3. Capitula disciform/ 头状花序盘状

        4. Petiole not sheathed at base/ 叶柄基部不具鞘

            5. Apex of style branches without a central appendage of fused papillae/ 花柱分枝顶端无合并的乳头状毛的中央附器 ....**4. *Senecio*/ 千里光属**

            5. Apex of style branches with a central appendage of fused papillae/ 花柱分枝顶端具合并的乳头状毛的中央附器 ....**5. *Crassocephalum*/ 野茼蒿属**

        4. Petiole broadly sheathed at base/ 叶柄基部宽，具鞘

6. Achenes hairy; leaf margin involute/ 瘦果多毛；叶缘内卷 .... **6. *Farfugium*/ 大吴风草属**
6. Achenes glabrous; leaf margin revolute/ 瘦果光滑；叶缘反卷 .... **7. *Ligularia*/ 橐吾属**

1. *Gynura formosana* Kitam./ 白凤菜；
2. *Emilia sonchifolia* (L.) DC./ 一点红；
3. *Erechtites valerianifolius* (Link ex Spreng.) DC./ 败酱叶菊芹；
4. *Senecio scandens* Buch.-Ham. ex D. Don/ 千里光；
5. *Crassocephalum crepidioides* (Benth.) S. Moore/ 野茼蒿；
6. *Farfugium japonicum* (L.) Kitam./ 大吴风草（photo by WeiLiang MA/ 马炜梁拍摄）；
7. *Ligularia intermedia* Nakai/ 狭苞橐吾（Showing glabrous achenes, not distributed here/ 仅展示光滑的瘦果，非本区域分布）

XII. Astereae/ 紫菀族

1. Capitula disciform or discoid/ 头状花序盘状（不具舌状花）
   2. Pappus absent, or of disk florets sometimes with 1 or 2 bristles/ 无冠毛，或盘花有时具 1 或 2 根刚毛状冠毛
   .... **1. *Dichrocephala*/ 鱼眼草属**
   2. Pappus of bristles/ 冠毛刚毛状
      3. Capitula discoid/ 头状花序杂性盘状,均为两性花 .... **2. *Aster*/ 紫菀属**
      3. Capitula disciform/ 头状花序盘状，花杂性
         4. Herbs glabrous/ 植株光滑 .... **3. *Symphyotrichum*/ 联毛紫菀属**
         4. Herbs glandular, hairy/ 植株具腺体，多毛
            5. Involucres urceolate or subcylindric/ 总苞片坛状或近圆柱状 .... **4. *Erigeron*/ 飞蓬属**
            5. Involucres campanulate to hemispheric-campanulate/ 总苞片钟状至半球状 ....

....**5. *Eschenbachia*/ 白酒草属**

1. Capitula radiate/ 头状花序辐射状（有舌状花）

    6. Ray floret lamina yellow, orange, red, or brown/ 舌状花黄色、橙色、红色或棕色

        7. Capitula numerous/ 头状花序多数 ....**6. *Solidago*/ 一枝黄花属**

        7. Capitula solitary/ 头状花序单生 ....**4. *Erigeron*/ 飞蓬属**

    6. Ray floret lamina white, pink, purple, or blue/ 舌状花白色、粉色、紫色或蓝色 ....**7. *Tripolium*/ 碱菀属**

1. *Dichrocephala integrifolia* (L. f.) Kuntze/ 鱼眼草；
2. *Aster trinervius* Roxb. ex D. Don subsp. *ageratoides* (Turcz.) Grierson/ 三脉紫菀；
3. *Symphyotrichum subulatum* (Michx.) G.L. Nesom/ 钻叶紫菀；
4. *Erigeron bonariensis* L./ 香丝草；
5. *Eschenbachia japonica* (Thunb.) J. Kost./ 白酒草；
6. *Solidago decurrens* Lour./ 一枝黄花；
7. *Tripolium pannonicum* (Jacq.) Dobrocz./ 碱菀

XIII. Inuleae/ 旋覆花族

1. Capitula radiate, disciform, or discoid; achenes with large oxalate crystals in epidermis cells/ 头状花序辐射状、杂性盘状或两性盘状；瘦果表皮细胞有大型草酸结晶

    2. Pappus absent/ 无冠毛 ....**1. *Carpesium*/ 天名精属**

    2. Pappus present/ 有冠毛

3. Marginal florets tubular or filiform/ 缘花细管状或丝状 ....**2. *Blumea*/ 艾纳香属**
3. Marginal florets radiate or absent/ 缘花舌状或无 ....**3. *Inula*/ 旋覆花属**
1. Capitula disciform; achenes without large epidermis crystals/ 头状花序杂性盘状；瘦果表皮细胞无大型结晶
 4. Pappus absent/ 无冠毛 ....**4. *Epaltes*/ 球菊属**
 4. Pappus of capillary bristles/ 冠毛为纤细刚毛 ....**5. *Pluchea*/ 阔苞菊属**

1. *Carpesium abrotanoides* L./ 天名精；
2. *Blumea megacephala* (Randeria) C.C. Chang & Y.Q. Tseng/ 东风草；
3. *Inula japonica* Thunb./ 旋覆花；
4. *Epaltes australis* Less./ 球菊；
5. *Pluchea indica* (L.) Less./ 阔苞菊

408 Adoxaceae 五福花科

Leaves opposite, lateral veins distinct / 叶对生，侧脉明显
Inflorescence often a large compound umbels / 多大型复伞形花序
Flowers small, bisexual, actinomorphic / 花小，两性，辐射对称
Ovary part-inferior to inferior / 子房半下位至下位

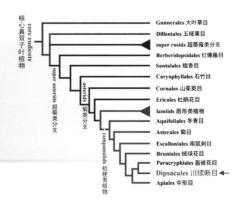

Shrubs, less often perennial **herbs**, or **small trees**. **Leaves** simple, trifoliolate to odd-pinnately compound; opposite or rarely whorled; margin entire to toothed; petioles present; stipules present or absent. **Inflorescence** terminal, panicles, umbels, spikes or head-like cymes. **Flowers** small (<5 mm in diam.), bisexual, rarely unisexual (gynodioecious), (3–)5-merous, actinomorphic. **Sepals** fused. **Petals** fused, often with a nectary at the base. **Stamens** 5, alternate corolla lobes, inserted on corolla tube; anthers basifixed, extrorse; staminodes 3–5. **Ovary** part-inferior to inferior; carpels fused; locules 1 or 3–5; placentation axile to apical; style short or reduced to absent; stigmas capitate or 2–3-fid. **Fruit** fleshy drupes.

灌木，偶为多年生**草本**或**小乔木**。**叶**为单叶，3 小叶至奇数羽状复叶；对生或稀轮生；全缘至有锯齿；具叶柄；托叶有或无。**花序**顶生，圆锥状、伞形、穗状或头状聚伞花序。花小（直径小于 5mm），两性，稀单性（雄全异株），(3~) 5 基数，辐射对称。**花萼**合生。**花瓣**合生，基部常具蜜腺。**雄蕊** 5 枚，与花冠裂片互生，贴生于冠筒；花药基着，外向；退化雄蕊 3~5 枚。**子房**半下位至下位；心皮合生；1 或 3~5 室；中轴胎座至顶生胎座；花柱短或退化至缺；柱头头状或 2~3 裂。**果实**为肉质核果。

World 4/220, worldwide, mainly in northern temperate regions and tropical montane habitats.
China 4/81.
This area 2/11.

全世界共 4 属 220 种，全球广布，主要分布于北温带和热带山地。
中国产 4 属 81 种。
本地区有 2 属 11 种。

Viburnum dilatatum Thunb./ 荚蒾

1. Leaves simple/ 单叶 .... 1. *Viburnum*/ 荚蒾属
1. Leaves compound/ 复叶 .... 2. *Sambucus*/ 接骨木属

1. *Viburnum propinquum* Hemsl./ 球核荚蒾
（1a. Inflorescence/ 花序；1b. Infructescence/ 果序）；

2. *Sambucus javanica* Blume/ 接骨草
（2a. Inflorescence/ 花序；2b. Infructescence/ 果序）

409 Caprifoliaceae 忍冬科

Piliferous / 常被毛
Bark peeling into strips / 老茎皮条状剥落
Leaves opposite / 叶对生
Stamens often 5, exserted / 雄蕊常 5 枚，伸出花冠筒

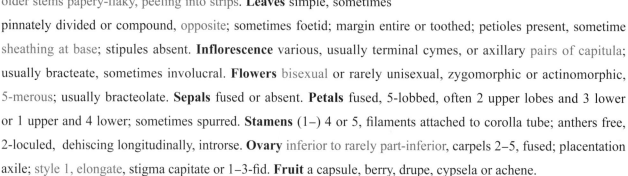

Herbs, shrubs, small trees, or woody climbers; hairs various; bark of older stems papery-flaky, peeling into strips. **Leaves** simple, sometimes pinnately divided or compound, opposite; sometimes foetid; margin entire or toothed; petioles present, sometime sheathing at base; stipules absent. **Inflorescence** various, usually terminal cymes, or axillary pairs of capitula; usually bracteate, sometimes involucral. **Flowers** bisexual or rarely unisexual, zygomorphic or actinomorphic, 5-merous; usually bracteolate. **Sepals** fused or absent. **Petals** fused, 5-lobbed, often 2 upper lobes and 3 lower or 1 upper and 4 lower; sometimes spurred. **Stamens** (1–) 4 or 5, filaments attached to corolla tube; anthers free, 2-loculed, dehiscing longitudinally, introrse. **Ovary** inferior to rarely part-inferior, carpels 2–5, fused; placentation axile; style 1, elongate, stigma capitate or 1–3-fid. **Fruit** a capsule, berry, drupe, cypsela or achene.

草本、灌木、小乔木或木质藤本；毛被多样；老茎茎皮纸质而易条状剥落。**叶**为单叶，有时羽裂或为复叶，对生；有时具臭味；全缘或具齿；具叶柄，有时基部鞘状；托叶缺。**花序**多样，常为顶生聚伞花序或腋生成对的头状花序；常具苞片，有时总苞状。**花**两性或稀单性，两侧对称或辐射对称，5 基数；常具小苞片。**花萼**合生或缺。**花瓣**合生，5 裂，上唇 2 枚下唇 3 枚或上唇 1 枚下唇 4 枚；有时具距。**雄蕊** (1~) 4 或 5 枚，花丝贴生于冠筒；花药离生，2 室，纵裂，内向。**子房**下位至稀半下位，心皮 2~5 枚，合生；中轴胎座；花柱 1 个，伸长，柱头头状或 1~3 裂。**果实**为蒴果、浆果、核果、连萼瘦果或瘦果。

World 36/810, worldwide, mainly in northern temperate regions.
China 20/ca. 144.
This area 3/9.

Lonicera fragrantissima Lindl. & Paxton/ 郁香忍冬

全世界共 36 属 810 种，全球广布，主要分布于北温带。
中国产 20 属约 144 种。
本地区有 3 属 9 种。

1. Herbs/ 草本 .... **1. *Patrinia*/ 败酱属**
1. Shrubs or trees or woody climbers/ 灌木、乔木或木质藤本
   2. Usually paired flowers with somewhat fused ovaries; often woody climbers/ 花常成对，两花子房多少合生；多木质藤本 .... **2. *Lonicera*/ 忍冬属**

2. Ovary not fused; often erect shrubs/ 子房离生；多直立灌木 ....**3. *Abelia*/ 糯米条属**

1. *Patrinia villosa* (Thunb.) Dufr./ 攀倒甑；
2. *Lonicera fragrantissima* Lindl. & Paxton/ 郁香忍冬；
3. *Abelia chinensis* R. Br./ 糯米条（photo by XinXin ZHU/ 朱鑫鑫拍摄）

413 Pittosporaceae 海桐科

Leaves clustered at branch ends / 叶常簇生于小枝顶
Flowers 5-merous and radial / 花 5 基数，辐射状
Capsule dehiscing by 2–5 valves / 蒴果 2~5 瓣裂
Seeds surrounded by sticky material / 种子为黏性物质包围

Trees, **shrubs** or woody climbers. **Leaves** simple, alternate, rarely opposite or whorled; usually clustered at branch ends; sometimes aromatic; margin entire or less often toothed; petioles present; stipules absent. **Inflorescence** panicles, cymes, thyrses or corymbs, or a solitary flower; bracteate and bracteolate. **Flowers** usually bisexual, usually actinomorphic, 5-merous. **Sepals** free or slightly fused, imbricate. **Petals** free or fused at base, imbricate, often clawed. **Stamens** opposite sepals; filaments filiform, free from petals; anther dorsifixed, introrse. **Ovary** superior, 2–5 carpels, fused, sometimes on a stalk; 1–2(–5)-loculed; ovules 5 to many per locule; placentation parietal or axile; style short, usually unlobed and persistent. **Fruit** a capsule dehiscing by 2–5 valves. **Seeds** usually surrounded by glutinous or greasy material.

乔木、**灌木**或木质藤本。**叶**为单叶，互生，稀对生或轮生；常簇生小枝顶；有时具芳香味；全缘或偶具齿；具叶柄；托叶缺。**花序**圆锥状、聚伞状、聚伞圆锥状或伞房状，或为单花；具苞片和小苞片。**花**常两性，多辐射对称，5 基数。**花萼**离生或稍合生，覆瓦状。**花瓣**离生或基部合生，覆瓦状，常具爪。**雄蕊**与花萼对生；花丝丝状，与花瓣分离；花药背着，内向。**子房**上位，2~5 枚心皮，合生，有时具柄；1~2(~5)室；胚珠每室 5 至多枚；侧膜胎座或中轴胎座；花柱短，常不分裂，宿存。**果实**为 2~5 瓣裂的蒴果。**种子**常有黏质或油质物包在外面。

World 6–9/200, Old World tropics and subtropics, especially in Australia, southwest Pacific Islands, Southeast Asia and subtropical East Asia.
China 1/44.
This area 1/2.

全世界共 6~9 属 200 种，分布于旧世界热带和亚热带，尤其是澳大利亚，西南太平洋群岛，东南亚及亚洲东部的亚热带地区。
中国产 1 属 44 种。
本地区有 1 属 2 种。

Pittosporum Banks. ex Gaertn. 海桐属

Pittosporum tobira (Thunb.) W.T. Aiton/ 海桐
（a. Inflorescence/ 花序；b. Capsules/ 蒴果）

414 Araliaceae 五加科

Often woody, stems with conspicuous leaf scars / 常木本，叶痕明显
Leaves palmately lobed or palmately compound / 叶掌状分裂或为掌状复叶
Inflorescence umbels or capitula / 伞形或头状花序
Ovary often inferior, drupe / 子房多下位，核果

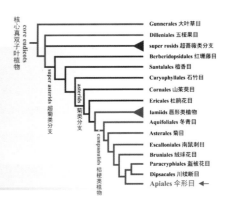

Trees or shrubs, sometimes woody climbers with aerial roots, rarely herbs; stems often with conspicuous leaf scars. **Leaves** sometimes aromatic; simple and palmately lobed, palmately compound, or pinnately compound; alternate, or rarely opposite; often clustered at branch ends; petioles present; stipules fused at base of petiole, often broad and sheathing. **Inflorescence** umbels or capitula, or arranged in various inflorescences; bracteate. **Flowers** small, bisexual or unisexual, actinomorphic, 5-merous, glandular disk present. **Sepals** free or fused, tube small. **Petals** 5–10; free or slightly fused, valvate or imbricate. **Stamens** usually as many as petals; opposite with the sepals; filaments inserted at edge of disk. **Ovary** inferior or rarely superior; carpels fused; ovule 1(–2) per locule; placentation apical; styles 1–5. **Fruit** a drupe, berry or schizocarp.

乔木或灌木，有时为木质藤本，具气生根，稀为草本；茎常具明显叶痕。**叶**有时具芳香味；单叶，掌状分裂，掌状复叶或羽状复叶；互生，或稀对生；常集生枝顶；具叶柄；托叶与叶柄基部合生，常宽大成鞘。**花序**伞形或头状，或排列成各式花序；具苞片。花小，两性或单性，辐射对称，5基数，具蜜腺盘。**花萼**离生或合生，萼筒小。**花瓣** 5~10 枚；离生或稍合生，镊合状或覆瓦状。**雄蕊**常与花瓣同数；与花萼对生；花丝嵌生于花盘边缘。**子房**下位或稀上位；心皮合生；胚珠每室 1(~2) 枚；顶生胎座；花柱 1~5 个。**果实**为核果、浆果或分果。

World 43/1,450, widely distributed in tropical and subtropical regions, a few extending into temperate regions.
China 22/ca. 192.
This area 6/16.

Aralia elata (Miq.) Seem./ 楤木

全世界共 43 属 1 450 种，主要分布于热带和亚热带地区，少数至温带。
中国产 22 属约 192 种。
本地区有 6 属 16 种。

1. Herbs/ 草本 .... **1. *Hydrocotyle*/ 天胡荽属**
1. Shrubs or trees/ 灌木或乔木
  2. Petals imbricate in bud/花瓣在芽中覆瓦状排列 .... **2. *Aralia*/ 楤木属**
  2. Petals valvate in bud/ 花瓣在芽中镊合状排列

3. Leaves palmately compound/ 叶为掌状复叶
    4. Plants with prickles/ 植株具刺 ....**3. *Eleutherococcus*/ 五加属**
    4. Plants unarmed/ 植株无刺 ....**4. *Heptapleurum*/ 鹅掌柴属**
3. Leaves simple or palmately lobed/ 叶为单叶或掌状分裂
    5. Woody climbers/ 木质藤本 ....**5. *Hedera*/ 常春藤属**
    5. Shrubs or trees, erect; stems with prickles/ 直立灌木或乔木；茎具刺 ....**6. *Kalopanax*/ 刺楸属**

1. *Hydrocotyle sibthorpioides* Lam./ 天胡荽；
2. *Aralia elata* (Miq.) Seem./ 楤木；
3. *Eleutherococcus nodiflorus* (Dunn) S.Y. Hu/ 细柱五加；
4. *Heptapleurum octophyllum* (Lour.) Benth. ex Hance/ 鹅掌柴；
5. *Hedera rhombea* (Miq.) Bean/ 菱叶常春藤；
6. *Kalopanax septemlobus* (Thunb.) Koidz./ 刺楸

416 Apiaceae 伞形科

Herbs, aromatic / 草本，具芳香味
Leaves decompound, petiole sheathing / 叶为多回复叶，叶柄具鞘
Compound umbels / 复伞形花序
Mericarps / 双悬果

Herbs, or rarely **shrubs** to **trees**; stem furrowed, hollow or solid; sometimes aromatic, spiny. **Leaves** alternate, pinnately or palmately compound to simple, then margin deeply dissected or lobed; petiole usually sheathing at base; stipules often absent. **Inflorescence** simple or compound umbels, rarely capitula, panicles, racemes or solitary flower; often subtended by bracts forming an involucre, sometimes showy. **Flowers** small, bisexual or unisexual; actinomorphic, marginal flowers often zygomorphic; bracteolate. **Sepals** tube wholly adnate to the ovary, 5-toothed. **Petals** 5, free, usually inflexed, valvate, often clawed. **Stamens** 5, alternate the petals; filaments free from petals; anthers introrse. **Ovary** inferior, carpels fused, 2-loculed; ovule 1 per locule; placentation apical; styles 2, usually swollen at the base. **Fruit** a dry schizocarp, of two mericarps united by their faces (commissure), and usually attached to a central axis (carpophore), from which the mericarps separate at maturity; each mericarp has 5 primary ribs, one down the back (dorsal rib), two on the edges near the commissure (lateral ribs), and two between the dorsal and lateral ribs (intermediate ribs), occasionally with 4 secondary ribs alternating with the primary, the ribs filiform to broadly winged; vittae (oil-tubes) usually present in the furrow and on the commissure face; each mericarp seed 1.

草本，或稀为**灌木至乔木**；茎具沟槽，中空或实心；有时具芳香味，具刺。**叶**互生，羽状复叶或掌状复叶至单叶，边缘深缺刻或分裂；叶柄常于基部成鞘状；托叶常缺。**花序**为单个或复合伞形花序，稀头状、圆锥状、总状或单花；常为总苞状苞片包围，有时显著。**花**小，两性或单性；辐射对称，缘花常两侧对称；具小苞片。**花萼**萼筒全部与子房合生，具5齿。**花瓣**5枚，离生，常反折，镊合状，常具爪。**雄蕊**5枚，与花瓣互生；花丝与花瓣分离；花药内向。**子房**下位，心皮合生，2室；胚珠每室1枚；顶生胎座；花柱2个，基部常膨大。**果实**为干燥的分果（双悬果），由两枚悬果爿合生（沿合生面），常附着于中央柄上（果瓣柄），成熟时分离；每个悬果爿具5条主棱，1枚在背部（背棱），2枚靠近合生面（侧棱），另2枚在背棱与侧棱之间（中棱），偶有4枚次棱与主棱间生，棱丝状至宽翅状；悬果爿沟槽及合生面上具油管；每个悬果爿含1粒种子。

World 300–440 (455)/3,000–3,750, widely distributed in temperate regions, mainly in Eurasia.
China ca. 99/ca. 616.
This area 11/14.

全世界共 300~440（455）属 3 000~3 750 种，广布温带，主要分布于欧亚大陆。

Glehnia littoralis F. Schmidt ex Miq./ 珊瑚菜

中国产约99属约616种。

本地区有11属14种。

1. Stems creeping, rarely erect/ 茎匍匐，稀直立 ....**1. *Centella*/ 积雪草属**
1. Stems usually erect, not creeping/ 茎常直立，非匍匐状
 2. Fruit with both primary and secondary ribs/ 果具主棱和次棱 ....**2. *Daucus*/ 胡萝卜属**
 2. Fruit with primary ribs, secondary ribs absent/ 果具主棱，无次棱
  3. Commissure face plane/ 分果合生面平直
   4. Fruit ribs equal/ 果棱等宽
    5. Primary ribs of fruit filiform, commissure narrow/ 主棱丝状，合生面狭窄 ....**3. *Cyclospermum*/ 细叶旱芹属**
    5. Primary ribs prominent to narrowly winged, commissure broad/ 主棱显著至狭翅状，合生面宽
     6. Semiaquatic or marshland herbs; outer petals of umbellule usually conspicuously radiant/ 半水生或沼生草本；小伞形花序外缘花的花瓣常呈显著的辐射瓣 ....**4. *Oenanthe*/ 水芹属**
     6. Terrestrial herbs; outer petals of umbellule not radiant/ 陆生草本；小伞形花序外缘花的花瓣常不为辐射瓣 ....**5. *Cnidium*/ 蛇床属**
   4. Lateral ribs broader than the dorsal ribs/ 果侧棱较背棱宽
    7. Lateral mericarp wings divergent at maturity/ 果侧翅成熟时分离
     8. White-pubescent throughout; coastal plants/ 全株被白色柔毛；滨海植物 ....**6. *Glehnia*/ 珊瑚菜属**
     8. Plant somewhat glabrous, not coastal plants/ 植株多少光滑，非滨海植物 ....**7. *Ostericum*/ 山芹属**
    7. Lateral mericarp wings adnate or closely appressed at maturity/ 果侧翅成熟时贴合
     9. Lateral mericarp wings not thickened/ 果侧翅不增厚 ....**8. *Peucedanum*/ 前胡属**
     9. Lateral mericarp wings thickened, margin rigid/ 果侧翅增厚，翅缘坚硬 ....**9. *Heracleum*/ 独活属**
  3. Commissure face deeply concave or sulcate/ 分果合生面深凹或中空
   10. Fruit beaked/ 果实具喙
    11. Fruit setulose or bristly, usually in longitudinal rows/ 果实常具直伸或钩状成行皮刺 ....**10. *Torilis*/ 窃衣属**
    11. Fruit glabrous or bristly, but not in longitudinal rows/ 果实光滑或有刺，但不明显成行 ....**11. *Anthriscus*/ 峨参属**
   10. Fruit not beaked/ 果实不具喙 ....**12. *Changium*/ 明党参属**

APIALES. Apiaceae. *Glehnia littoralis* F. Schmidt ex Miq./ 伞形目 伞形科 珊瑚菜

A. Umbel/ 伞形花序
B1. Flower segregated/ 花离析
B2. Cross section of ovary/ 子房横切面
B3. Vertical section of flower/ 花纵切面
B4. Stamens/ 雄蕊
B5. Vertical section of flower/ 花纵切面
B6. Dorsal view of flower/ 花背面观
B7. Ventral view of flower/ 花腹面观

C1. Vertical section of fruit (parallel to cotyledon)/ 果实纵切面 (平行于子叶)
C2. Fruit/ 果实
C4. Mericarp/ 双悬果分果爿
C5. Vertical profile of infructescence/ 果序纵剖面
C6. Vertical section of mericarp/ 双悬果纵切面
C7. Cross section of mericarp/ 双悬果横切面
(Scale/ 标尺：1 mm)

1. *Centella asiatica* (L.) Urb./ 积雪草；
2. *Daucus carota* L./ 野胡萝卜；
3. *Cyclospermum leptophyllum* (Pers.) Sprague ex Britton & P. Wilson/ 细叶旱芹；
4. *Oenanthe benghalensis* (Roxb.) Benth. & Hook. f./ 短辐水芹；
5. *Glehnia littoralis* F. Schmidt ex Miq./ 珊瑚菜；
6. *Ostericum citriodorum* (Hance) C.Q. Yuan & R.H. Shan/ 隔山香；
7. *Peucedanum japonicum* Thunb./ 滨海前胡；
8. *Heracleum sphondylium* L. var. *nipponicum* (Kitag.) H. Ohba/ 日本独活；
9. *Torilis scabra* (Thunb.) DC./ 窃衣；
10. *Anthriscus sylvestris* (L.) Hoffm./ 峨参；
11. *Changium smyrnioides* Fedde ex H. Wolff/ 明党参
12. *Cnidium japonicum* Miq./ 滨蛇床

Index to Scientific Names 拉丁学名索引

A

Abelia	353
Abelia chinensis	353
Abelmoschus	228
Abelmoschus esculentus	229
Abutilon	228
Abutilon theophrasti	229
Acacia farnesiana	139
Acacia	139
Acacieae	135, 139
Acalypha	202
Acalypha australis	202
Acanthaceae	313
Acer	220
Acer tataricum subsp. *ginnala*	220
Achillea	344
Achillea millefolium	344
Achnatherum coreanum	99
Achyranthes	247
Achyranthes bidentata	247
Acmella	342
Acmella paniculata	343
Acoraceae	22
Acorus	22
Acorus gramineus	22
Acorus calamus	22
Actinidia	279
Actinidia chinensis	279
Actinidia chinensis var. *deliciosa*	279
Actinidiaceae	279
Actinostemma	183
Actinostemma tenerum	184
Adenophora	332
Adenophora petiolata subsp. *hunanensis*	332
Adina	284
Adina pilulifera	285
Adinandra	267
Adinandra millettii	267
Adoxaceae	350
Aeginetia	327
Aeginetia indica	328
Aeschynomene	142
Aeschynomene indica	142
Aeschynomeneae	135, 136, 142
Agave	57
Agave americana	58
Ageratum	340
Ageratum conyzoides	340
Agrimonia	154
Agrimonia pilosa	158
Agrostis	95, 96
Aidia	284
Aidia cochinchinensis	285
Ailanthus altissima	223
Ailanthus	223
Ainsliaea	339
Ainsliaea kawakamii	339
Aizoaceae	249
Ajania	344
Ajania pallasiana	345
Ajuga	319
Ajuga nipponensis	321
Akebia	102
Akebia quinate	103
Alangium	263
Alangium platanifolium	264
Albizia	140
Albizia julibrissin	137, 140
Aletris	32
Aletris scopulorum	32
Alisma	26
Alisma orientale	27
Alismataceae	26
ALISMATALES	24, 29
Allium	52
Allium tuberosum	54

361

Allocasuarina	180	*Aphananthe aspera*	168
Alopecurus	95	Apiaceae	357, 359
Alopecurus aequalis	96	*Apluda*	87
Alpinia	70	*Apluda mutica*	89
Alpinia japonica	71	Apocynaceae	291
Alpinia zerumbet	71	Aquifoliaceae	329
Alternanthera	247	Araceae	23, 24
Alternanthera philoxeroides	247	*Aralia*	355
Althaea	228	*Aralia elata*	355, 356
Althaea officinalis	229	Araliaceae	355
Altingiaceae	118	*Archidendron*	140
Alysicarpus	145	*Archidendron lucidum*	140
Alysicarpus vaginalis	146	*Ardisia*	271
Alyxia	292	*Ardisia crenata*	273
Alyxia sinensis	291, 293	Arecaceae	60
Amana	38	*Arenaria*	244
Amana edulis	40	*Arenaria serpyllifolia*	245
Amaranthaceae	246	*Arisaema erubescens*	25
Amaranthus	247	*Arisaema heterophyllum*	25
Amaranthus viridis	247	*Aristolochia*	9
Amaryllidaceae	52, 53, 56	*Aristolochia dabieshanense*	11
Ambrosia	341	Aristolochiaceae	9, 10
Ambrosia artemisiifolia	342	*Armeniaca*	155
Amischotype	62	*Armeniaca mume*	159
Amischotype hispida	64	*Artemisia*	344
Ammannia	207	*Artemisia annua*	345
Ammannia multiflora	208	*Arthraxon*	88
Amorpha	148	*Arthraxou hispidus*	89
Amorpheae	136, 148	*Arum*	23
Ampelopsis	130	Arundineae	81, 94
Ampelopsis glandulosa	130	*Arundinella hirta*	85
Amygdalus	155	*Arundinella*	84
Amygdalus persica	159	Arundinelleae	80, 81, 84
Anacardiaceae	217	*Arundo*	94
Anagallis	270	*Arundo donax* 'Versicolor'	94
Anagallis arvensis	270, 272	*Asarum*	9, 10
Andropogoneae	81, 87	*Asarum ichangense*	11
Androsace	271	*Asarum*	10
Androsace umbellata	272	Asparagaceae	55
Anodendron	292	ASPARAGALES	42, 47, 50, 53, 56
Anodendron affine	293	*Asparagus*	57
Anredera	255	*Asparagus cochinchinensis*	58
Anredera cordifolia	255	Asphodelaceae	49, 50
Anthemideae	335, 344	*Aster*	347
Anthriscus	358	*Aster trinervius* subsp. *ageratoides*	348
Anthriscus sylvestris	360	Astereae	336, 347
Aphananthe	167	*Astragalus*	147

Astragalus sinicus	147	Boraginaceae	294
Athroismeae	336, 346	*Bothriochloa*	88
Aucuba	282	*Bothriochloa ischaemum*	89
Aucuba japonica	282	*Bothriospermum*	295
AUSTROBAILEYALES	2	*Bothriospermum zeylanicum*	296
Avena	95	*Bougainvillea*	252
Avena fatua	95	*Bougainvillea glabra*	253
Aveneae	82, 95	Brachypodieae	81, 92
		Brachypodium	92
		Brachypodium sylvaticum	92
		Brassicaceae	233
B		*Breynia*	204
Bacopa	306	*Breynia rostrata*	204
Bacopa monnieri	307	*Briza*	92
Balsaminaceae	265	*Briza minor*	93
Bambusa	84	Bromeae	82, 97
Bambusa ventricosa	84	*Bromus catharticus*	97
Bambuseae	80, 84	*Bromus remotiflorus*	97
Barnardia japonica	57	*Broussonetia*	170
Basella	255	*Broussonetia kazinoki*	170
Basella alba	255	*Buddleja*	308
Basellaceae	255	*Buddleja lindleyana*	309
Bassia	247	*Bulbophyllum*	41
Bassia scoparia	248	*Bulbophyllum chondriophorum*	43
Bauhinia	140	Buxaceae	116
Bauhinia championii	140	*Buxus*	116
Beckmannia	95	*Buxus sinica*	117
Beckmannia syzigachne	96	*Buxus rugulosa*	117
Begonia grandis	185		
Begonia	185		
Begoniaceae	185	**C**	
Berberidaceae	106	Cactaceae	258
Berchemia	164	*Caesalpinia decapetala*	140
Berchemia floribunda	165	*Caesalpinia*	140
Betulaceae	181	Caesalpinieae	135, 140
Bidens	341	*Calamagrostis*	95
Bidens pilosa	343	*Calamagrostis epigeios*	96
Bletilla	41	*Callerya*	147
Bletilla striata	43	*Callerya reticulata*	148
Blumea	349	*Callicarpa*	319
Blumea megacephala	349	*Callicarpa cathayana*	321
Boehmeria	172	*Callitriche*	306
Boehmeria macrophylla	172	*Callitriche palustris*	307
Boerhavia	252	Calycanthaceae	15, 16
Boerhavia diffusa	253	*Calystegia*	298
Bolboschoenus	77	*Calystegia soldanella*	299
Bonnaya	310	*Camellia*	274
Bonnaya antipoda	311		

Camellia fraterna	275	*Celosia*	246
Camellia japonica	275	*Celosia argentea*	247
Campanulaceae	331	*Celtis*	167
Camptotheca	259	*Celtis sinensis*	168
Camptotheca acuminata	259	*Centaurium*	288
Campylotropis	145	*Centaurium pulchellum* var. *altaicum*	287, 288
Campylotropis macrocarpa	145	*Centella*	358
Canavalia	143	*Centella asiatica*	360
Canavalia rosea	134, 143	*Centipeda*	346
Cannabaceae	167	*Centipeda minima*	346
Capparaceae	232	Centotheceae	81, 82, 94
Capparis cantoniensis	232	*Cerastium*	244
Capparis	232	*Cerastium glomeratum*	244
Caprifoliaceae	352	*Cerasus*	155
Capsella	234	*Cerasus discoidea*	159
Capsella bursa-pastoris	234	Ceratophyllaceae	100
Caragana	146	*Ceratophyllum demersum*	100
Caragana sinica	146	*Ceratophyllum*	100
Cardamine	234	Cercideae	135, 140
Cardamine impatiens	234	*Cercis*	140
Cardueae	335, 345	*Cercis glabra*	140
Carex	77	*Chamaecrista*	141
Carex pumila	78	*Chamaecrista mimosoides*	141
Carpesium	348	*Changium*	358
Carpesium abrotanoides	349	*Changium smyrnioides*	360
Carpinus	181	*Celtis chekiangensis*	168
Carpinus putoensis	181	*Chenopodium*	247
Caryophyllaceae	243	*Chenopodium album*	248
Caryopteris	318	*Chimonanthus*	15
Caryopteris incana	321	*Chimonanthus praecox*	15, 16
Casearia	199	*Chionanthus*	302
Casearia glomerata	200	*Chionanthus retusus*	303
Cassieae	135, 141	Chloranthaceae	21
Cassytha	17	*Chloranthus*	21
Cassytha filiformis	19	*Chloranthus fortunei*	21
Castanea	173	*Chloris*	98
Castanea seguinii	175	*Chloris formosana*	99
Castanopsis	173	*Choerospondias*	218
Castanopsis carlesii	175	*Choerospondias axillaris*	218
Casuarina	180	*Christia*	145
Casuarina glauca	180	*Christia* obcordata	146
Casuarinaceae	179	*Chrysanthemum*	344
Cayratia	130	*Chrysanthemum indicum*	344
Cayratia japonica	130	*Chukrasia*	226
Celastraceae	187	*Chukrasia tabularis*	226
Celastrus	188	Cichorieae	334, 338
Celastrus gemmatus	189	*Cinnamomum*	17

Cinnamomum burmannii	18	*Crossostephium*	344
Cinnamomum japonicum 'Chenii'	20	*Crossostephium chinense*	345
Cirsium	345	*Crotalaria*	144
Cirsium japonicum	345	*Crotalaria sessiliflora*	144
Citrus	222	Crotalarieae	135, 144
Citrus japonica	222	*Croton*	202
Cladium	77	*Croton cascarilloides*	202
Cleisostoma	42	Cucurbitaceae	182
Cleisostoma arietium	44	*Cuscuta*	297
Clematis	108	*Cuscuta campestris*	299
Clematis apiifolia	110	*Cyanthillium*	339
Clematis lanuginosa	110	*Cyanthillium cinereum*	339
Clerodendrum	319	*Cyclospermum*	358
Clerodendrum kaichianum	321	*Cyclospermum leptophyllum*	360
Cleyera	267	*Cymbopogon*	88
Cleyera japonica	267	*Cymbopogon goeringii*	90
Clinopodium	320	*Cynanchum*	292
Clinopodium chinense	322	*Cynanchum fordii*	293
Cnidiuml	358	*Cynodon*	98
Cocculus	105	*Cynodon dactylon*	98
Cocculus orbiculatus	105	Cynodonteae	82, 98
Coix	87	*Cynoglossum*	295
Coix lacryma-jobi	88	*Cynoglossum lanceolatum*	296
Commelina	62	Cyperaceae	76
Commelina communis	63, 64	*Cyperus*	77
Commelinaceae	62, 63	*Cyperus cyperoides*	79
COMMELINALES	63	*Cyrtococcum*	85
Convolvulaceae	297	*Cyrtococcum patens*	86
Corchoropsis	228		
Corchoropsis crenata	229		
Coreopsis	341	**D**	
Coreopsis lanceolata	343	*Dactyloctenium*	90
Cornaceae	263	*Dactyloctenium aegyptium*	91
Cornus	263	*Dalbergia*	148
Cornus walteri	264	Dalbergieae	136, 148
Corydalis	101	*Damnacanthus*	284
Corydalis heterocarpa	101	*Damnacanthus indicus*	286
Corydalis incisa	101	*Daphne*	230
Cosmos	341	*Daphne kiusiana* var. *atrocaulis*	231
Cosmos bipinnatus	343	Daphniphyllaceae	122
Cotula	344	*Daphniphyllum*	122
Cotula anthemoides	344	*Daphniphyllum oldhamii*	122
Crassocephalum	346	*Datura*	301
Crassocephalum crepidioides	347	*Datura stramonium*	301
Crassulaceae	126	*Daucus*	358
Crataegus	154	*Daucus carota*	360
Crataegus pinnatifida	157	*Delphinium*	109

Delphinium anthriscifolium var. *calleryi*	110
Desmodieae	135, 136, 145
Desmodium	145
Desmodium heterophyllum	146
Deutzia	260
Deutzia crenata	262
Deyeuxia	95
Deyeuxia effusiflora	95
Dianella	49
Dianella ensifolia	51
Dianthus	244
Dianthus plumarius	243
Dianthus superbus	245
Dichondra	298
Dichondra micrantha	299
Dichrocephala	347
Dichrocephala integrifolia	348
Digitaria	85
Digitaria ciliaris	86
Dioscorea	33
Dioscorea cirrhosa	33
Dioscorea tenuipes	35
Dioscoreaceae	33, 34
DIOSCOREALES	34
Diospyros	268
Diospyros kaki var. *silvestris*	268, 269
Diospyros morrisiana	269
Diploclisia	105
Diploclisia affinis	105
Diplospora	284
Diplospora dubia	285
Distylium	119
Distylium myricoides	121
Dodonaea	220
Dodonaea viscosa	220
Drosera	242
Drosera peltata	242
Droseraceae	242
Duchesnea	155
Duchesnea indica	159
Dunbaria	142
Dunbaria villosa	143
Duranta	317
Duranta erecta	317
Dysphania	247
Dysphania ambrosioides	248

E

Ebenaceae	268
Echinochloa	86
Echinochloa colona	87
Echinopeae	335, 340
Echinops grijsii	340
Echinops	340
Eclipta	342
Eclipta prostrata	343
Edgeworthia	231
Edgeworthia chrysantha	231
Ehretia	295
Ehretia acuminata	296
Eichhornia	66
Eichhornia crassipes	67
Elaeagnaceae	160
Elaeagnus argyi	162
Elaeagnus	160
Elaeocarpaceae	192
Elaeocarpus glabripetalus	192
Elaeocarpus	192
Elaeagnus pungens	162
Eleocharis	77
Eleocharis geniculata	78
Elephantopus	339
Elephantopus scaber	339
Eleusine	90
Eleusine indica	91
Eleutherococcus	356
Eleutherococcus nodiflorus	356
Elsholtzia	320
Elsholtzia splendens	322
Elymus	92
Elymus kamoji	92
Embelia	271
Embelia vestita	273
Emilia	346
Emilia sonchifolia	347
Epaltes	349
Epaltes australis	349
Epilobium	209
Epilobium pyrricholophum	211
Eragrostideae	81, 82, 90
Eragrostis	90
Eragrostis cumingii	91
Eragrostis ferruginea	91

Erechtites	346
Erechtites valerianifolius	347
Eremochloa	88
Eremochloa ophiuroides	89
Ericaceae	280
Erigeron	347, 348
Erigeron bonariensis	348
Eriocaulaceae	73
Eriocaulon	73
Eriocaulon buergerianum	73
Eriochloa	86
Eriochloa villosa	87
Erythrina	143
Erythrina crista-galli	143
Eschenbachia	348
Eschenbachia japonica	348
Eulalia	87
Eulalia wightii	89
Euonymus	187
Euonymus alatus	187, 188
Euonymus centidens	189
Eupatorieae	335, 340
Eupatorium	340
Eupatorium cannabinum	340
Euphorbia	202
Euphorbia maculata	201
Euphorbia sieboldiana	202
Euphorbiaceae	201
Eurya	266
Eurya japonica	267
Euscaphis japonica	216
Euscaphis	216
Evolvulus	298
Evolvulus alsinoides	299
Exochorda	154
Exochorda racemosa	157

F

Fabaceae	134
Fabeae	136, 149
Fagaceae	173
Fagopyrum	241
Fagopyrum dibotrys	241
Fallopia	240, 241
Fallopia multiflora	241
Farfugium	347

Farfugium japonicum	347
Fatoua	169
Fatoua villosa	170
Festuca	92
Festuca parvigluma	93
Ficus	169
Ficus pumila var. *awkeotsang*	170
Fimbristylis	77
Fimbristylis sericea	79
Firmiana	227
Firmiana simplex	228
Flemingia	142
Flemingia prostrata	143
Fontanesia	302
Fontanesia phillyreoides subsp. *fortunei*	303
Fraxinus	302
Fraxinus chinensis	303
Fritillaria	38
Fritillaria thunbergii	39, 40
Fuirena	77
Fuirena ciliaris	79

G

Gaillardia	341
Gaillardia pulchella	342
Galegeae	136, 147
Galinsoga	341
Galinsoga quadriradiata	343
Galium	284
Galium spurium	286
Gamochaeta	345
Gamochaeta pensylvanica	346
Gardenia	284
Gardenia jasminoides	283, 285
Gardneria	290
Gardneria multiflora	289, 290
Garryaceae	282
Gentiana	288
Gentiana zollingeri	288
Gentianaceae	287
Geraniaceae	205
Geranium carolinianum	205
Geranium	205
Gesneriaceae	304
Glebionis	344
Glebionis coronaria	344

Glechoma	319
Glechoma longituba	321
Glehnia	358
Glehnia littoralis	357, 359, 360
Glochidion	203
Glochidion puber	204
Glossocardia	341
Glossocardia bidens	343
Glycine	143
Glycine tomentella	144
Gnaphalieae	335, 345
Gnaphalium	345
Gnaphalium japonicum	346
Gonocarpus	128
Gonocarpus micranthus	128
Gonostegia	172
Gonostegia hirta	172
Goodeniaceae	333
Goodyera nankoensis	42
Grewia	228
Grewia biloba var. *parviflora*	229
Grossulariaceae	123
Gueldenstaedtia	147
Gueldenstaedtia verna	147
Gymnema	292
Gymnema sylvestre	293
Gymnosporia	188
Gymnosporia diversifolia	189
Gymnostoma	179
Gynostemma	183
Gynostemma pentaphyllum	184
Gynura	346
Gynura formosana	347

H

Haloragaceae	128
Hamamelidaceae	119
Hedera	356
Hedera rhombea	356
Hedyotis	284
Hedyotis strigulosa	285
Hedysareae	136, 146
Heliantheae	335, 341
Helianthus	342
Helianthus annuus	343
Helicia	114
Helicia cochinchinensis	115
Helicteres	228
Helicteres angustifolia	228
Hemerocallis	49
Hemerocallis fulva	50, 51
Hemisteptia	345
Hemisteptia lyrata	345
Heptapleurum	356
Heptapleurum octophyllum	356
Heracleum	358
Heracleum sphondylium var. *nipponicum*	360
Herminium	42
Herminium lanceum	44
Hibiscus	228
Hibiscus mutabilis	229
Hosta	57
Hosta ventricosa	56, 59
Houttuynia	4
Houttuynia cordata	5
Hoya	291
Hoya carnosa	293
Humulus	167
Humulus scandens	168
Hydrangea	260
Hydrangea robusta	262
Hydrangeaceae	260
Hydrilla	28
Hydrilla verticillata	30
Hydrocharitaceae	28, 29
Hydrocotyle	355
Hydrocotyle sibthorpioides	356
Hylodesmum	145
Hylodesmum podocarpum	146
Hylotelephium	126
Hylotelephium erythrostictum	127
Hypericaceae	194
Hypericum ascyron	194
Hypericum	194
Hypericum japonicum Thunb.	194
Hypericum sampsonii Hance	194
Hypoxidaceae	45
Hypoxis	45
Hypoxis aurea	45

I

Ilex crenata	329

Ilex rotunda	329
Ilex	329
Ilex integra	329
Illicium lanceolatum	2
Impatiens	265
Impatiens balsamina	265
Imperata	87
Imperata cylindrica var. *major*	88
Indigofera amblyantha	147
Indigofera	146
Indigofereae	136, 146
Ingeae	135, 140
Inula	349
Inula japonica	349
Inuleae	348
Ipomoea	297
Ipomoea nil	299
Iridaceae	46, 47
Iris	46
Iris lactea	47
Iris proantha var. *valida*	48
Isachne	84
Isachne albens	84
Isachneae	80, 81, 82, 84
Ischaemum	87, 89
Isodon	320
Isodon inflexus	322
Ixeridium	338
Ixeridium laevigatum	338
Ixeris	338
Ixeris repens	338

J

Jasminum	302
Jasminum lanceolarium	303
Juglandaceae	177
Juncaceae	74
Juncus	74
Juncus setchuensis	75
Justicia	313
Justicia hayatae	314

K

Kadsura	1
Kadsura longipedunculata	3
Kali	247
Kali tragus	248
Kalopanax	356
Kalopanax septemlobus	356
Kandelia	193
Kandelia obovata	193
Kerria	154
Kerria japonica	158
Kochia	247
Koelreuteria	220
Koelreuteria bipinnata	220
Korthalsella	235
Korthalsella japonica	237
Kummerowia	145
Kummerowia striata	145
Kyllinga	77
Kyllinga brevifolia var. *leiolepis*	79

L

Lablab	142
Lablab purpureus	143
Lagerstroemia	207
Lagerstroemia indica	208
Lamiaceae	318
Lamium	319
Lamium amplexicaule	322
Lantana	316
Lantana camara	317
Lapsanastrum	338
Lapsanastrum apogonoides	338
Lardizabalaceae	102
Lathyrus	149
Lathyrus japonicus	149
Lauraceae	17, 18
LAURALES	16, 18
Leersia	97
Leersia sayanuka	97
Lemna	23
Lemna aequinoctialis	25
Lentibulariaceae	315
Leonurus	319
Leonurus japonicus	322
Lepidium	234
Lepidium virginicum	234
Leptochloa	90
Leptochloa chinensis	90

Lespedeza	145
Lespedeza virgata	146
Leucaena	139
Leucas	319
Leucas chinensis	322
Ligularia	347
Ligularia intermedia	347
Ligustrum	302
Ligustrum obtusifolium subsp. *microphyllum*	303
Liliaceae	38, 39
LILIALES	39
Lilium	38
Lilium lankongense	40
Limonium	238
Limonium sinense	239
Lindera	17
Lindera erythrocarpa	19
Lindernia	310
Lindernia procumbens	311
Linderniaceae	310
Lipocarpha	77
Lipocarpha microcephala	78
Liquidambar formosana	118
Liquidambar	118
Liriodendron	12
Liriodendron chinense	14
Liriodendron × *sinoamericanum*	14
Liriope	57
Liriope spicata	58
Lithocarpus	173
Lithocarpus glaber	175
Lithospermum	294
Lithospermum zollingeri	296
Litsea	17
Litsea cubeba	19
Lobelia	331
Lobelia chinensis	331, 332
Loganiaceae	289
Lolium	92, 93
Lonicera	352
Lonicera fragrantissima	352, 353
Lophatherum gracile	94
Loropetalum	119
Loropetalum chinense	120, 121
Ludwigia	209
Ludwigia prostrata	211
Luisia	42
Luisia hancockii	43
Luzula	74
Luzula campestris	75
Luzula multiflora	75
Lycianthes	301
Lycianthes biflora	301
Lycium	301
Lycium chinense	301
Lycopus	320
Lycopus lucidus	322
Lycoris sprengeri	54
Lysimachia	270
Lysimachia capillipes	272
Lysionotus	304
Lysionotus pauciflorus	304, 305
Lythraceae	207

M

Machilus tricuspidata	170
Machilus	18
Machilus thunbergii	20
Maclura	170
Maclura tricuspidata	170
Maesa	271
Maesa japonica	273
Magnoliaceae	12, 13
MAGNOLIALES	13
Mahonia	106
Mahonia bealei	107
Malus	154, 158
Malva	228
Malva pusilla	229
Malvaceae	227
Malvastrum	228
Malvastrum coromandelianum	229
Matsumurella	319
Matsumurella chinensis	322
Mazaceae	323
Mazus	323
Mazus pumilus	323, 324
Medicago	144
Medicago polymorpha	145
Melampyrum	326
Melampyrum roseum	328
Melanthera	342
Melanthera prostrata	343

Melanthiaceae	36
Melastoma	215
Melastoma dodecandrum	214, 215
Melastomataceae	214
Melia	226
Melia azedarach	225, 226
Meliaceae	225
Meliceae	81, 91
Melicope	221
Melicope pteleifolia	222
Melilotus	144
Melilotus officinalis	145
Melinis	85
Melinis repens	86
Meliosma	111
Meliosma rigida	112
Melochia	228
Melochia corchorifolia	228
Menispermaceae	104
Mentha	320
Mentha crispata	322
Merremia	298
Merremia sibirica	299
Metaplexis	292
Metaplexis japonica	292, 293
Michelia	12
Michelia figo	14
Microtis	41
Microtis unifolia	43
Millettia	147
Millettieae	136, 147
Mimosa	139
Mimoseae	135, 139
Mirabilis	252
Mirabilis jalapa	253
Miscanthus	87
Miscanthus floridulus	89
Mitrasacme	290
Mitrasacme pygmaea	290
Molluginaceae	254
Mollugo	254
Mollugo stricta	254
Monochasma	327
Monochasma savatieri	327, 328
Monochoria	66
Monochoria vaginalis	67
Moraceae	169
Morella	176
Morella rubra	176
Morinda	284
Morinda umbellata subsp. *obovata*	286
Morus	170
Morus australis	170
Mosla	320
Mosla scabra	322
Mucuna	143
Mucuna sempervirens	143
Murdannia	62
Murdannia spirata	64
Murraya	222
Murraya exotica	222
Musa	68
Musaceae	68
Mussaenda	284
Mussaenda pubescens	285
Mutisieae	335, 339
Myosoton	244
Myosoton aquaticum	244
Myricaceae	176
Myriophyllum	128
Myriophyllum aquaticum	128
Myrsine	271
Myrsine seguinii	273
Myrtaceae	212

N

Nandina	106
Nandina domestica	107
Nanocnide	172
Nanocnide japonica	172
Narcissus	52
Narcissus tazetta var. *chinensis*	52, 53, 54, 56
Nartheciaceae	32
Nelumbo	113
Nelumbo nucifera	113
Nelumbonaceae	113
Neofinetia	42
Neofinetia falcata	44
Neolitsea aurata var. *chekiangensis*	19
Neyraudia	90
Neyraudia reynaudiana	90
Nicandra	300
Nicandra physalodes	301

Nyctaginaceae	252
Nyssaceae	259

O

Oenanthe	358
Oenanthe benghalensis	360
Oenothera	209
Oenothera laciniata	211
Ohwia	145
Ohwia caudata	146
Oleaceae	302
Onagraceae	209
Ophiopogon	57
Ophiopogon japonicus	58
Oplismenus	85
Oplismenus undulatifolius var. *binatus*	86
Opuntia	258
Opuntia dillenii	258
Orchidaceae	41, 42
Oreocharis	304
Oreocharis maximowiczii	305
Oreocnide	172
Oreocnide frutescens	171, 172
Orobanchaceae	326
Orobanche	327
Orobanche brassicae	328
Orostachys	126
Orostachys fimbriata	127
Orychophragmus	234
Orychophragmus violaceus	234
Oryzeae	82, 97
Osbeckia	215
Osbeckia stellata	215
Osmanthus	302
Osmanthus fragrans	303
Osteomeles	154
Osteomeles subrotunda	157
Ostericum	358
Ostericum citriodorum	360
Oxalidaceae	190
Oxalis corniculata	190, 191
Oxalis	190

P

Paris polyphylla var. *chinensis*	36
Panicum repens	86
Pachysandra	116
Pachysandra terminalis	117
Padus	155
Padus obtusata	159
Paederia	284
Paederia foetida	286
Paliurus	164
Paliurus ramosissimus	165
Paniceae	81, 85
Panicum	85
Panicum sumatrense	86
Papaveraceae	101
Parapholis	92
Parapholis incurva	93
Paris	36
Paris polyphylla var. *stenophylla*	36
Parthenium	341
Parthenium hysterophorus	342
Parthenocissus	130
Parthenocissus tricuspidata	130
Paspalum	86
Paspalum scrobiculatum var. *orbiculare*	87
Passiflora edulis	198
Passiflora	198
Passifloraceae	198
Patrinia	352
Patrinia villosa	353
Paulownia	325
Paulownia tomentosa	325
Paulowniaceae	325
Pedaliaceae	312
Pennisetum	85
Pennisetum alopecuroides	86
Pentaphylacaceae	266
Peperomia	6
Peperomia blanda	8
Perilla	320
Perilla frutescens	322
Peristrophe	313
Peristrophe japonica	314
Persicaria	241
Peucedanum	358
Peucedanum japonicum	360
Phacelurus	88, 89
Phaenosperma globosum	98
Phaenospermateae	82, 98

Phalaris	95	*Plantago*	306
Phalaris arundinacea	96	*Plantago virginica*	307
Phaseoleae	135, 136, 142	*Platycarya*	178
Phedimus	126	*Platycarya strobilacea*	177, 178
Phedimus aizoon	127	*Platycodon*	332
Philydraceae	65	*Platycodon grandiflorus*	332
Philydrum	65	*Pleioblastus*	84
Philydrum lanuginosum	65	*Pleioblastus amarus*	84
Phoebe	18	*Ploygala hongkongensis* var. *stenophylla*	152
Phoebe sheareri	20	*Ploygala japonica*	152
Phoenix	60	*Pluchea*	349
Phoenix roebelenii	61	*Pluchea indica*	349
Photinia	154	*Plumbago*	238
Photinia parvifolia	157	*Plumbago zeylanica*	239
Phragmites	94	*Poa*	92
Phragmites australis	94	*Poa annua*	83, 93
Phtheirospermum	327	Poaceae	80, 83
Phtheirospermum japonicum	328	POALES	83
Phyla	316	Poeae	81, 82, 92
Phyla nodiflora	317	*Pogonatherum*	87
Phyllanthaceae	203	*Pogonatherum crinitum*	88
Phyllanthus	204	*Polycarpon*	244
Phyllanthus leptoclados	204	*Polycarpon prostratum*	244
Phyllanthus urinaria	203	*Polygala*	150
Phyllostachys	84	Polygalaceae	150
Phyllostachys nidularia	84	Polygonaceae	240
Physalis	301	*Polygonatum*	57
Physalis angulata	301	*Polygonatum cyrtonema*	58
Phytolacca	251	*Polygonum*	240, 241
Phytolacca americana	251	*Polygonum aviculare*	241
Phytolaccaceae	251	*Polypogon*	95
Pilea	172	*Polypogon fugax*	96
Pilea angulata	172	Pontederiaceae	66
Pinellia	23	*Portulaca*	257
Pinellia pedatisecta	25	*Portulaca oleracea*	257
Pinellia ternata	24	Portulacaceae	257
Piper	6	*Potamogeton*	31
Piper kadsura	7	*Potamogeton distinctus*	31
Piper	8	Potamogetonaceae	31
Piperaceae	6, 7	*Potentilla*	155, 159
PIPERALES	7, 10	*Pouzolzia*	172
Pistacia	217	*Pouzolzia zeylanica*	172
Pistacia chinensis	218	*Praxelis*	340
Pittosporaceae	354	*Praxelis clematidea*	340
Pittosporum	354	*Premna*	319
Pittosporum tobira	354	*Premna microphylla*	321
Plantaginaceae	306	*Primula*	271

Primula cicutariifolia	272
Primulaceae	270
Proteaceae	114
Prunella	319
Prunella vulgaris	322
Pseudognaphalium	345
Pseudognaphalium affine	346
Psychotria	284
Psychotria asiatica	286
Pterocarya	178
Pterocarya stenoptera	177, 178
Pueraria	143
Pueraria montana var. *lobata*	144
Pycreus	77
Pycreus flavidus	79
Pyracantha	154
Pyracantha fortuneana	157
Pyrus	154
Pyrus calleryana	158

Q

Quercus	174
Quercus acutissima	175
Quercus ciliaris	175
Quercus phillyreoides	173

R

Ranunculaceae	108
Ranunculus	109
Ranunculus ternatus	110
Raphanus	234
Raphanus raphanistrum	233, 234
Reineckea	57
Reineckea carnea	59
Reynoutria	241
Reynoutria japonica	241
Rhamnaceae	163
Rhamnella	164
Rhamnella franguloides	165
Rhamnus	164
Rhamnus crenata	165
Rhaphiolepis	154
Rhaphiolepis indica	158
Rhizophoraceae	193
Rhododendron	280

Rhododendron ovatum	280
Rhodomyrtus	213
Rhodomyrtus tomentosa	213
Rhus	218
Rhus chinensis	218
Rhynchosia	142
Rhynchosia volubilis	143
Rhynchospora	77
Rhynchospora rubra	78
Ribes fasciculatum var. *chinense*	123
Ribes glaciale	123
Ribes	123
Ricinus	202
Ricinus communis	202
Robinia	149
Robinia pseudoacacia	149
Robinieae	137, 149
Rorippa	234
Rorippa indica	234
Rosa	154
Rosa henryi	158
Rosaceae	153
Rotala	207
Rotala rotundifolia	208
Rottboellia	88
Rottboellia cochinchinensis	89
Rubia	284
Rubia argyi	286
Rubiaceae	283
Rubus	154
Rubus adenophorus	159
Rubus buergeri	158
Rubus hirsutus	159
Rumex	240, 241
Rumex japonicus	241
Rutaceae	221

S

Sabia	111
Sabia discolor	112
Sabiaceae	111
Saccharum	87
Saccharum spontaneum	89
Sacciolepis	85
Sacciolepis indica	86
Sageretia	164

Sageretia thea	163, 165	*Scutellaria indica*	321
Sagina	244	*Sedum*	126
Sagina japonica	244	*Sedum formosanum*	127
Sagittaria	26	*Semiaquilegia*	109
Sagittaria lancifolia	27	*Semiaquilegia adoxoides*	110
Salicaceae	199	*Senecio*	346
Salicornia	247	*Senecio scandens*	347
Salicornia europaea	248	Senecioneae	336, 346
Salix	199	*Senna*	141
Salix babylonica	200	*Senna occidentalis*	141
Salvia	320	*Serissa*	284
Salvia japonica	322	*Serissa serissoides*	286
Sambucus	350	*Sesamum*	312
Sambucus javanica	351	*Sesamum indicum*	312
Sanguisorba	154	*Sesbania cannabina*	138. 149
Sanguisorba officinalis	158	*Sesbania*	149
Santalaceae	235	Sesbanieae	137, 149
Sapindaceae	219	*Sesuvium*	249
Sapindus	220	*Sesuvium portulacastrum*	250
Sapindus saponaria	219, 220	*Setaria*	85
Saururaceae	4	*Setaria faberi*	86
Saururus	4	*Shizhenia*	42
Saururus chinensis	5	*Shizhenia pinguicula*	44
Saxifraga	124	*Sida*	228
Saxifraga stolonifera	125	*Sida rhombifolia*	229
Saxifragaceae	124	*Sigesbeckia*	341
Scaevola	333	*Sigesbeckia orientalis*	343
Scaevola taccada	333	*Silene*	244
Schima	274	*Silene aprica*	245
Schima superba	275	Simaroubaceae	223
Schisandra	1	*Sindechites*	292
Schisandra sphenanthera	3	*Sindechites henryi*	293
Schisandraceae	1, 2	*Sinomenium*	105
Schizachyrium delavayi	91	*Sinomenium acutum*	105
Schoenoplectus	78	*Siphonostegia*	326
Schoenoplectus tabernaemontani	79	*Siphonostegia chinensis*	328
Scirpus	77	*Sisyrinchium*	46
Scirpus karuisawensis	79	*Sisyrinchium palmifolium*	48
Scleria	77, 78	Smilacaceae	37
Scolopia	199	*Smilax*	37, 237
Scolopia chinensis	200	*Smilax glabra*	37
Scoparia	306	*Smilax nipponica*	37
Scoparia dulcis	307	Solanaceae	300
Scrophularia	308	*Solanum*	300
Scrophularia ningpoensis	309	*Solanum lyratum*	301
Scrophulariaceae	308	*Solena*	183
Scutellaria	319	*Solena heterophylla*	184

Solidago	348
Solidago decurrens	348
Soliva	344
Soliva anthemifolia	344
Sonchus	338
Sonchus oleraceus	338
Sophoreae	135, 141
Sorghum	88
Sorghum nitidum	89
Sorghum propinquum	89
Spartina	98
Spartina alterniflora	98
Spergularia	244
Spergularia marina	244
Spermacoce	284
Spermacoce remota	286
Sphaerocaryum	84
Sphaerocaryum malaccense	84
Sphagneticola	342
Sphagneticola calendulacea	343
Spinifex	85
Spinifex littoreus	86
Spiraea	154
Spiraea chinensis	157
Spirodela	23
Spirodela polyrhiza	25
Spodiopogon	87
Spodiopogon sibiricus	88
Sporobolus	90
Sporobolus fertilis	90
Stachys	319, 322
Staphyleaceae	216
Stauntonia	102
Stauntonia leucantha	103
Stellaria	244
Stellaria media	244
Stephanandra	154
Stephanandra chinensis	157
Stephania japonica	105
Stimpsonia	271
Stimpsonia chamaedryoides	272
Stipeae	82, 99
Styphnolobium	141
Styphnolobium japonicum	141
Styracaceae	277
Styrax	277
Styrax confusus	277
Suaeda	247
Suaeda glauca	248
Swertia	288
Swertia hickinii	288
Symphyotrichum	347
Symphyotrichum subulatum	348
Symphytum	294
Symphytum officinale	296
Symplocaceae	276
Symplocos	276
Symplocos paniculata	276
Syzygium	213
Syzygium buxifolium	213

T

Tacca chantrieri	34
Talinaceae	256
Talinum	256
Talinum paniculatum	256
Taraxacum	338
Taraxacum mongolicum	338
Tarenna	284
Tarenna mollissima	286
Ternstroemia	266
Ternstroemia gymnanthera	267
Tetradium	221
Tetradium glabrifolium	222
Tetragonia	249
Tetragonia tetragonoides	250
Tetrastigma	130
Tetrastigma hemsleyanum	130
Teucrium	319
Teucrium viscidum	321
Thalictrum	109
Thalictrum fortunei	110
Theaceae	274
Themeda	88
Themeda triandra	90
Thermopsideae	135, 141
Thermopsis chinensis	141
Thermopsis	141
Thesium	235
Thesium chinense	236, 237
Thymelaeaceae	230
Thyrocarpus	295
Thyrocarpus sampsonii	296

Tilia	228
Tilia henryana var. *subglabra*	229
Tithonia	342
Tithonia diversifolia	343
Toddalia	222
Toddalia asiatica	222
Toona	226
Toona sinensis	226
Torenia	311
Torenia violacea	311
Torilis	358
Torilis scabra	360
Toxicodendron	218
Toxicodendron vernicifluum	218
Trachelospermum	292
Trachelospermum jasminoides	293
Trachycarpus	60
Trachycarpus fortunei	61
Trapa	207
Trapa natans	208
Trema	167
Trema cannabina var. *dielsiana*	168
Triadica	202
Triadica sebifera	202
Tribulus	131
Tribulus terrestris	132, 133
Trichosanthes	182
Trichosanthes	184
Tricyrtis	38
Tricyrtis chinensis	40
Trifolieae	135, 136, 144
Trifolium	144
Trifolium repens	145
Trigonotis	295
Trigonotis peduncularis	296
Triodanis	332
Triodanis perfoliata subsp. *biflora*	332
Tripolium	348
Tripolium pannonicum	348
Triticeae	81, 92
Tubocapsicum	301
Tubocapsicum anomalum	301
Typha	72
Typha angustifolia	72
Typhaceae	72

U

Ulmaceae	166
Ulmus	166
Ulmus parvifolia	166
Uncaria	284
Uncaria rhynchophylla	285
Urena	228
Urena lobata	229
Urticaceae	171
Utricularia	315
Utricularia aurea	315

V

Vaccinium	280
Vaccinium bracteatum	280
Vallisneria	28
Vallisneria natans	29, 30
Vandellia	311
Vandellia anagallis	311
Verbena	317
Verbena officinalis	316, 317
Verbenaceae	316
Vernicia	202
Vernicia fordii	202
Vernonieae	334, 335, 339
Veronica	307
Veronica polita	307
Veronicastrum	307
Veronicastrum axillare	307
Viburnum	350
Viburnum dilatatum	350
Viburnum propinquum	351
Vicia	149
Vicia sativa	149
Vigna	142
Vigna unguiculata	143
Viola betonicifolia	196
Viola cornuta	197
Viola diffusa	196
Viola	196
Violaceae	196
Vitaceae	129
Vitex	319
Vitex rotundifolia	318, 320, 321
Vitis	129, 130
Vitis wilsoniae	130

Vulpia	92
Vulpia myuros	93

W

Wahlenbergia	332
Wahlenbergia marginata	332
Waltheria	228
Waltheria indica	229
Wikstroemia	231
Wikstroemia indica	230, 231
Wisteria	147
Wisteria sinensis	148
Wolffia	23
Wolffia globosa	25

X

Xanthium	341
Xanthium strumarium	342
Xylosma	199
Xylosma congesta	200

Y

Ycoris	52
Youngia	338
Youngia japonica	338
Yulania	12
Yulania biondii	14
Yulania liliiflora	13

Z

Zanthoxylum	222
Zanthoxylum armatum	221
Zanthoxylum schinifolium	222
Zehneria	183
Zehneria japonica	184
Zelkova	166
Zelkova serrata	166
Zingiberaceae	70
Zinnia	341
Zinnia elegans	342
Zizania	97
Zizania latifolia	97
Zornia	142
Zornia gibbosa	142
Zoysia	98
Zoysia japonica	98
Zygophyllaceae	131

Index to Chinese Names 中文名称索引

A

阿福花科	49, 50
矮生薹草	78
矮水竹叶	64
艾纳香属	349
爱玉子	170
安息香科	277, 278
安息香属	277

B

八宝	127
八宝属	126
八角枫属	263
巴豆属	202
巴戟天属	284
芭蕉科	68
芭蕉属	68
菝葜科	37
菝葜属	37, 237
白背黄花稔	229
白背牛尾菜	37
白车轴草	145
白凤菜	347
白花丹科	238
白花丹属	238
白花丹	239
白花苦灯笼	286
白花柳叶箬	84
白花马蔺	47
白及	43
白及属	41
白酒草	348
白酒草属	348
白鹃梅	157
白鹃梅属	154
白蜡树	303
白马骨	286
白毛鹿茸草	327
白茅属	87
白檀	276
白羊草	89
白英	301
百齿卫矛	189
百合科	38, 39
百合目	39
百合属	38
百金花	287, 288
百金花属	288
百日菊	342
百日菊属	341
百蕊草	236, 237
百蕊草属	235
败酱属	352
败酱叶菊芹	347
稗荩	84
稗荩属	84
稗属	86
斑地锦	201
斑鸠菊族	334, 335, 339
斑种草属	295
板凳果属	116
半边莲	331, 332
半边莲属	331
半夏	24
半夏属	23
棒头草	96
棒头草属	95
宝盖草	322
报春花科	270
报春花属	271
北美车前	307
北美独行菜	234
北鱼黄草	299
贝母属	38
荸荠属	77
秕壳草	97
笔龙胆	288
笔罗子	112
蓖麻	202
蓖麻属	202
蝙蝠草属	145
扁担杆属	228
扁豆	143
扁豆属	142
扁莎属	77
扁穗雀麦	97
变叶裸实	189
蔗草属	77
滨海白绒草	322
滨海前胡	360
薄荷属	320
补血草	239
补血草属	238

C

菜豆族	135, 136, 142
苍耳	342
苍耳属	341
糙叶树	168
糙叶树属	167
草海桐	333
草海桐科	333
草海桐属	333
草胡椒属	6
草木犀	145
草木犀属	144
粗枝绣球	262
箣柊	200
箣柊属	199
梣属	302
叉唇角盘兰	44
茶藨子科	123
茶藨子属	123
茶条枫	220
钗子股属	42

379

菖蒲	22	刺桐属	143	地苏	214, 215
菖蒲科	22	刺子莞	78	地笋	322
菖蒲属	22	刺子莞属	77	地笋属	320
长柄山蚂蝗	146	葱属	52	地桃花	229
长柄山蚂蝗属	145	葱叶兰	43	地杨梅	75
长萼母草	311	葱叶兰属	41	地杨梅属	74
长萼母草属	311	粗毛牛膝菊	343	地榆	158
长叶冻绿	165	粗枝木麻黄	180	地榆属	154
长籽柳叶菜	211	粗壮小鸢尾	48	棣棠花	158
常春藤属	356	簇花庭菖蒲	48	棣棠属	154
常春油麻藤	356	翠雀属	109	点地梅	271, 272
常夏石竹	243			点地梅属	271
车前科	306			吊石苣苔	304, 305
车前属	306	**D**		吊石苣苔属	304
车桑子	220	打碗花属	298	丁癸草	142
车桑子属	220	大白茅	88	丁癸草属	142
车轴草属	144	大别山马兜铃	11	丁香蓼	211
车轴草族	135, 136, 144	大豆属	143	丁香蓼属	209
城口卷瓣兰	43	大狗尾草	86	顶花板凳果	117
橙桑属	170	大花石上莲	305	东风草	349
秤钩风	105	大戟科	201	冬青科	329, 330
秤钩风属	105	大戟属	202	冬青属	329
匙羹藤	293	大麻科	167	东方泽泻	27
匙羹藤属	292	大麻叶泽兰	340	东南茜草	286
匙叶合冠鼠曲草	346	大青属	319	东亚女贞	303
齿叶冬青	329	大吴风草	347	豆腐柴	321
齿叶溲疏	262	大吴风草属	347	豆腐柴属	319
赤楠	213	大叶直芒草	99	豆科	134, 137, 138
稠李属	155	大芽南蛇藤	189	豆梨	158
臭草族	81, 91	大油芒	88	独活属	358
臭椿	223	大油芒属	87	独行菜属	234
臭椿属	223	单叶蔓荆	318, 320, 321	杜茎山	273
楮	170	弹裂碎米荠	234	杜茎山属	271
穿鞘花	64	淡竹叶	94	杜鹃花科	280
穿鞘花属	62	刀豆属	143	杜鹃花属	280
垂柳	200	稻槎菜	338	杜英科	192
垂序商陆	251	稻槎菜属	338	杜英属	192
春黄菊族	335, 344	稻族	82, 97	短柄草	92
唇形科	318, 320	灯台兔儿风	339	短柄草族	81, 92
慈姑属	26	灯芯草科	74	短柄粉条儿菜	32
刺槐	149	灯芯草属	74	短辐水芹	360
刺槐属	149	地胆草	339	短绒野大豆	144
刺槐族	137, 149	地胆草属	339	椴属	228
刺葵属	60	地耳草	194	盾果草	296
刺楸	356	地肤	248	盾果草属	295
刺楸属	356	地肤属	247	钝药野木瓜	103
刺沙蓬	248	地锦	130	多花地杨梅	75

多花勾儿茶	165	枫香树	118	菰属	97
多花黄精	58	枫香树属	118	谷精草	73
多花木蓝	147	枫杨	177, 178	谷精草科	73
多花水苋菜	208	枫杨属	178	谷精草属	73
多荚草	244	凤仙花	265	瓜本	264
多荚草属	244	凤仙花科	265	瓜子金	151, 152
		凤仙花属	265	栝楼	184
		凤眼莲	67	栝楼属	182
E		凤眼莲属	66	观音草属	313
峨参	360	佛肚竹	84	光高粱	89
峨参属	358	腹水草属	307	光头稗	87
鹅肠菜	244	拂子茅	96	光叶丰花草	286
鹅肠菜属	244	拂子茅属	95	光叶子花	253
鹅耳枥属	181	芙兰草属	77	光药列当	328
鹅绒藤属	292	浮萍属	23	鬼针草	343
鹅掌柴	356	芙蓉菊	345	鬼针草属	341
鹅掌柴属	356	芙蓉菊属	344	过江藤	317
鹅掌楸	14	附地菜	295, 296	过江藤属	316
鹅掌楸属	12	附地菜属	295		
耳草属	284	复羽叶栾树	220	**H**	
				海滨山黧豆	149
		G		海刀豆	134, 143
F		甘蔗属	87	海马齿	250
番杏	250	柑橘属	222	海马齿属	249
番杏科	249	刚竹属	84	海桐	354
番杏属	249	笄子梢	145	海桐科	354
番薯属	297	笄子梢属	145	海桐属	354
繁缕	244	高粱属	88	海州香薷	322
繁缕属	244	高粱族	81, 87	含笑花	14
梵天花属	228	隔距兰属	42	含笑属	12
方木麻黄属	179	隔山香	360	含羞草属	139
防己科	104	葛麻姆	144	含羞草族	135, 139
飞廉族	335, 345	葛属	143	寒莓	158
飞龙掌血	222	弓果黍	86	韩信草	321
飞龙掌血属	222	弓果黍属	85	蔊菜	234
飞蓬属	347, 348	勾儿茶属	164	蔊菜属	234
费菜	127	构属	170	旱茅	91
费菜属	126	钩藤	285	杭州石荠苎	322
粉绿狐尾藻	128	钩藤属	284	蒿属	344
粉条儿菜属	32	钩腺大戟	202	禾本科	80, 83
风兰	44	狗骨柴	285	禾本目	83
风兰属	42	狗骨柴属	284	合冠鼠曲草属	345
风龙	105	狗尾草属	85	合欢	137, 140
风龙属	105	狗牙根	98	合萌	142
风轮菜	322	狗牙根属	98	合萌属	142
风轮菜属	320	菰	97	合萌族	135, 136, 142
风藤	7				

何首乌	241	虎尾草属	98	喙果黑面神	204
何首乌属	240, 241	虎尾草族	82, 98	活血丹	321
盒子草	184	虎掌	25	活血丹属	319
盒子草属	183	虎杖	241	火棘	157
褐冠小苦荬	338	虎杖属	241	火棘属	154
黑麦草属	92, 93	互花米草	98	霍州油菜	141
黑面神属	204	花点草	172	藿香蓟	340
黑藻	30	花点草属	172	藿香蓟属	340
黑藻属	28	花椒属	222		
黑籽荸荠	78	华东蘘草	79		
红淡比	267	华东蓝刺头	340	**J**	
红淡比属	267	华东唐松草	110	鸡冠刺桐	143
红果山胡椒	19	华空木	157	鸡桑	170
'红茎'泽泻慈姑	27	华蔓茶藨子	123	鸡蛋果	198
红毛草	86	华中五味子	3	鸡矢藤	286
红楠	20	华重楼	36	鸡矢藤属	284
红树科	193	华紫珠	321	鸡血藤属	147
红丝线	301	化香树	177, 178	鸡眼草	145
红丝线属	301	化香树属	178	鸡眼草属	145
猴耳环属	140	画眉草属	90	积雪草	360
篌竹	84	画眉草族	81, 82, 90	积雪草属	358
厚壳树	296	桦木科	181	吉祥草	59
厚壳树属	295	槐	141	吉祥草属	57
厚皮香	267	槐属	141	蕺菜	5
厚皮香属	266	槐族	135, 141	蒺藜	132, 133
狐尾藻属	128	换锦花	54	蒺藜科	131, 132
胡椒科	6, 7	黄鹌菜	338	蒺藜属	131
胡椒目	7, 10	黄鹌菜属	338	戟菜属	4
胡椒属	6, 8	黄背草	90	戟叶堇菜	196
胡萝卜属	358	黄海棠	194	蓟	345
胡桃科	177	黄花蒿	345	蓟属	345
胡颓子	162	黄花狸藻	315	檵木	120, 121
胡颓子科	160	黄花棯属	228	檵木属	119
胡颓子属	160	黄金茅属	87	夹竹桃科	291, 292
胡枝子属	145	黄精属	57	荚蒾	350
湖瓜紫荆	140	黄连木	218	荚蒾属	350
湖瓜草属	77	黄连木属	217	假臭草	340
蝴蝶草属	311	黄芪属	147	假臭草属	340
葫芦科	182, 183	黄芩属	319	假淡竹叶族	81, 82, 94
虎刺	286	黄檀属	148	假稻属	97
虎刺属	284	黄檀族	136, 148	假俭草	89
虎耳草	125	黄细心	253	假俭草属	88
虎耳草科	124	黄细心属	252	假连翘	317
虎耳草属	124	黄杨	117	假连翘属	317
虎皮楠	122	黄杨科	116	假马齿苋	307
虎皮楠科	122	黄杨属	116	假马齿苋属	306
虎皮楠属	122	灰背清风藤	112	假牛鞭草	93

假牛鞭草属	92	金鱼藻	100	柯属	173
假婆婆纳	272	金鱼藻科	100	壳斗科	173, 174
假婆婆纳属	271	金鱼藻属	100	刻叶紫堇	101
假酸浆	301	筋骨草属	319	孔颖草属	88
假酸浆属	300	堇菜科	196, 197	苦草	29, 30
尖帽草属	290	堇菜属	196	苦草属	28
菅属	88	锦鸡儿	146	苦苣菜	338
剪股颖属	95, 96	锦鸡儿属	146	苦苣菜属	338
碱蓬	248	锦葵科	227	苦苣苔科	304
碱蓬属	247	锦葵属	228	苦荬菜属	338
碱菀	348	荩草	89	苦木科	223
碱菀属	348	荩草属	88	苦蘵	301
箭根薯	34	江浙獐牙菜	288	苦竹	84
姜科	70	景天科	126	苦竹属	84
豇豆	143	景天属	126	阔苞菊	349
豇豆属	142	九节	286	阔苞菊属	349
角盘兰属	42	九节属	284	阔叶十大功劳	107
绞股蓝	184	九里香	222		
绞股蓝属	183	九里香属	222		
脚骨脆属	199	九头狮子草	314	**L**	
接骨草	351	救荒野豌豆	149	拉拉藤属	284
接骨木属	350	菊苣族	334, 338	蜡梅	15, 16
节节菜属	207	菊芹属	346	蜡梅科	15, 16
结缕草	98	菊三七属	346	蜡梅属	15
结缕草属	98	菊属	344	兰科	41, 42
结香	231	橘草	90	兰香草	321
结香属	231	枸杞	301	蓝刺头属	340
桔梗	332	枸杞属	301	蓝刺头族	335, 340
桔梗科	331	榉属	166	蓝果树科	259
桔梗属	332	榉树	166	蓝花参	332
金发草属	87	聚合草	296	蓝花参属	332
金柑	222	聚合草属	294	狼尾草	86
金合欢	139	瞿麦	245	狼尾草属	85
金合欢属	139	卷耳属	244	榔榆	166
金合欢族	135, 139	绢毛飘拂草	79	老鹳草属	205
金鸡菊属	341	决明属	141	老鼠芳	86
金锦香属	215	决明族	135, 141	老鸦瓣	40
金缕梅科	119, 120	爵床科	313	老鸦瓣属	38
金钮扣	343	爵床属	313	簕竹属	84
金钮扣属	342			簕竹族	80, 84
金钱蒲	22			了哥王	230, 231
金荞	241	**K**		类芦	90
金丝草	88	咖啡黄葵	229	类芦属	90
金丝桃科	194, 195	看麦娘	96	冷水花属	172
金丝桃属	194	看麦娘属	95	狸藻科	315
金粟兰科	21	柯	175	狸藻属	315
金粟兰属	21	柯孟披碱草	92	梨属	154

藜	248	龙舌兰	58	马齿苋	257
藜芦科	36	龙舌兰属	57	马齿苋科	257
藜属	247	龙须藤	140	马齿苋属	257
鳢肠	343	龙芽草	158	马兜铃科	9, 10
鳢肠属	342	龙芽草属	154	马兜铃属	9
栎属	174, 237	龙珠	301	马儿属	183
栗寄生	237	龙珠属	301	马甲子	165
栗寄生属	235	龙爪茅	91	马甲子属	164
栗属	173	龙爪茅属	90	马铃苣苔属	304
联毛紫菀属	347	芦苇	94	马钱科	289
莲	113	芦苇属	94	马松子	228
莲科	113	芦竹'花叶'	94	马松子属	228
莲属	113	芦竹属	94	马唐属	85
莲子草属	247	芦竹族	81, 94	马蹄金	299
链荚豆	146	卤地菊	343	马蹄金属	298
链荚豆属	145	卤地菊属	342	马银花	280
链珠藤	291, 293	鹿藿	143	马缨丹	317
链珠藤属	292	鹿藿属	142	马缨丹属	316
楝	225, 226	鹿角草	343	麦冬	58
楝科	225	鹿角草属	341	曼陀罗	301
楝属	226	鹿茸草属	327	曼陀罗属	301
楝叶吴萸	222	卵瓣还亮草	110	芒属	87
亮叶猴耳环	140	栾树属	220	牻牛儿苗科	205, 206
蓼科	240	罗浮柿	269	猫乳	165
蓼属	241	萝卜属	234	猫乳属	164
列当科	326, 327	萝藦	292, 293	猫爪草	110
列当属	327	萝藦属	292	毛芙兰草	79
裂叶月见草	211	裸柱菊	344	毛秆野古草	85
鬣刺属	85	裸柱菊属	344	毛茛科	108, 109
柃木	267	络石	293	毛茛属	109
柃属	266	络石属	292	毛花连蕊茶	275
凌风草属	92	落葵	255	毛棶	264
菱属	207	落葵科	255	毛茛叶报春	272
菱叶常春藤	356	落葵属	255	毛泡桐	325
流苏树	303	落葵薯	255	毛瑞香	231
流苏树属	302	落葵薯属	255	毛叶铁线莲	110
琉璃草属	295	葎草	168	毛药藤	293
琉璃繁缕	270, 272	葎草属	167	毛药藤属	292
琉璃繁缕属	270			茅膏菜属	242
柳属	199	**M**		茅膏菜科	242
柳叶菜科	209, 210	麻栎	175	茅膏菜	242
柳叶菜属	209	麻楝	226	茅瓜	184
柳叶箬属	84	麻楝属	226	茅瓜属	183
柳叶箬族	80, 81, 82, 84	马㼎儿	184	茅栗	175
六月雪属	284	马鞭草	316, 317	梅	159
龙胆科	287	马鞭草科	316	美登木属	188
龙胆属	288	马鞭草属	317	美味猕猴桃	279

苎麻属	172
蒙古蒲公英	338
猕猴桃科	279
猕猴桃属	279
米草属	98
米口袋属	147
米槠	175
密齿酸藤子	273
密花树	273
蜜茱萸属	221
绵枣儿	57
绵枣儿属	57
明党参	360
明党参属	358
陌上菜	311
陌上菜属	310
母草科	310
牡荆属	319
木防己	105
木防己属	105
木芙蓉	229
木荷	275
木荷属	274
木姜子属	17
木槿属	228
木兰科	12, 13
木兰目	13
木兰藤目	2
木蓝属	146
木蓝族	136, 146
木麻黄科	179
木麻黄属	180
木通	103
木通科	102
木通属	102
木樨	303
木樨科	302
木樨属	302
苜蓿属	144

N

南湖斑叶兰	42
南苜蓿	145
南蛇藤属	188
南酸枣	218
南酸枣属	218
南天竹	107
南天竹属	106
南五味子	3
南五味子属	1
南烛	280
楠属	18
囊颖草	86
囊颖草属	85
内折香茶菜	322
泥胡菜	345
泥胡菜属	345
泥花草	311
泥花草属	310
拟高粱	89
拟漆姑	244
拟漆姑属	244
牛角隔距兰	44
牛筋草	91
牛膝	247
牛膝菊属	341
牛膝属	247
纽扣草属	284
农吉利	144
糯米椴	229
糯米条	353
糯米团	172
糯米团属	172
女娄菜	245
女萎	110
女贞属	302

O

欧菱	208

P

爬山虎属	130
爬岩红	307
攀倒甑	353
泡花树属	111
泡桐科	325
泡桐属	325
蓬莱葛	289, 290
蓬莱葛属	290
蓬蘽	159
蟛蜞菊	343

蟛蜞菊属	342
萹蓄	241
萹蓄属	240, 241
飘拂草属	77
苹果属	154, 158
婆婆纳	307
婆婆纳属	307
朴属	167
朴树	168
铺地蝙蝠草	146
铺地黍	86
匍茎百合	40
葡萄科	129
葡萄属	129, 130
蒲公英属	338
蒲桃属	213
普陀鹅耳枥	181
普陀樟	20

Q

槭属	220
七星莲	196
漆姑草	244
漆姑草属	244
漆树	218
漆树科	217
漆树属	218
荠	234
荠菜属	234
千斤拔	143
千斤拔属	142
千金藤	105
千金藤属	104
千金子	90
千里光	347
千里光属	346
千里光族	336, 346
千屈菜科	207
牵牛	299
前胡属	358
茜草科	283
茜草属	284
茜树	285
茜树属	284
蔷薇科	153, 156
蔷薇属	154

荞麦属	241	软叶刺葵	61	山罗花	328
茄科	300	瑞香科	230	山罗花属	326
茄属	300	瑞香属	230	山蚂蝗属	145
窃衣	360	润楠属	18	山蚂蝗族	135, 136, 145
窃衣属	358			山麦冬	58
青花椒	222			山麦冬属	57
青木	282	**S**		山芹属	358
青萸	247	赛葵	229	山羊豆族	136, 147
青萸属	246	赛葵属	228	山油麻	168
清风藤科	111	赛山梅	278	山芫荽属	344
清风藤属	111	三白草	5	山楂	157
清香藤	303	三白草科	4	山楂属	154
苘麻	229	三白草属	4	山芝麻	228
苘麻属	228	三棱草属	77	山芝麻属	228
秋海棠	185	三脉紫菀	348	山茱萸科	263
秋海棠科	185, 186	三色堇品种	197	山茱萸属	263
秋海棠属	185	三叶崖爬藤	130	珊瑚菜	357, 360
秋葵属	228	三桠苦	222	珊瑚菜属	358
秋茄树	193	伞形科	357, 359	穇属	90
秋茄树属	193	桑科	169	鳝藤	293
秋英	343	桑属	170	鳝藤属	292
秋英属	341	沙冰藜属	247	商陆科	251
求米草属	85	沙参属	332	商陆属	251
球核荚蒾	351	沙苦荬菜	338	少花米口袋	147
球花脚骨脆	200	莎草科	76	佘山羊奶子	162
球菊	349	莎草属	77	蛇床属	358
球菊属	349	山白前	293	蛇莓	159
球兰	293	山扁豆	141	蛇莓属	155
球兰属	291	山扁豆属	141	蛇婆子	229
球穗扁莎	79	山茶	275	蛇婆子属	228
球序卷耳	244	山茶科	274	蛇葡萄	130
全缘冬青	329	山茶属	274	蛇葡萄属	130
雀稗属	86	山矾科	276	肾叶打碗花	299
雀麦族	82, 97	山矾属	276	省沽油科	216
雀梅藤	163, 165	山柑科	232	薯	344
雀梅藤属	164	山柑属	232	薯属	344
		山胡椒属	17	十大功劳属	106
R		山黄菊族	336, 346	十字花科	233
荛花属	231	山黄麻属	167	湿鼠曲草属	345
忍冬科	352	山鸡椒	19	石斑木	158
忍冬属	352	山菅	51	石斑木属	154
日本独活	360	山菅属	49	石蝉草	8
日本薯蓣	35	山姜	71	石豆兰属	41
榕属	169	山姜属	70	石胡荽	346
柔弱斑种草	296	山蚂蟥属	149	石胡荽属	346
肉叶耳草	285	山龙眼科	114	石楠属	154
软条七蔷薇	158	山龙眼属	114	石荠苎属	320

石蒜科	52, 53, 56	水蔗草	89	天南星	25
石蒜属	52	水蔗属	87	天南星科	23, 24
石竹科	243	水竹叶属	62	天南星属	23
石竹属	244	水烛	72	天人菊	342
时珍兰	44	丝穗金粟兰	21	天人菊属	341
时珍兰属	42	丝缨花科	282	田葱	65
柿科	268	松蒿	328	田葱科	65
柿属	268	松蒿属	327	田葱属	65
疏花雀麦	97	楤木	355, 356	田菁	138, 149
疏穗野青茅	95	楤木属	355	田菁属	149
黍属	85	溲疏属	260	田菁族	137, 149
黍族	81, 85	素馨属	302	田麻	229
鼠李科	163, 164	粟米草	254	田麻属	228
鼠李属	164	粟米草科	254	甜根子草	89
鼠茅	93	粟米草属	254	铁冬青	329
鼠茅属	92	酸浆属	301	铁线莲属	108
鼠曲草	346	酸模属	240, 241	铁苋菜	202
鼠曲草属	345	酸藤子属	271	铁苋菜属	202
鼠曲草族	335, 345	算盘子	204	铁仔属	271
鼠尾草	322	算盘子属	203	庭菖蒲属	46
鼠尾草属	320	碎米荠属	234	通泉草	323, 324
鼠尾粟	90			通泉草科	323, 324
鼠尾粟属	90			通泉草属	323
薯蓣	33	**T**		茼蒿	344
薯蓣科	33, 34	台湾佛甲草	127	茼蒿属	344
薯蓣目	34	台湾虎尾草	99	筒轴茅	89
薯蓣属	33	薹草属	77	筒轴茅属	88
束尾草属	88, 89	檀香科	235, 236	秃瓣杜英	192
双穗求米草	86	唐松草属	109	土丁桂	299
水鳖科	28, 29	糖蜜草属	85	土丁桂属	298
水葱	79	桃	159	土茯苓	37
水葱属	78	桃金娘	213	土荆芥	248
水晶花	21	桃金娘科	212	土人参	256
水马齿	307	桃金娘属	213	土人参科	256
水马齿属	306	桃属	155	土人参属	256
水苎麻	172	桃叶珊瑚属	282	兔儿风属	339
水芹属	358	天胡荽	356	菟丝子属	297
水蛇麻	170	天胡荽属	355	豚草	342
水蛇麻属	168, 169	天葵	110	豚草属	341
水苏属	319, 322	天葵属	109	橐吾属	347
水田白	290	天门冬	58		
水团花	285	天门冬科	55	**W**	
水团花属	284	天门冬目	42, 47, 50, 53, 56		
水蜈蚣属	77	天门冬属	57	瓦松	127
水仙	52, 53, 54, 56	天名精	349	瓦松属	126
水仙属	52	天名精属	348	网络鸡血藤	148
水苋菜属	207	天目朴树	168	网脉葡萄	130

莴草	96	喜树	259	小果山龙眼	115
莴草属	95	喜树属	259	小花扁担杆	229
望春玉兰	14	细柄黍	86	小花琉璃草	296
望江南	141	细齿稠李	159	小槐花	146
委陵菜属	155, 159	细梗胡枝子	146	小槐花属	145
卫矛	187, 188	细梗香草	272	小金梅草	45
卫矛科	187, 188	细叶旱芹	360	小金梅草属	45
卫矛属	187	细叶旱芹属	358	小苦荬属	338
魏氏金茅	89	细叶青冈	175	小麦族	81, 92
蚊母树属	119	细叶湿鼠曲草	346	小石积属	154
乌冈栎	173	细辛属	9, 10	小叶马蹄香	11
乌桕	202	细枝叶下珠	204	小叶石楠	157
乌桕属	202	细柱五加	356	小叶野决明	141
乌口树属	284	狭苞囊吾	347	小野芝麻	322
乌蔹莓	130	狭叶香港远志	152	小野芝麻属	319
乌蔹莓属	130	狭叶重楼	36	小颖羊茅	93
无刺鳞水蜈蚣	79	夏枯草	322	叶下珠	203
无根萍	25	夏枯草属	319	叶下珠科	203
无根萍属	23	仙茅科	45	叶下珠属	204
无根藤	19	仙人掌	258	叶子花属	252
无根藤属	17	仙人掌科	258	血见愁	321
无患子	219, 220	仙人掌属	258	新木姜子属	17
无患子科	219	纤毛马唐	86	星毛金锦香	215
无患子属	220	纤叶钗子股	43	杏属	155
无心菜	245	显子草	98	杏叶沙参	332
无心菜属	244	显子草族	82, 98	绣球防风属	319
吴茱萸属	221	线叶金鸡菊	343	绣球科	260
梧桐	228	苋科	246	绣球属	260, 261
梧桐属	227	苋属	247	绣线菊属	154
五福花科	350	腺毛藜属	247	萱草	50, 51
五加科	355	腺毛莓	159	萱草属	49
五加属	356	香茶菜属	320	玄参	309
五节芒	89	香椿	226	玄参科	308
五列木科	266	香椿属	226	玄参属	308
五味子科	1, 2	香科科属	319	旋覆花	349
五味子属	1	香茅属	88	旋覆花属	349
雾水葛	172	香蒲科	72	旋覆花族	336, 348
雾水葛属	172	香蒲属	72	悬钩子属	154
		香薷属	320	旋花科	297, 298
		香丝草	348	雪柳	303
X		向日葵	343	雪柳属	302
西番莲科	198	向日葵属	342	荨麻科	171
西番莲属	198	向日葵族	335, 341	薯树科	118
稀脉浮萍	25	小檗科	106		
豨莶	343	小二仙草	128		
豨莶属	341	小二仙草科	128	**Y**	
喜旱莲子草	247	小二仙草属	128	鸭舌草	67

鸭跖草	63, 64	野蕉	69	银胶菊	342
鸭跖草科	62, 63	野菊	344	银胶菊属	341
鸭跖草目	63	野决明属	141	银叶巴豆	202
鸭跖草属	62	野决明族	135, 141	印加树族	135, 140
鸭嘴草属	87, 89	野老鹳草	205, 206	罂粟科	101
崖豆藤属	147	野萝卜	233, 234	樱属	155
崖豆藤族	136, 147	野牡丹科	214	迎春樱桃	159
崖爬藤属	130	野牡丹属	215	蝇子草属	244
亚菊	345	野木瓜属	102	油点草	40
亚菊属	344	野青茅属	95	油点草属	38
岩黄芪族	136, 146	野柿	268, 269	油麻藤属	143
沿阶草属	57	野黍	87	油桐	202
盐麸木	218	野黍属	86	油桐属	202
盐麸木属	218	野茼蒿	347	莸属	318
盐角草	248	野茼蒿属	346	鱼黄草属	298
盐角草属	247	野豌豆属	149	鱼腥草	5
广州山柑	232	野豌豆族	136, 149	鱼眼草	348
眼子菜	31	野鸦椿	216	鱼眼草属	347
眼子菜科	31	野鸦椿属	216	榆科	166
眼子菜属	31	野燕麦	95	榆属	166
艳山姜	71	野芝麻属	319	雨久花科	66
燕麦属	95	野珠兰属	154	雨久花属	66
燕麦族	82, 95	夜香牛	339	玉兰属	12
羊角藤	286	夜香牛属	339	玉叶金花	285
羊茅属	92	一把伞南星	25	玉叶金花属	284
羊蹄	241	一本芒	78	玉簪属	57
羊蹄甲属	140	一本芒属	77	郁香忍冬	352, 353
杨柳科	199	一点红	347	鸢尾科	46, 47
杨梅	176	一点红属	346	鸢尾属	46
杨梅科	176	一枝黄花	348	元宝草	194
杨梅属	176	一枝黄花属	348	圆瓣冷水花	172
杨梅蚊母树	121	异果黄堇	101	圆果雀稗	87
杨桐	267	异木麻黄属	180	芫荽菊	344
杨桐属	267	异叶山蚂蝗	146	圆叶节节菜	208
药葵	229	异檐花	332	圆叶锦葵	229
药葵属	228	异檐花属	332	圆叶小石积	157
野百合	144	益母草	322	原野菟丝子	299
野扁豆	143	益母草属	319	远志科	150, 151
野扁豆属	142	薏苡	88	远志属	150
野灯芯草	75	薏苡属	87	月见草属	209
野甘草	307	蘮草	96	越橘属	280
野甘草属	306	蘮草属	95	云实	140
野菰	328	阴香	18	云实属	140
野菰属	327	阴行草	328	云实族	135, 140
野古草属	84	阴行草属	326	芸香科	221
野古草族	80, 81, 84	银合欢属	139		
野胡萝卜	360	银鳞茅	93		

Z

早熟禾	83, 93		砖子苗	79
早熟禾属	92		锥属	173
早熟禾族	81, 82, 92		紫背金盘	321
早田氏爵床	314		紫草科	294, 295
泽兰属	340		紫草属	294
泽兰族	335, 340		紫萼	56, 59
泽泻科	26		紫萼蝴蝶草	311
泽泻目	24, 29		紫金牛属	271
泽泻属	26		紫堇属	101
芝麻属	312		紫荆属	140
獐牙菜属	288		紫荆族	135, 140
樟科	17, 18		紫麻	171, 172
樟目	16, 18		紫麻属	172
樟属	17		紫茉莉	253
沼金花科	32		紫茉莉科	252
柘	170		紫茉莉属	252
浙贝母	39, 40		梓木草	296
浙江大青	321		紫楠	20
浙江新木姜子	19		紫萍	25
针茅族	82, 99		紫萍属	23
珍珠菜属	270		紫苏	322
珍珠茅属	77, 78		紫苏属	320
知风草	91		紫穗槐属	148
猪毛菜属	247		紫穗槐族	136, 148
栀子	283, 285		紫藤	148
栀子属	284		紫藤属	147
芝麻	312		紫菀属	347
芝麻科	312		紫菀族	336, 347
中华猕猴桃	279		紫薇	208
中华绣线菊	157		紫薇属	207
肿柄菊	343		紫玉兰	13
肿柄菊属	342		紫云英	147
重楼属	36		紫珠属	319
帚菊木族	335, 339		棕榈	61
皱果苋	247		棕榈科	60
皱叶黄杨	117		棕榈属	60
皱叶留兰香	322		钻叶紫菀	348
朱砂根	273		醉鱼草	309
诸葛菜	234		醉鱼草属	308
诸葛菜属	234		柞木	200
珠芽画眉草	91		柞木属	199
猪毛菜属	247		酢浆草	190, 191
猪屎豆属	144		酢浆草科	190, 191
猪屎豆族	135, 144		酢浆草属	190
猪殃殃	286			
竹叶花椒	221			